HANDBOOK OF ORGANIC WASTE CONVERSION

HANDBOOK OF ORGANIC WASTE CONVERSION

Michael W. M. Bewick, Ph.D, M.I. Biol.
Holder of the Leaf Fellowship,
Trinity College, University of Cambridge.

Van Nostrand Reinhold Environmental Engineering Series

VNR **VAN NOSTRAND REINHOLD COMPANY**
NEW YORK **CINCINNATI** **ATLANTA** **DALLAS** **SAN FRANCISCO**
LONDON **TORONTO** **MELBOURNE**

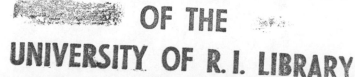

Van Nostrand Reinhold Company Regional Offices:
New York   Cincinnati   Atlanta   Dallas   San Francisco

Van Nostrand Reinhold Company International Offices:
London   Toronto   Melbourne

Library of Congress Catalog Card Number: 79-18568
ISBN: 0-442-20679-8

Manufactured in the United States of America

Published by Van Nostrand Reinhold Company
135 West 50th Street, New York, N.Y. 10020

Published simultaneously in Canada by Van Nostrand Reinhold Ltd.

15 14 13 12 11 10 9 8 7 6 5 4 3 2 1

Library of Congress Cataloging in Publication Data

Main entry unter title:
Handbook of organic waste conversion.

(Van Nostrand Reinhold environmental engineering
series)
    Includes index.
    1. Organic wastes as fertilizer.   2. Organic wastes
as feed.   3. Organic wastes—Recycling.   I. Bewick,
Michael W. M.
S654.H36   628'.445   79-18568
ISBN 0-442-20679-8

Van Nostrand Reinhold Environmental Engineering Series

THE VAN NOSTRAND REINHOLD ENVIRONMENTAL ENGINEERING SERIES is dedicated to the presentation of current and vital information relative to the engineering aspects of controlling man's physical environment. Systems and subsystems available to exercise control of both the indoor and outdoor environment continue to become more sophisticated and to involve a number of engineering disciplines. The aim of the series is to provide books which, though often concerned with the life cycle—design, installation, and operation and maintenance—of a specific system or subsystem, are complementary when viewed in their relationship to the total environment.

The Van Nostrand Reinhold Environmental Engineering Series includes books concerned with the engineering of mechanical systems designed (1) to control the environment within structures, including those in which manufacturing processes are carried out, and (2) to control the exterior environment through control of waste products expelled by inhabitants of structures and from manufacturing processes. The series includes books on heating, air conditioning and ventilation, control of air and water pollution, control of the acoustic environment, sanitary engineering and waste disposal, illumination, and piping systems for transporting media of all kinds.

Van Nostrand Reinhold Environmental Engineering Series

ADVANCED WASTEWATER TREATMENT, by Russell L. Culp and Gordon L. Culp

ARCHITECTURAL INTERIOR SYSTEMS—Lighting, Air Conditioning, Acoustics, John E. Flynn and Arthur W. Segil

SOLID WASTE MANAGEMENT, by D. Joseph Hagerty, Joseph L. Pavoni and John E. Heer, Jr.

THERMAL INSULATION, by John F. Malloy

AIR POLLUTION AND INDUSTRY, edited by Richard D. Ross

INDUSTRIAL WASTE DISPOSAL, edited by Richard D. Ross

MICROBIAL CONTAMINATION CONTROL FACILITIES, by Robert S. Rurkle and G. Briggs Phillips

SOUND, NOISE, AND VIBRATION CONTROL (Second Edition), by Lyle F. Yerges

NEW CONCEPTS IN WATER PURIFICATION, by Gordon L. Culp and Russell L. Culp

HANDBOOK OF SOLID WASTE DISPOSAL: MATERIALS AND ENERGY RECOVERY, by Joseph L. Pavoni, John E. Heer, Jr., and D. Joseph Hagerty

ENVIRONMENTAL ASSESSMENTS AND STATEMENTS, by John E. Heer, Jr. and D. Joseph Hagerty

ENVIRONMENTAL IMPACT ANALYSIS: A New Dimension in Decision Making by R. K. Jain, L. V. Urban and G. S. Stacey

CONTROL SYSTEMS FOR HEATING, VENTILATING, AND AIR CONDITIONING (Second Edition), by Roger W. Haines

WATER QUALITY MANAGEMENT PLANNING, edited by Joseph L. Pavoni

HANDBOOK OF ADVANCED WASTEWATER TREATMENT (Second Edition), by Russell L. Culp, George Mack Wesner and Gordon L. Culp

HANDBOOK OF NOISE ASSESSMENT, edited by Daryl N. May

NOISE CONTROL: HANDBOOK OF PRINCIPLES AND PRACTICES, edited by David M. Lipscomb and Arthur C. Taylor

AIR POLLUTION CONTROL TECHNOLOGY, by Robert M. Bethea

POWER PLANT SITING, by John V. Winter and David A. Conner

DISINFECTION OF WASTEWATER AND WATER FOR REUSE, by Geo. Clifford White

LAND USE PLANNING: Techniques of Implementation, by T. William Patterson

BIOLOGICAL PATHS TO SELF-RELIANCE, by Russell E. Anderson

HANDBOOK OF INDUSTRIAL WASTE DISPOSAL, by Richard A. Conway and Richard D. Ross

HANDBOOK OF ORGANIC WASTE CONVERSION, edited by Michael W. M. Bewick

LAND APPLICATIONS OF WASTE (Volume 1), by Raymond C. Loehr, William J. Jewell, Joseph D. Novak, William W. Clarkson and Gerald S. Friedman

LAND APPLICATIONS OF WASTE (Volume 2), by Raymond C. Loehr, William J. Jewell, Joseph D. Novak, William W. Clarkson and Gerald S. Friedman

STRUCTURAL DYNAMICS: Theory and Computation, by Mario Paz

HANDBOOK OF MUNICIPAL WASTE MANAGEMENT SYSTEMS: Planning and Practice, by Barbara J. Stevens

Foreword

During three decades of professional activity, this writer has witnessed a sequence of upsurges and subsequent and inevitable declines in fascinations with special concerns or subjects by both the public at large and professionals. This phenomenon, which has taken on the proportions and characteristics of fads, continues today. Most often such fads arise from genuine concerns and bring attention and corrective actions to areas in need of improvement. As sometimes happens with fads, the movements may be accompanied by excesses perpetrated by the overzealous. Happily, repercussions from the excesses usually are limited to a degree of alienation on the part of the more moderately inclined of the populace. Correlative to the fad is an intensification of related research activity. This should not be surprising, since magnitude of research money is a function of the dimensions of the prevailing fad. Modern research being an expensive activity, a major share of research activity would naturally be devoted to the subject of the fad.

Because an effective means for spreading the message is important in the development of a fad, it was not until the 1960's that fads began to take on the broad proportions that they have now. With the advent of the late 1950's and early 1960's, the variety of media for spreading a message expanded from the newspaper and periodical to include the radio, and even more importantly, television. Thus, while in the early 1950's there was something that might loosely be construed as the beginnings of a fad with regard to composting, it never reached the scale of the "ecology" fad that made its appearance in the 1960's. In this latter movement, the term "ecology" was stretched to cover everything that even remotely pertained to the environment, although it was mostly reserved for the restoration, or at least the preservation, of a reasonably good quality of the environment. It reached its peak with the institution of "earth day." But along with the disappearance of the unrest of the 1960's, the popularity of "ecology" began to wane and eventually was replaced to some extent by "resource conservation" (that is, conservation not only of those resources inherent in the environment, but of all resources). Eventually, resource conservation also suffered the fate common to fads, and now its place is held by energy. The onset of the energy fad was triggered by the artificial interruption in the supply of petroleum-based fuels, and is fostered by the long deferred realization of the finite nature of our energy resources.

While fads may come and go, there is one field of endeavor which it behooves us to advance to the proportions of a fad—albeit without its ephemerality—and that is the expansion of the world's food supply to at least a point at which no one need suffer want. After all, what good would it do mankind to have an abundance of energy and be surrounded by an environment of the highest quality, if its food supply were insufficient? It follows, therefore, that the provision of an adequate food supply must be the primary concern not only of those developing nations in which food shortages may be chronic, but also of more fortunate nations which presently enjoy an abundance of food. The problem in the latter nations is the failure of their citizenry to realize the vulnerability (not only to the forces of nature, but also to human folly) of food production. In the absence of precautionary measures, the factors that eventually will lead to the ending of the abundance are a shortage of plant nutrients and the disappearance of good quality land. Prime agricultural land is rapidly being taken up either for more crop production or, catastrophically, for urban expansion. Overuse and erosion through neglect exacerbate the problem. As for plant nutrients, the dependence upon petroleum-derived plant nutrients (for example, the use of natural gas in the synthesis of ammonia) places a ceiling on the supply of nutrients which will soon be reached. Therefore, unless remedial measures are taken, the magnitude of the food supply will level off, and perhaps even decline. The tragedy is that the world's population will not likewise become stationary, and instead will expand. The inevitable result is a worldwide food shortage.

The message to be found in these forebodings is that it is essential to conserve those resources which pertain to food production. Conservation implies not only the application of sound management to the prevention of the loss of good land, but also to the extraction of maximum use from the food that is produced. Maximum use involves several courses, among which are the following:

1. The type and the available supply of individual feedstuffs should be fitted to the type of consumer, animal or human, that would derive the greatest nutritional benefit from it, and because of this benefit ultimately serve the greatest use in human nutrition. For example, there is debate as to whether feeding grain to large meat-animals (beef cattle) rather than to poultry or small mammals (rabbits) or even solely to humans is a luxury rather than a necessity.

2. Maximum use of the plant must be obtained. The ultimate but rarely practical goal here would be the utilization of the entire plant as a feedstuff. For example, the fruit might be used for human consumption, while the remainder of the plant would be converted into an animal feedstuff.

3. The impracticality, in most cases, of the second course leads to a third course, which is to return non-edible crop residues to the soil, where they can serve as a source of plant nutrients as well as enhance future crop production by improving the quality of the soil.

4. The fourth course is related to the third one. It consists in putting to good use the wastes generated in food processing and preparation. This can be done either by converting them into a feedstuff or by returning them to the soil in a controlled manner. In the past, all to often, such wastes were disposed of in a completely unsatisfactory manner. Now, with the concern over the environment and the search for new food sources, the practice of refeeding wastes is becoming increasingly more common.

5. A fifth and eminently practical approach is to extract the ultimate utility of the food crop by returning to the food chain the nutrients contained in the wastes excreted by the consumers of the crop. This can be done by incorporating the digestive wastes of one animal into the feedstuff of another, or by utilizing them as a fertilizer. The first method is commonplace in nature, as is witnessed by birds feeding on the undigested grain in the dung of ruminants. Within the past decade, extensive studies have been conducted on the conversion of cattle manure into food adjuncts. The second method, the use of animal wastes as fertilizer, is as ancient as agriculture.

It is hard to assign an hierarchical order to the five courses, and in fact it would serve no purpose to do so, since *all* are needed to ensure the continuation of a supply of food adequate to meet the needs of an expanding world population. However, of the approaches, the fourth and fifth are the most readily followed or applied, since the technology and practices involved are fairly well developed. However, a practical difficulty in setting up a program for applying the two approaches is encountered when one searches the literature for needed information. While much has been written on the subject, specially within the past few years, the writing has appeared in a wide assortment of periodicals, journals, and reports that are difficult to obtain. Thus, this handbook serves an extremely useful purpose. It represents a distillation of the existing literature plus the summation of the research activities of a group of very competent specialists. Moreover, the wide range of subjects covered in the book is a distinct asset. Included here are discussions of both agricultural and municipal wastes, as well as several wastes that hitherto have received only scant attention in the literature, such as wood and parkland wastes and special industrial wastes.

CLARENCE G. GOLUEKE

Preface

The aim of this book is to examine the whole spectrum of organic wastes, assessing their present and potential uses as feedstuffs and fertilizers.

The contributors have been selected for their detailed knowledge of particular waste materials, and it is hoped that the reader will be able to select specific chapters on the waste in which he is interested, or to read the book as a whole to obtain an overall picture of the uses of organic wastes. Since the book was prepared with these two purposes in mind, the reader may find some overlap between certain chapters. Different wastes may be treated using the same methods, or particular waste materials may fall on the borderline between two chapters. Such repetition ensures that specialists in particular waste materials may obtain all the information on that material from the relevant chapter.

The last two chapters in the book are somewhat different from those preceding, as they examine the use of one particular method, fermentation, in the reclamation of waste materials from many sources. Liquid wastes, those containing only 1 or 2% solids, tend to be overlooked in most reclamation schemes, since it is not practical in most cases to produce an economic feedstuff or fertilizer directly from them. Chapter 14 presents a review of virtually the only method of producing feedstuffs from these wastes, as single cell protein via fermentation. In Chapter 15, the use of fermentation is again reviewed as a method of converting organic wastes to feedstuffs and/or fertilizers. In this process, anaerobic fermentation is used to produce a feedstuff or a fertilizer as the final product, together with another vital commodity, energy. This method of producing energy in the process of organic waste conversion may point the way to yet another aspect of the utilization of organic wastes: as a source of energy.

M. W. M. BEWICK

Contents

HANDBOOK OF ORGANIC WASTE CONVERSION

Agricultural wastes as fertilizers

H. Tunney, Ph.D.

The Agricultural Institute, Johnstown Castle Research Centre, Wexford, Ireland

INTRODUCTION

It is appropriate to start by outlining what is meant by agricultural wastes. In this context, agricultural wastes include all organic wastes produced and disposed of or used in primary agricultural production. These wastes are exclusively animal manures and crop residues or mixtures of the two. Of course, these materials are essentially by-products of production rather than wastes in the strict sense of the word.

Historical Perspective

For thousands of years, agricultural wastes have been used by farmers as the major source of maintaining and improving soil fertility. Indeed, on a world scale this is still true today. As long ago as 400 B.C. Xenophon referred to the importance of animal manures (Flaig *et al*, 1978). There are references in the

Bible to the use of manures for improving crop growth. Almost 2000 years ago, Virgil, in his book of verse on agricultural practice and advice, stressed the importance of wastes for good crop production. The importance of manure is pointed out in a translation from the original Latin text: ". . . what e'er the sets thou plantest in the fields thereon strew refuse rich" (Virgil, c 79 A.D.).

There have been many publications and scientific papers, particularly over the past 100 years, on the value of farmyard manure in crop production. A number of long-term experiments to study the benefits of farm wastes and chemical fertilizers were started in the last century, and some of these are still in progress. The long-term experiments at Rothamsted in England (Johnston, 1969) and at Askov in Denmark (Kofoed, 1976) are but two examples.

Important changes in agricultural production have been taking place over the past half century or so, and have had an important influence on the use of agricultural wastes as fertilizers. Therefore, we will take a brief look at some of the these changes.

Recent Changes in Agricultural Production. In recent years, a revolutionary change has been taking place in agricultural production. This change is most noticeable in developed countries and has been accelerating over the past 30 years, since the end of World War II. The most obvious effect of this ongoing revolution has been the rapid decrease in the number of people engaged in agricultural production along with a significant increase in agricultural production itself. In short, a large agricultural labor force and draught animals are being replaced by machinery powered by fossil fuels. New technology, including fertilizers, herbicides, pesticides, and improved crop varieties, is playing an important part in the changes.

It is estimated that the agricultural labor force in the developed countries will be 30 million in 1985, in comparison with 70 million in 1950 (Loehr, 1974). The agricultural work force in the United States decreased from 8.4 to 4.6 million between 1955 and 1969. A similar trend has occurred in other developed countries; for example, in the United Kingdom, the male agricultural work force decreased from 611,000 to 315,000 between 1947 and 1967 (S.F.B.I.U., 1969). In developed countries, agriculture is employing in the region of 10% of the total labor force (Anon, 1972). The situation in developing countries is different, and it is estimated that the agricultural labor force will increase from 344 million in 1950 to 500 million in 1985 (Loehr, 1974).

Of course, it is not possible to predict future trends with confidence. Changes in the availability of energy or in the price or availability of fossil fuels could have pronounced effects on the world food supply and how it is produced. Food supply in the developed countries could be particularly vulnerable if fossil fuel or an alternative is not available. It would take many years to get a skilled labor force and horses to cultivate the soil to produce food for the population with the traditional methods that were used up to a few decades ago.

There is, of course, an increased demand for food because of an increasing world population. However, equally important are the changes in human diets that have taken place in recent years. In developed countries with higher living standards, there is an increased consumption of livestock products and a decrease in consumption of cereals products. The nature of these changes for the United States is illustrated in Fig. 1.1 (Stout, 1972), in which the steady and continuing increase in the consumption of animal products and the corresponding decrease in the use of starchy foods is evident. A similar trend, on a wider scale, is suggested by Barreveld (1977), who predicts that world demand for meat and animal products will almost double between 1970 and 1990.

A distinction can be made between developed and developing countries. The changes discussed are most evident in the developed countries. In many of the poorer developing countries, a high proportion of the labor force is engaged in

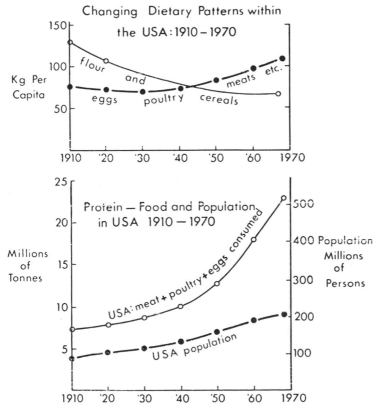

Fig. 1.1. Relationship of changing dietary patterns in the United States. *Source*: Stout. 1972. Reproduced from the Western Hemisphere Nutritional Congress III, Symposia Specialists, Inc., Miami, Florida.

agriculture. The capital for machinery, fertilizers, and other aids to increased agricultural production is often lacking, as is the expertise to increase production.

Changes in Use of Manures and Fertilizers. Over the past century, many of the basic scientific principles of plant nutrition and soil fertility have been explained. In developed countries, inorganic chemical fertilizers have been widely accepted as a major source of improving and maintaining soil fertility.

With the advent of abundant and relatively inexpensive chemical fertilizers, agricultural manures became agricultural wastes, and what was formerly an asset is now often regarded as a disposal problem. In addition to these events, rapid and radical changes have taken place in methods of livestock production, partic-

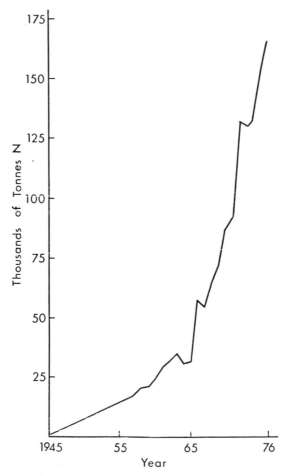

Fig. 1.2. Increased nitrogen fertilizer use for the Republic of Ireland.

ularly over the last two decades. Traditionally, cattle, pigs, and poultry were fed on free range or housed in relatively small numbers. Today, in most developed countries, intensive animal production is now being practiced where increasing numbers of livestock are being fed on fewer larger farms. Technological know-how and economic pressures are facilitating this trend. It is now possible—and increasingly found in practice—to have over 1000 cattle or 10,000 pigs or a 100,000 poultry in one single intensive farm unit. On a somewhat different scale, many thousands of beef animals are being produced in large open feedlots in semi-arid regions such as the southwestern United States.

These changes have been accompanied by dramatic increases in the use of chemical fertilizers. In the region of 70 million tonnes of plant nutrients were used as chemical fertilizer worldwide in 1976 (F.A.O., 1977a). There was ap-

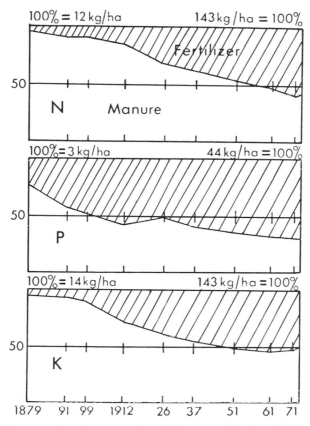

Fig. 1.3. Nitrogen, phosphorus, and potassium nutrients applied in fertilizer and animal manures in Germany over the past 100 years. *Source*: Flaig *et al*. 1978.

proximately a doubling in fertilizer use between 1960 and 1970 and an eight-fold increase between the end of World War II and 1970 (Loehr, 1974). This trend in increased chemical fertilizer consumption is illustrated by the change in nitrogen use for the Republic of Ireland (shown in Fig. 1.2). Similar trends are evident for many developed countries (U.S.D.A., 1970). Nitrogen use has shown the greatest increase, for example, in the United States, there was a ten-fold increase in nitrogen use in the quarter century from 1945 to 1970.

The increase in the use of chemical fertilizers and the relative decrease in the use of animal manures in Germany over the past 100 years is described by Flaig *et al* (1978). This change is illustrated in Fig. 1.3, where it can be seen that up to about 1890, over 90% of the nitrogen and potassium was applied as animal manures, whereas in 1970, over 50% was in the form of chemical fertilizers. It is important to note that there has been an overall increase in nutrients supplied as manure, despite a decrease relative to chemical fertilizers.

Chemical fertilizers play a very important role in increased agricultural production. However, it should be noted that on a world scale, the total quantity of plant nutrients recycled to agricultural land in organic wastes is greater than that applied as artificial fertilizer. Voorhoeve (1974) estimated that in 1971 the production of plant nutrients from organic wastes was about eight times greater than the consumption of chemical fertilizers in developing countries.

It is clear that the fertilizer value of agricultural wastes has a greater potential for agricultural production if properly used and this is particularly true for developing countries (Flaig *et al*, 1978).

HANDLING

In general, the handling of agricultural wastes—and here we are dealing principally with animal manures—includes collection, storage, and spreading or disposal on land.

Traditionally, animals were bedded with straw or other litter and the manure was a mixture of urine, feces, and straw. This farmyard manure was normally stored and spread as a solid. A high proportion of agricultural waste is still handled with straw, as farmyard manure. However, an increasing proportion is being stored and handled as a liquid or slurry where no bedding is used. This is particularly true on most new and large intensive animal enterprises.

Animal wastes can be divided into three main categories:

1. Farmyard manure, where straw or other bedding is used;
2. Slurry, which is a mixture of feces and urine, often containing variable quantities of rainwater and washwater; and
3. Separated manure, where the waste is separated into two fractions—solid and liquid—which are handled separately.

The third category is receiving increased attention with the use of mechanical slurry separators, where the liquid fraction is spread or irrigated to land, and the solid fraction is handled separately as a solid. Other categories of manure in which non-mechanical separation takes place can be included here. In arid or semi-arid regions, manure quickly dries out by evaporation after it is produced in feedlots or corrals. This manure is handled as a solid and is sometimes stacked into huge "manure mountains" (Flaig et al, 1978). With this method of storage there is no liquid for disposal; however, there can be runoff problems after heavy rainfall. In temperate regions with relatively high rainfall, such as Northern Europe, slurry is sometimes pushed into compounds and solid/liquid separation takes place during storage. A solid crust forms on the top of the compound and sludge on the bottom with liquid in the center (Robertson, 1977). This method is sometimes described as dungstead storage and is essentially non-mechanical solid/liquid separation. The liquid fraction can be drawn off and spread first, and then the solid fraction remaining can be spread.

Crop residues could be considered a source of agricultural wastes. Residues from most root crops, however, are returned directly to the soil during harvesting (for example, potato haulms). These residues could not be considered as wastes and are really in the same category as plant roots. The residues do, however, supply considerable quantitites of essential nutrients for the next crops after mineralization by soil micro-organisms.

Straw on soil after harvesting cereal crops is in a somewhat different category, as it can interfere with subsequent mechanical cultivation. To avoid this problem, straw is often burned or mechanically chopped into small pieces after harvesting. Straw left on the soil in this manner supplies the small quantities of phosphorus, potassium, and other nutrients present for subsequent crops. Due to the high carbon to nitrogen ratio of straw, the mineralizing organisms may reduce the available soil nitrogen, and extra should usually be applied—about 30 kg N/ha, if straw is not burned or removed. In recent years, straw is increasingly being used as source of winter feed for cattle. Research has shown that the feeding value and digestibility of straw can be increased by treatment with an alkali, such as sodium hydroxide, or by mechanical treatment, such as very fine grinding (which will be dealt with in more detail in Chapter 2).

Collection and Storage

Beef and dairy cattle are sometimes housed in straw bedded sheds. The manure is usually allowed to accumulate to a depth of about 1 meter before cleaning. The manure is then either spread directly to land or stored in a heap for subsequent spreading.

Increasingly, manure is being collected and stored as slurry. The concept of handling manure as slurry containing a mixture of feces, urine, and water is not a new one. It was described by Mechi (1859) in his book on farming. Different methods of slurry handling were practiced over the past century, principally in the Alpine regions of Switzerland, Austria, and Germany. This material, called Gulle, was highly regarded for its fertilizer value (Gisinger, 1950).

Modern methods of handling animal wastes as slurry were greatly influenced by the development, in Scandinavia, of slatted floors for animal rearing (Hoibs, 1960). This development has revolutionized cattle and pig housing over the past two decades. Along the same lines is the "slope slot flush," which was recently developed and is currently being studied in the United States (Natwick and Goodrich, 1975). With this system, manure passes through slots into pipes in the floor and flows to storage tanks outside the building.

Free stall cubicle houses are a popular method of housing cattle, particularly dairy cows. With this system, the manure is usually cleaned by tractor scraper off concrete yards and passages into storage tanks. The storage tanks may be underground or overground and can be concrete or steel, glass lined, or made from other materials. Tractor-driven pumps are sometimes used to transfer the manure to an overground storage tank.

Slatted floors are extensively used for beef animals, and in the past two years, slatted passages for dairy cows have been more widely adopted. With the slatted floor house, the manure is usually stored in 2-meter deep tanks beneath the house. There is usually adequate manure storage for the winter feeding period, and there is no labor involved in manure handling while animals are in the house. It passes directly as it is produced through the floor into the storage tanks below.

For pigs, straw is still sometimes used for bedding, but the majority of intensive units collect and store manure as a slurry. Part of the floor area is normally slatted and the slurry flows in manure channels to outside storage.

In the case of poultry, broilers are usually reared on deep litter. The litter is usually wood shavings, chopped straw, or similar materials. The house is cleaned out after about 11 weeks, when the broilers are sold. Fresh litter is then put in for a new batch of chicks. This manure usually has about 75% dry matter and is either stored in a heap or spread straight from the house. Laying hens are now usually housed in cages and the droppings pass through the cage floor. The droppings are usually taken by conveyor belt to a spreader or storage outside the house. The hen droppings have about 25% dry matter. Of all agricultural wastes, poultry manure contains the highest levels of plant nutrients and therefore has the highest fertilizer value.

Spreading and Treatment

Traditionally, farmyard manure was transported by horse and cart or spread by horse-drawn tanks, and an 1869 patent depicts a horse-drawn tank and plough for liquid manure (Robertson, 1977).

There are now over 100 different types of tractor-drawn machines for spreading manures. Farmyard manure is spread by muck spreaders and slurry wagons or tankers. Vacuum tankers have been most widely used for spreading slurry. Filling is by atmospheric pressure when the tank is under vacuum. Spreading is by reversing the system and pumping air into the tankers. Helical rotor pumps and other systems are also used for spreading slurry.

Many machines for spreading manure have been designed for disposal, so that manure is often spread unevenly. To get maximum fertilizer value from manure it is important that machines should be capable of spreading evenly, as with fertilizer distributors. Most manure is spread on the soil surface. However, the possibility of injecting slurry into the soil is receiving attention. Slurry injection into cultivated soil has been used for many years, and more recently, injecting tines suitable for grassland are being used. Injecting is now carried out primarily to reduce smell after spreading, but it can also increase the effectiveness of slurry nitrogen and reduce contamination of grass.

Treatment of slurry along the lines used for urban sewage has been studied extensively over the past decade. The consensus of opinion at present is that application to land is the only economically practical method of disposing of animal manures. An intensive pig enterprise with 5,000 pig places would have a waste treatment problem equivalent to a town with about 12,000 people. Conventional treatment on this scale would make animal products very expensive. Veal calf wastes are very dilute, by comparison with other animal slurries and are being treated successfully in Holland. Some pretreatment of slurry prior to spreading on land may be desirable; for example, solid/liquid separation may reduce some of the problems involved in pumping high dry matter slurry. To avoid spread of disease, slurry can be treated with lime before spreading on land (Thunegard, 1975).

The foregoing material on handling of agricultural wastes and manures is intended as a brief outline. For more detailed information on handling and treatment of animal wastes, the following authors should be consulted: Taiganides (1977), Loehr (1974), Miner and Smith (1975), Robertson (1977), Anon (1976), Strauch *et al* (1977).

WASTE CHARACTERISTICS

Agricultural wastes differ from most municipal and industrial wastes in that they are usually highly concentrated and relatively low volume wastes. Biochemical oxygen demands can vary from about 1000 parts per million for dilute slurry or dirty water to over 50,000 parts per million for silage effluent and concentrated pig slurry. The biochemical oxygen demand is a laboratory estimate of the quantity of oxygen required to oxidize part of the organic matter in waste in a specified period of time, usually five days (Anon, 1965). Details of characteristics of food processing wastes—including meat, dairy, and vegetable processing—are adequately described in later chapters. Here we will deal mainly with the quantities and composition of animal manures, with some details on straw, silage effluent, compost, and other wastes. There are many other agricultural wastes, including dead animals, waste fuel and chemicals, plastic and other packaging materials, old machinery, and building materials; however, they are not relevant to the purpose of this chapter, as they have little or no fertilizer value and would not normally be recommended for application to agricultural land.

The animal wastes are primarily undigested plant materials with a high proportion of bacterial and intestinal cells.

Quantities of Wastes Produced

The quantity of wastes produced daily by an animal is a function of the type of animal, its size, its feed, and the temperature and humidity of the environment (Loehr, 1974). Quantities quoted in the literature often refer to total weight or volume of fresh waste produced, including variable quantities of litter or water.

The nature of the relationship between manure production and liveweight gain of the different farm animals is shown in Table 1.1 (Miner and Smith, 1975).

The figures in Table 1.1 indicate that on average, daily undiluted fresh manure production is equivalent to between 5 and 8% of the animal liveweight. The daily manure dry matter produced is usually between 0.6 and 1.7% of the animal liveweight. These figures are an oversimplification in that there is considerable variation in manure produced as a percentage of animal liveweight depending on the type and age of the animal and other factors. For example, a dairy cow will produce more manure during lactation than when she is not milking (Tietjen, 1966). The values in Table 1.1 do, however, indicate a relationship between manure production and weight of animal.

Studies by Hendrick (1915) indicated the importance of diet on manure production. Cows fed 68 kg of mangolds produced 19 kg feces and 40 kg urine per day. When the same cows were fed on 12 kg hay and 30 kg water, they produced

TABLE 1.1. Daily Manure Production Expressed on a Fresh Manure and a Dry Matter Basis, as a Percentage of Animal Liveweight. Adapted From Miner and Smith (1975).

	Dairy Cow	Beef Feeder	Swine Feeder	Poultry	
				Layer	Broiler
Fresh manure	8.2	6.0	6.5	5.3	7.1
Dry matter	1.0	0.7	0.6	1.3	1.7

22 kg feces and 6.5 kg urine. (Urine production tends to show a greater variation in quantity than does feces production.)

Intensive livestock rearing tends toward standardization of diet and type of animal. With this tendency, the quantities and composition of manure should be more uniform. The greatest source of variation will probably be from varying quantities of water or litter mixed with the manure.

Cattle. The slurry production rate most often quoted for cattle is in the region of 40 to 45 kg per day per livestock unit (Tunney, 1977a). A livestock unit (L.U.) is equivalent to a 540 kg dairy cow (Attwood and Heavey, 1964). Experience at Johnstown Castle Research Centre over the past three years indicates that 500 kg beef cattle produce 40 kg of undiluted slurry per animal per day.

There is less information in recent literature on the quantities of farmyard manure and dungstead manure produced. It has been estimated that about 3 tonnes of dungstead manure or 6 tonnes of farmyard manure would be produced by one livestock unit during a 140 day winter period (Tunney, 1977a).

Pigs. The most often quoted slurry production for meal-fed pigs is 4.5 kg per pig per day on the average over the fattening period (Tunney, 1977a). Sauerlandt and Tietjen, (1970) studied the quantities of manure produced by fattening pigs of different weights. At an average weight of 87 kg per pig, 4.35 kg slurry composed of 1.73 kg feces and 2.62 kg urine was produced. In a recent study in Scotland (Anon, 1976) pigs at 68 kg mean weight produced 4.5 kg slurry per day with 10% dry matter. One sow and piglets produces about 14.3 kg slurry per day (Christensen, 1976).

Poultry. Loehr (1974) indicated that daily fresh poultry manure represents about 5% of bird liveweight. A 2 kg mature bird would produce about 0.1 kg manure daily with 25% dry matter. Broiler droppings are mixed with litter and normally contain about 75% dry matter when removed from the house.

Silage Effluent. The quantity of silage effluent from ensiled grass and other materials is usually related to the dry matter of the ensiled materials: the lower the dry matter, the greater the quantity of effluent. A tonne of ensiled material with 10 to 20% dry matter will produce 260 to 450 kg of effluent (Harrison, 1973). At 20% dry matter, approximately 100 kg effluent per tonne is produced. At over 25% dry matter, there is generally no effluent. Wilting before ensiling not only improves silage quality, but also has the advantage of increasing dry matter and therefore reducing or preventing effluent disposal problems. Straw yield can vary considerably but is usually in the region of 5 tonnes/ha for high yielding crops.

Spent compost from growing mushrooms, along with other composts, are spread on land and can have a significant fertilizer value. The quantities are small by comparison with other agricultural wastes; however, they can be important in localized areas.

Composition of Agricultural Wastes

Many of the factors already mentioned as influencing quantities also influence the composition of manures. The diet, perhaps more than any other single factor, has an important bearing on composition. Type and age of the animal, as well as storage conditions also have important effects on composition. The degree of dilution with water will greatly influence composition of manure expressed on a fresh weight basis; however, degree of dilution will have little effect on results expressed on manure dry matter.

The plant nutrient composition of agricultural wastes usually has some relationship to the nutrient content of plant materials. This is understandable, as most agricultural wastes originate directly or indirectly from plant materials. There can be considerable variation in the nitrogen (N), phosphorus (P), and potassium (K) content of plant material. For example, animal wastes in developed countries now have a higher content of N, P, K, and other nutrients than they did 50 years ago. This is due to the higher soil nutrient status with the resultant increase in plant uptake due mainly to the increased use of chemical fertilizers. The improved utilization of animal feed (for example, higher digestability) also contributes the higher nutrient content of manures. Legumes such as clover or beans normally have a higher N content than do cereal grains, reflecting their higher protein content. This is partly due to their ability to fix atmospheric nitrogen gas in symbiosis with rhizobium bacteria in nodules on the legume roots.

Agricultural crops normally contain 1.5 to 4% N, 0.2 to 0.4% P, and 1 to 3% K in the plant dry matter. During digestion or decomposition, much of the organic matter is used up; however, most of the plant nutrients are retained. For this reason, the nutrient content of agricultural wastes may be four times higher than in the original plant material. In the more extreme situation, where the waste is burned and only the ash remains, the nutrient concentration on a dry matter basis will be at a maximum. It should be noted, however, that some nutrients, such as nitrogen, may be partly lost during burning or decomposition. Inorganic nutrients (phosphorus, for example) are often added to animal rations and will increase levels in the manure.

It is estimated that between 80 and 95% of the inorganic nutrients ingested by animals in their feed is excreted in the feces and urine. Growing and lactating animals retain a higher proportion of the nutrients from their feed than store or fattening animals. For example, Gisinger (1960) showed that dairy cattle excreted 80%, 80%, and 95% of ingested N, P, and K, respectively. Fattening cattle, on the other hand, excreted 94%, 98%, and 98% of the N, P, and K, respectively. The proportion of organic matter excreted will depend on digestability, but on the average, organic matter excreted is equivalent to about 40% of organic matter intake. This explains the higher inorganic nutrient content of manures.

It is evident that a high proportion of the plant nutrients in animal feed are excreted in the manure and are potentially available for recycling to cropland for production of subsequent crops.

The proportion of feces to urine in manure will influence composition. Urine is normally low in phosphorus and high in potassium, whereas about equal parts of nitrogen may be excreted in the feces and urine of cattle (Tietjen, 1976). Therefore, manure in which a proportion of the urine was allowed to drain away would be relatively low in nitrogen and potassium.

Cattle, Pig, and Poultry Manures.   A recent review of the literature on the composition of manures shows there is considerable variation in manure composition (Tunney, 1977a). However, in view of the wide geographical diversity of origin of the manures there is reasonable agreement between the values reported. The results of the values from the literature are summarized in Table 1.2.

In general, the higher the dry matter, the higher the nutrient content. For example, in Table 1.2, the higher nutrient content of the poultry manure is due mainly to the higher dry matter content. There is usually about the same level of nitrogen and potassium in cattle slurry, and phosphorus content is usually about 25% of the nitrogen and potassium level.

Pig manure usually has about the same level of nitrogen as does cattle manure, although pig manure is usually higher in phosphorus and lower in potassium than is cattle slurry. The difference in the phosphorus to potassium ratio between cattle and pig manure is principally a reflection of the diet of these animals. The relatively low potassium and high phosphorus in pig slurry reflects a mainly cereal diet, and the high potassium in cattle manure reflects the high potassium

TABLE 1.2.   Range of Values from Literature on Composition of Cattle, Pig, and Poultry Manures. Adapted from Tunney (1977a).

| | | Nutrients (kg per 10 tonnes fresh manure) | | | |
	Dry Matter (%)	N	P	K	Mg
Cattle manure	4–23	24–65	4–18	20–58	2–6
Pig manure	5–25	16–68	6–21	17–36	3–7
Poultry manure	23–68	96–230	24–120	38–116	12–22

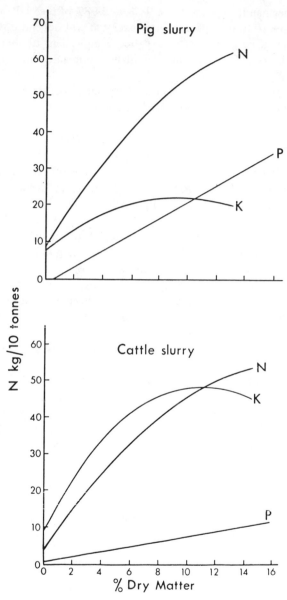

Fig. 1.4. The relationships between dry matter and nutrient content of cattle and pig slurry. *Source*: Tunney. 1977a.

content of herbage which often makes up a high proportion of the ruminant diet. Further evidence of the differences between cattle and pig manure is illustrated in Fig. 1.4.

The nutrient ratio of poultry manure is somewhat similar to pig manure and probably reflects the similarity between the diets of pigs and poultry. The variation in slurry composition can be greater than for solid manures. The variations in composition of cattle slurry from 33 farms and pig slurry from 25 farms were studied in Ireland (Tunney, 1977a). These results showed a very wide variation. In fact, there was over a ten-fold variation in dry matter between the highest and lowest farms. The results of this study are summarized in Table 1.3.

It can be seen from Table 1.3 that there can be a very wide difference in slurry composition between farms. This variation presents a problem when trying to integrate slurry into a farm fertilizer program. In farm practice, slurry is usually applied on a volume per area basis (m^3/ha or gals/acre) with little attention to slurry quality. From these results it is clear that variation between farms is such that an estimate of quality is necessary before deciding on application rates to supply adequate nutrients for crop needs. This study also showed a positive correlation between dry matter and plant nutrient content of slurry. The relationship between dry matter and nutrient content of cattle and pig slurry is shown in Fig. 1.4.

A mixture of feces and urine as produced by pigs would normally contain about 9% dry matter (O'Callaghan et al, 1971). It appears that the variation in dry matter levels in Fig. 1.4 are due principally to dilution with rain or wash water. The high dry matter samples are probably the result of storage tanks being emptied by vacuum tanker, where the more liquid fraction is drawn off and spread, and the more solid fraction tends to accumulate in the storage tank.

TABLE 1.3.   Summary of Analyses of Cattle and Pig Slurry. Adapted from Tunney (1977a).

| | Dry Matter (%) | Nutrients (kg per 10 tonnes manure) | | | |
		N	P	K	Mg
		Cattle slurry (33 farms)			
Mean	8	28	6	42	4
Range	1–14	8–56	1–12	8–64	1–11
		Pig slurry (25 farms)			
Mean	4	30	9	15	4
Range	1–13	4–70	1–34	2–33	1–20

With greater care in storage to avoid dilution with water, slurry could be much more uniform in composition, and it would be much easier to use it as a fertilizer.

A simple field test to estimate the dry matter and fertilizer value of slurry based on the relationship between dry matter and specific gravity of slurry has been developed (Tunney *et al*, 1975). The relationship between dry matter and specific gravity for cattle and pig slurry is shown in Fig. 1.5. Based on this relationship, a hydrometer called a Slurry Meter, which is calibrated in percent dry matter, can be used to get a quick estimate of the value of slurry. Fig. 1.6 shows the Slurry Meter being used to estimate the dry matter of two samples of pig slurry.

Methods of intensive animal rearing are becoming standardized. For example, the diet of pigs is very similar in most developed countries. In these circumstances it appears likely that the nutrient composition of pig manure on a dry matter basis will be more predictable and this should facilitate its use as a fertilizer. There is a need to ensure that losses of nutrients during storage, spreading, and dilution of slurry are minimized. In addition to plant nutrients and other inorganic chemicals, manures can contain residues of organic chemicals used in

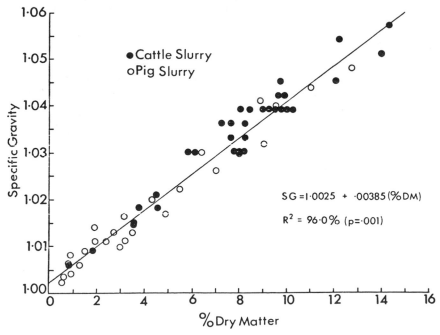

Fig. 1.5. The relationship between dry matter and specific gravity of cattle and pig slurry. *Source*: **Tunney. 1977a.**

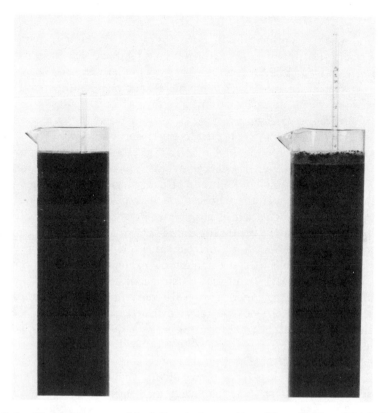

Fig. 1.6.   Slurry Meter used to estimate dry matter of low (left) and high (right) dry matter pig slurries.

animal production. Residues of chemicals for disease control and hormone based growth regulators may occur in manure and they can influence plant growth (Flaig *et al*, 1978). Cereal straw dry matter contains amounts of nutrients in the region of 1.4% N, 0.4% P, and 0.7% K. Sawdust and wood shavings used as litter for broilers has a very low plant nutrient content. An average sample of silage effluent would contain about 5% dry matter, and 10 tonnes of effluent would contain about 30, 6, and 30 kg of N, P, and K, respectively (Stewart and McCullough, 1974). Dirty runoff water from cattle yards and milking parlors would usually have the composition of dilute slurry, and normally would have less than 2% dry matter.

For values on the composition of other agricultural wastes, the reader may wish to consult work by Berryman (1970) and Loehr (1974).

FERTILIZER VALUE

Agricultural wastes are principally of plant origin, and therefore contain all the nutrients essential for plant growth. It has been established that at least 15 inorganic elements are essential for healthy plant growth (Sailsbury and Ross, 1969). These essential elements are classified into marco- and micro-nutrients.

The macro-nutrients normally required in the percentage range are carbon, hydrogen, nitrogen, potassium, phosphorus, calcium, magnesium and sulfur. The micro-nutrients include iron, manganese, copper, zinc, boron, molybdenum, and perhaps others such as chlorine. In addition to the essential nutrients, plants can contain more than 50 other inorganic elements, most of which are present in only trace amounts, and are not known to be necessary for healthy plant growth and development. There are other nutrients, such as selenium, cobalt, and iodine, which are essential for animal nutrition and must be obtained from the diet.

Chemical fertilizers usually contain only three macro-nutrients: nitrogen, phosphorus, and potassium. Calcium carbonate as ground limestone is the other fertilizer commonly applied for the purpose of maintaining soil pH at optimum level for crop growth. Most crops can obtain adequate quantities of the other nutrients for healthy plant growth from soil reserves and from the atmosphere. Certain soils can show deficiencies of these nutrients, and additions are necessary for healthy plant growth. For example, some peat soils are often copper deficient, and boron must be added to many soils for healthy sugar beet production.

The major agricultural value of animal manures and other wastes is that they supply plant nutrients essential for plant growth. They supply both macro- and micro-nutrients. Claims are sometimes made that organic manures have hidden benefits. There is increased interest in "organic foods" grown without the aid of chemicals and where organic manure is used instead of artificial fertilizers, but no evidence exists in scientific literature to support the view that either crop yield or quality can be significantly and consistently improved by the use of organic manure rather than chemical fertilizers. Of course, it is important to supply the correct quantity and balance of nutrients to get maximum crop yield and quality. This applies whether the nutrients are applied as organic manure or chemical fertilizers.

There has been considerable controversy for many years as to the uptake of organic humic substances by plants and the possible effects on plant growth (Flaig et al, 1978). It is well established that plants can grow and develop satisfactorily in a water-based nutrient solution containing only the essential inorganic nutrients. No organic compound is known to be essential for plant growth. In the presence of sunlight, water, and the essential nutrients, plants can synthesize all the organic compounds necessary for life. This synthesis, which is the basis of all life in the world, can even take place in asceptic conditions where all micro-organisms have been excluded. Thus it is indicated that organic substances or manures *per se* are not essential for plant growth.

It is not surprising that the nutrients in animal wastes are available to plants when it is considered that they were recently taken up from the soil by the plants that made up the animal feed. Portions of these nutrients are present as simple inorganic compounds that are readily available for uptake by plants. The remainder is present in organic molecules, and must usually be mineralized by micro-organisms in the soil before they become available for plant uptake.

We will now look in more detail at the fertilizer efficiency of nutrients in manure.

Nitrogen

The nitrogen availability from wastes is complex and it is influenced by several factors, as is evident from the wide range of nitrogen efficiency from manures reported in the literature.

Efficiency levels of nitrogen ranging from over 100% as effective as chemical fertilizer to zero efficiency and even negative responses have been reported in the literature. The efficiency of nitrogen in manure is complicated by the fact that only part of the total nitrogen is available in the first year after application, and there may be a residual effect over a number of years (Lande Cremer, 1976). It has been estimated that about 50% of organic matter in manures will be mineralized in the first year after application (Sluijsmans and Kolenbrander, 1976).

A recent report from England (Anon, 1976) states "approximately half to three-quarters of the nitrogen from both pig and cattle slurry is available for plant growth in the year of application; the corresponding figure for poultry excreta is approximately four-fifths." The report stated that manure at 34 tonnes/ha caused no damage to grass sward, and when applied in late winter or early spring, the nitrogen was about 50% effective. When applied in mid-winter, only 25% of the nitrogen was effective, and in early winter, only 16%. A dressing of 68 tonnes/ha did not increase yield over the 34 tonnes/ha application.

It has been estimated that 50 to 60% of applied fertilizer nitrogen is usually recovered in the crop. Another 10 to 20% may volatilize or be denitrified, and the remaining nitrogen may be leached down the soil profile out of reach of plant roots, or be washed off the soil surface as runoff.

The proportion of nitrogen utilized by crops is influenced by many factors. Usually the higher the rate of nitrogen applied, the lower the percentage uptake by the plant. A study of grass response to slurry and fertilizer nitrogen showed that increased rates of cattle slurry gave very little increase in yield (Tunney, 1977b). The grass yield response to fertilizer and slurry nitrogen is summarized in Fig. 1.7. There was a very low nitrogen efficiency at the high rates of cattle slurry. The results in Fig. 1.7 also show the higher yield response obtained with pig slurry relative to cattle slurry. All slurry treatments improved yield over the no slurry treatments.

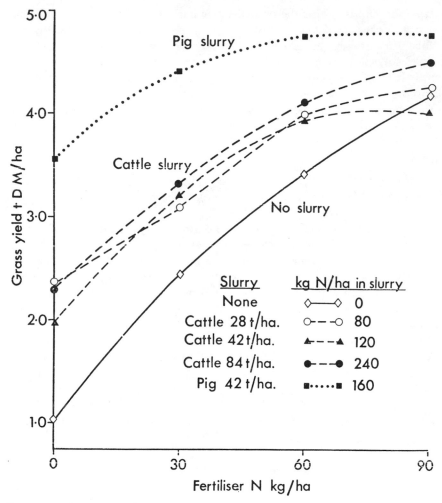

Fig. 1.7. Response of grass yield to fertilizer and slurry nitrogen. *Source*: Tunney. 1977b.

Results of experience in America (Viets, 1974) suggest that wastes with a carbon to nitrogen ratio (C:N ratio) of about 25 (i.e., 37.5% C and 1.5% N in waste dry matter) will not contribute nitrogen for increased crop production in the year of application. It may—as in the case of straw—depress the yield, if extra nitrogen is not added to meet microbial demands for decomposition of the extra carbon. For a waste to release significant amounts of nitrogen, it should have 2% nitrogen or greater on a dry matter basis. Viets' study suggests that organic waste might be expected to release for a subsequent crop in available form about 1 kg N per tonne of manure dry matter for each 0.1% N over 2.0%. Thus, one tonne of manure dry matter with 2.5% N would be expected to yield 5 kg of available nitrogen. To take another example, 12.5 tonnes of an 8% dry matter cattle slurry (1 tonne dry matter) containing 4.8% nitrogen on a dry matter basis would have 28 kg available nitrogen or a little over 50% of the total nitrogen. In other words, the 12.5 tonnes of slurry would be expected to have a nitrogen benefit equivalent to about 100 kg of calcium ammonium nitrate (27.5% N) fertilizer. This clearly illustrates the high volumes and low nutrient concentrations of wastes in comparison with artificial fertilizers.

Work in Ireland suggests that two-thirds of the nitrogen in pig slurry is available on grassland in the year of application (McAllister, 1976). This work also indicates that the soluble nitrogen in slurry is comparable with fertilizer nitrogen. Soluble nitrogen levels are similar to ammonia N content and this accounts for about 50% of total N in anaerobically stored cattle or pig manure.

Experience in Denmark suggests that 40% of the N in animal manures is available (Kofoed, 1976), while it is also indicated that the mean carbon contents for cattle, pig, and poultry manures are 34.4%, 33.8%, and 30.1%, respectively. The mean C:N ratio for cattle, pig, and poultry manures were 6.1, 4.3, and 6.4 respectively. The maximum carbon content was 40% and the maximum C:N ratio was 9.0 from over 60 samples.

Tietjen (1976), reporting on German work, outlines the importance of manure composition on nitrogen availability. He indicates that urine nitrogen is as effective as fertilizer nitrogen and that farmyard manure nitrogen can be from 0 to 30% effective. He estimates cattle slurry or full-gulle (i.e., a mixture of feces and urine) nitrogen to be 50% as effective as fertilizer nitrogen. A recent publication from Norway indicates that nitrogen in cattle slurry mixed with an equal volume of water is 60 to 80% as effective as fertilizer nitrogen on grassland, (Naess and Myhr, 1975). Chumley (1975), in England, suggests that 60% of the nitrogen in animal manures is available in the year of application.

Experiments on sweet corn in the Coachella Valley of California by Tyler *et al* (1964) determined that nitrogen in farmyard manure was only 20% as effective as fertilizer nitrogen. Another American report suggests that poultry manure nitrogen is 50% as effective as fertilizer nitrogen.

Most of these estimates of nitrogen efficiency refer only to the year of application. However, work in California (Pratt *et al*, 1973) attempted to quantify the residual effect of manures, and decay series have been developed to show the proportion of nitrogen applied in manure that can become available in years subsequent to application. For example, for fresh cattle manure with 3.5% N, the decay series is 0.75, 0.15, 0.10, 0.05. This can be interpreted as follows: 75% of the nitrogen is effective in first year of application, 15% of the remaining nitrogen is effective in the second year, and so on. This means that the apparent effectiveness of manure could be higher in the year of application on land that had been receiving appreciable quantities of manure for a number of years previously.

There is a very large reserve of nitrogen in soil organic matter, and this is the main natural source of nitrogen. There can be up to 10 tonnes of nitrogen per hectare in the top 20 cm of soil (Flaig *et al*, 1978), but only a small proportion of this total is available for plant uptake in any one year. Most of the nitrogen is present in stable humic substances which are very slowly mobilized by micro-organisms. Broadbent (1968) discusses the immobilization and mineralization of soil nitrogen and there is a dynamic equilibrium between organic and inorganic soil nitrogen that controls the proportion available for plant growth. Scarsbrook (1965) reports that 2 to 4% of soil organic matter is mineralized per year.

In practice, the efficiency of nitrogen in manure is influenced by many factors, such as:

- Loss of nitrogen by volatilization or by leaching
- Fixation of nitrogen by legumes and soil micro-organisms
- Release of nitrogen from soil organic matter
- The position on the nitrogen response curve where efficiency is being studied
- Time of year
- Crop being studied
- Rate of application.

Losses of Nitrogen.   Many of the efficiency levels for nitrogen discussed above were recorded for tillage crops. Work on the use of cattle slurry on grassland shows that the yield response can be very low (see Fig. 1.7).

Ammonia loss by volatilization after spreading slurry may account for a significant proportion of the total nitrogen present. One experiment indicates that 50 to 75% of the ammonia nitrogen in slurry may be lost within one week of application (Sherwood, 1968). In Fig. 1.8, the nature of ammonia loss after pig slurry application under Irish conditions is shown. Most of the reduction in ammonia shown in Fig. 1.8 can be explained by volatilization, and results suggest that only a relatively small proportion is taken up by the plant, converted to nitrate in the soil, or lost by denitrification. These results are in good agreement with work in Denmark (Iversen, 1934), where about 90% of the ammonia was volatilized within four days of spreading; however, when the manure was ploughed in immediately after spreading, the loss was small. Vanderholm (1975) estimates that from 30 to 65% of nitrogen from slurry can be lost during storage and shortly after spreading. These losses include surface runoff as well as ammonia volatilization.

Recent developments in soil injection of slurry on both tillage and grassland hold the promise of reducing or even eliminating ammonia loss after slurry application. Other approaches to reduce ammonia loss, such as the addition of acid to reduce slurry pH, may be worth further study. Ammonia volatilization is

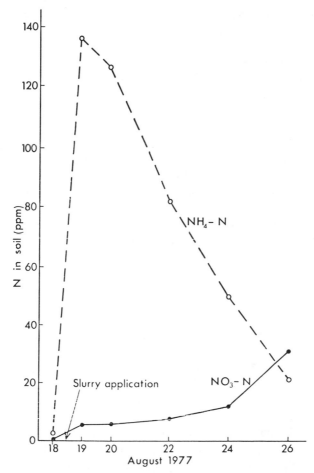

Fig. 1.8. Loss of ammonia after slurry application. *Source:* **Sherwood. 1968.**

likely to be greatest under conditions where the slurry dries out on the grass surface after application. There is often a greater yield response to slurry when it is diluted with water or washed off the grass surface shortly after application. When the slurry is washed down into the soil, the ammonia becomes attached to the soil exchange complex and will not be volatilized to the atmosphere.

High rates of slurry can reduce crop yield. Cattle slurry at 65 tonnes/ha in two applications gave higher grass yields than 110 tonnes/ha (Tunney, 1975). Results such as these suggest that in addition to the positive effect on grass yield of the nitrogen in slurry, there may also be a negative effect, particularly at high rates. The negative effect may be due, at least in part, to physical effects of slurry on the grass surface. There is also some evidence that cattle slurry may contain chemicals that are phytotoxic and therefore reduce crop yield. Negative effects of slurry would, of course, further complicate estimates of nitrogen efficiency.

Volatilization of ammonia from agricultural wastes accounts for a large loss of valuable nitrogen from agriculture. It is estimated that half the nitrogen in manures in the United States is lost by volatilization of ammonia (Porter, 1975). This is estimated at about 4 million tonnes of nitrogen, or half the amount of nitrogen fertilizer used annually in American agriculture.

Bremner (1977) states that denitrification and loss of nitrogen are aided by the presence of readily decomposable organic matter and anaerobic conditions. These conditions favor the volatilization of N_2O and N_2 gases from the soil, and the relative proportions are influenced by the type of easily decomposable organic matter. The volatilization of nitrogen from urea fertilizer is also aided by the presence in the soil of easily decomposable organic matter, and the activity of urease in the soil is prolonged by the presence of organic matter. Though in theory it seems possible, in practice there is little or no information indicating that slurry contributes to the loss of nitrogen from soil.

Phosphorus, Potassium, and other Nutrients

Some authors suggest that phorphorus in wastes is 50% as effective as fertilizer phosphorus (Anon, 1976). However, most research indicates that manure phosphorus is equally as effective as fertilizer phosphorus (May and Martin, 1966; Kolenbrander and Lande Cremer, 1967; Vetter, 1973).

On soils with adequate phosphorus fertility, where maintenance dressings only are required, the manure phosphorus can probably be regarded as 100% effective. On phosphorus deficient soils, it can be argued that only the inorganic phosphorus and the portion of organic phosphorus that will be mineralized will become available in the first year. Some results, however, indicate that even on deficient soils manure is as effective a source of phosphorus as is an inorganic fertilizer.

It is generally accepted in the literature that potassium in agricultural wastes is as effective as fertilizer potassium. This is understandable, considering that all potassium in wastes is soluble and is available without mineralization of the organic matter.

Other plant nutrients in wastes, which do not normally limit plant growth under farming conditions, become part of the soil reserves, and can be regarded as an effective source for plants. Magnesium in manure, for example, is regarded to be 100% as effective as fertilizer magnesium (Kolenbrander and Lande Cremer, 1967).

Organic wastes may help to mobilize nutrients such as copper and other heavy metals in soil and increase their availability to plants (Haan et al, 1976). This may be due to the chelating power of some organic molecules in wastes. It is interesting to note that copper in pig slurry is more readily available to plants than is copper from inorganic salts (McGrath, 1977).

Organic Matter

It is well established that organic wastes, in addition to supplying essential nutrients, can have beneficial effects on soil physical properties (Azevedo and Stout, 1974). Under most soil conditions, the physical benefits are probably secondary to those of the nutrients present. However, under certain soil conditions, manures can improve soil structure, with the results of an improved soil environment for root growth and nutrient uptake (Flaig et al, 1978). The value of organic wastes for improving soil structure and water holding capacity is often cited. When attempting to evaluate the fertilizer value of wastes in money terms, it is necessary to base the value on the available nutrients in the manure, and it is, of course, more difficult to put a value on physical benefits that may be of value in some soils and not in others. It is sometimes stated that organic wastes have a beneficial effect on soil by building up soil organic matter. However, most scientific evidence indicates that normal levels of organic waste application have minimal effects on soil organic matter content (Broadbent, 1947; Russell, 1961).

UTILIZATION

Ideally, agricultural wastes should be recycled to supply nutrients for subsequent crop growth. But this is not always possible. For example, the food for pigs and poultry is usually transported over long distances, often from other countries. In this situation, the bulky animal wastes cannot be recycled back to the land that produced the crop. The land where the pigs were reared may be inadequate or unsuitable for spreading the manure produced. Manure can then become a disposal problem and be a potential source of pollution.

With cattle, the situation is usually better. The land on which the feed is produced is normally available for spreading manure. Forage crops make up a high proportion of cattle feed, which is usually fed on or near the farm where it is produced. There are, of course, some exceptions to this: for example, the feed lots in the United States where feed for fattening animals is transported long distances and the resulting manure is disposed of rather than utilized. On pasture, grazing animals recycle the waste as it is produced. With winter feed for cattle, there is often a simple recycling relationship where the manure can be spread on land to supply the nutrients for the next crop. Manure from cattle and other ruminants is the greatest single source of agricultural waste for most countries. Poultry and pig manure is usually small by comparison. For this reason, most of the agricultural wastes on a world scale can be recycled to the land from which the crop was produced.

When and Where to Spread

When and where to spread wastes are important practical questions for agriculture. Wastes cannot be spread at some stages of crop growth; it is usually not possible to spread wastes on mature crops near harvesting time. It is also true that some crops are more suitable for waste application than are others.

From a fertilizer point of view, spring application usually gives best results, supplying maximum benefit from the nitrogen in the waste. However, it is not always possible to spread all waste in spring. In pig production, for example, the manure is produced throughout the year, so adequate storage for twelve months would be necessary if all the manure was spread in spring. This would be very costly. In practice, it is customary that manure produced during winter be spread spring and summer, and that produced in autumn be spread during late autumn. In some situations, where soil conditions are suitable, manure is sometimes spread throughout the year so that very little storage is required.

Grassland is perhaps the most versatile crop, in that waste can be spread for most of the year. For most tillage crops it is normally not practical to spread waste during most of the growing season. The pattern of agriculture will therefore determine when and where wastes can be spread.

Traditionally, most waste was spread on tilled land. With intensive agriculture, in areas like the British Isles—where most of the land is under grass—a high proportion of the waste is spread on grassland. In countries such as Denmark, where most of the land is tilled, a high proportion of the waste is spread for tillage crops.

On grassland, wastes are normally spread on the grass surface. On tilled land, the wastes are either ploughed or tilled into the soil after spreading. In this way a higher proportion of the nitrogen can be retained for crop production. Spreading wastes on a grass surface can cause problems by reducing palatability of the grass and increasing the danger of spread of disease to grazing animals. Soil conditions will also influence time of spreading. Time of application of waste is often determined by when soil conditions are dry enough to carry manure spreading equipment. Under some soil and climatic conditions, there may be six months of the year when manure spreading equipment could cause serious damage to the soil surface. Under other conditions, it is possible to spread manure throughout the year without serious soil damage. Applying slurry to grassland in winter, when the soil is wet, can cause more damage to grass than the same quantity applied in spring or summer. It appears that slurry can deplete the already low oxygen levels in the soil near the grass roots, killing the better yielding grasses, which will be replaced by docks and other weeds. It is therefore advisable that if slurry has to be spread on grassland in winter, low rates (say not more than 20 tonnes/ha) should be used.

Rates of Application

The quantities of plant nutrients required for different crop and soil conditions are reasonably well established for most countries. Where soil fertility is adequate, maintenance dressings of phosphorus and potassium are applied annually. Maintenance dressings are rates that will maintain soil fertility over the years and should equal the difference between losses and the nutrient supplying power of the soil. The main nutrient losses are crop removal and loss in drainage water. Nitrogen is more complex because there are many sources of losses and inputs and it is the primary nutrient influencing yield in conditions of high soil fertility. With low soil fertility, extra rates of nutrients, above maintenance dressings, may have to be applied initially for maximum crop yield. Losses of phosphorus and potassium in water from agricultural land are small in comparison with crop uptake and are usually in the region of a few kilograms per hectare per annum (Kolenbrander, 1973). Removal by the crop is related to crop yield and composition. Intensive grass production for winter food could produce dry matter in the region of 15 tonnes/ha/annum. This crop could remove over 400 kg N, 50 kg P, and 400 kg K/ha/annum. The maintenance dressing for this crop would vary with soils and other conditions but would probably be in the region of 50 kg P and 200 kg K/ha/annum for Irish conditions. Most agricultural soils can supply large quantities of potassium annually from soil minerals. From 50 to 100 kg K/ha can normally be supplied annually under Irish conditions. Rates of nitrogen fertilization for agricultural land are normally less than 300 kg N/ha/annum. Intensive grassland has the highest nitrogen requirement. Cereals usually require in the region of 100 kg N/ha/annum for maximum production. Leguminous crops that fix nitrogen from the air can give good yields without fertilizer nitrogen.

To get maximum benefit from the nutrients in agricultural wastes, they should be applied at rates to supply adequate nutrients for crop needs. Rates of application greatly in excess of crop needs are undesirable since that means nutrients are not being used effectively and the risk of soil and water pollution is increased. In the past, when manures were valued for their plant nutrients, the normal rate of application was between 25 and 50 tonnes/ha/annum or 5 to 10 tonnes of manure dry matter. Now manures are often applied in much higher quantities of disposal rates of up to ten times the traditional rates. However, such rates could not be recommended either from an agricultural or environmental viewpoint. Approximately 100 tonnes of 4% dry matter slurry would be required to supply the same quantity of nutrients as 50 tonnes 8% dry matter slurry. At 100 tonnes of cattle slurry per hectare, based on composition shown in Table 1.3, 380 kg N, 60 kg P, and 420 kg K/ha would be applied. It is clear that the potassium level at this rate is over double the maintenance requirement of most agricultural crops, while the nitrogen and phosphorus rates are high but not excessive. In fact, if only 50% of the nitrogen is available, supplementary fertilizer would be necessary for maximum yield of many crops. If 100 tonnes/ha of pig slurry (see Table 1.3) were used, then 300 kg N, 90 kg P, and 150 kg K would be applied. In this case, the phosphorus level is in excess of the needs of most agricultural crops but the nitrogen and potassium levels are not excessive. The 100 tonnes of cattle slurry would supply 8 tonnes dry matter and the pig slurry would supply 4 tonnes dry matter per hectare based on dry matter shown in Table 1.3.

From the above values, it is evident that the amounts of nitrogen, phosphorus, and potassium in manure are not in the correct proportion for many crops. There is generally too much potassium in cattle slurry and too much phosphorus in pig slurry relative to the other nutrients. To maximize the value of nutrients in manure, cattle slurry should be applied at rates to give adequate potassium for crop needs and be supplemented with fertilizer nitrogen and phosphorus. Pig slurry, on the other hand, should be applied at a rate to supply adequate phosphorus and be supplemented with fertilizer nitrogen and potassium. The same approach can be used to estimate the application rate of other wastes so as to maximize the fertilizer benefit of the nutrients present.

Where more wastes are available than are required for the crop nutient needs, most crops will tolerate higher application rates. For example, cattle slurry could be applied at a rate to supply adequate phosphorus for the crop, and the excess potassium may not cause serious problems. It may be possible to get maximum yield when relying on nitrogen from cattle slurry alone, as high rates of application may depress yield, as discussed earlier in this chapter.

For Irish conditions, 15 kg P and 25 kg K are recommended per hectare per annum as a maintenance dressing for intensive pastures. This would be supplied by 15 to 25 tonnes pig slurry per hectare. So pig slurry can be regarded as a balanced source of P and K for pastures. For grassland conserved for silage, the requirements are much higher and suggested rates of slurry and fertilizer are summarized in Table 1.4. It can be seen from this table that all the potassium is supplied by cattle slurry and all the phosphorus by pig slurry. The best combination of waste and fertilizer can be calculated for any crop or any waste based on the requirements of the crop and the nutrient content of the waste. Table 1.4 also reflects the higher nitrogen availability obtained with pig slurry relative to cattle slurry.

A dressing of 40 to 50 tonnes/ha of good slurry is suitable for most root crops. As for silage, cattle slurry normally needs to be supplemented with phosphorus and pig slurry with potassium fertilizer. Nitrogen fertilizer may also be necessary, and the rate will depend on crop requirements. Care is needed

TABLE 1.4.   Recommendations of Slurry and Fertilizer Per Hectare for Grass Silage Production, One Cut System.

	Spring	Autumn
Cattle Slurry	40 tonnes + 60 kg N*	10 kg P*
Pig Slurry	40 tonnes + 30 kg N*	60 kg K*

*Nutrients supplied as artificial fertilizer supplement.

when using waste for cereals because of their sensitivity to excess nitrogen and the difficulty of applying accurate rates of nitrogen due to the variation in waste composition.

Timing of Application

To get a good response from waste application, good management practices should be followed. No more than 50 tonnes/ha should be applied in any one application, and there should be at least 30 days between applications. Lower application rates would be advisable for wet soil conditions or when applying in the wintertime on grassland. Heavy rates of wastes, particularly slurry, can block soil pores (Stevens and Cornforth, 1974). This can lead to anaerobic conditions and reduced infiltration capacity, leading in turn, to increased surface runoff during rainfall. The blocked soil pores will usually be cleared within 30 days by micro-organisms breaking down the organic matter. However, if waste is applied at too frequent intervals, soil micro-organisms will not be able to clear the blocked soil pores.

On grassland, wastes should be applied when the grass is short, either after cutting or after close grazing. Waste application can knock down grass when it is at an advanced stage of growth. Recovery may be slow and the yield may be reduced as a result. Furthermore, the grass is likely to remain contaminated with the waste for a longer period of time. Silage should not be cut for at least six weeks after waste application to reduce the risk of contamination and spoilage of the silage. Figure 1.9 shows the difference in appearance between farmyard manure and cattle slurry applied to grassland. Slurry spreaders distribute slurry more finely, and much of the slurry adheres to the grass surface; farmyard manure spread with muck spreaders falls in small pieces and is not atomized like slurry.

Cattle or pig slurry at rates of 45 tonnes/ha applied six weeks before cutting did not reduce silage quality or animal performance (Tunney, 1976a). In order to reduce surface contamination and disease risks, it is advisable that wastes should be applied at least four weeks before pasture is grazed.

Fig. 1.9. Appearance of farmyard manure (foreground) and cattle slurry (top center) on grass following application.

Effects on Composition of Soil and Crop

Wastes can have considerable effects on the composition of soils and crops. Soil analyses on experimental plots indicated that land receiving cattle slurry had high soil potassium levels and land receiving pig slurry had high levels of phosphorus (Tunney, 1975). Effects of fertilizer, cattle slurry, and pig slurry on soil phosphorus and potassium levels is illustrated in Table 1.5. In Tunney's work, slurry was applied in two dressings in both 1973 and 1974. It can be seen from Table 1.5 that pig slurry had a large effect on soil phosphorus levels and soils using pig or poultry manure usually had high phosphorus levels. Recent unpublished work at Johnstown Castle Research Centre indicates that phosphorus in pig slurry gives higher available soil phosphorus levels than does the equivalent amount of fertilizer phosphorus. Soil receiving pig slurry shows a small but significant increase in pH and a decrease in available manganese (Tunney, 1977a). Regular soil analyses on land receiving wastes is advisable to ensure that correct level and balance of nutrients is maintained.

Grass on soil treated with pig slurry had higher calcium and magnesium levels and lower manganese levels than on artificial fertilizer treatments (Tunney, 1977a). The manganese content of grass from cattle slurry treated plots was significantly lower than with fertilizer treatments (Tunney, 1976b). This suggests that the risk of grass tetany may be higher on pasture receiving cattle slurry. The effect of slurry on magnesium content of grass is summarized in Table 1.6.

Copper sulfate is added to the diet of fattening pigs and the slurry may contain several hundred parts per million on a dry matter basis. Crops receiving pig slurry can contain increased levels of copper. There can also be a higher total removal of phosphorus from pig slurry than from fertilizer or cattle treatments, as shown in Fig. 1.10 (Tunney, 1975). The higher phorphorus removal is probably due to the higher soil phosphorus levels on pig slurry treatments (see Table

TABLE 1.5. Effects of Slurry and Fertilizer on "Available" Soil Phosphorus and Potassium with Morgan's Extractant. Adapted from Tunney (1975).

		Soil P ppm		Soil K ppm	
		Feb. 1973*	Oct. 1974*	Feb. 1973*	Oct. 1974*
Fertilizer	Low	5.5	8.0	135	103
	High		6.4		72
Pig slurry	70 tonnes/ha	5.8	18.4	155	106
	112 tonnes/ha		34.4		84
Cattle slurry	70 tonnes/ha	5.5	11.8	162	174
	112 tonnes/ha		9.8		176

*Date sampled.

TABLE 1.6. Effects of Slurry on Percent Magnesium in Grass Dry Matter. Adapted from Tunney (1976b).

| | 1974 | | | 1975 | | |
	Cut 1	Cut 2	Cut 3	Cut 1	Cut 2	Cut 3
Cattle slurry	.19	.19	.18	.18	.19	.22
Pig slurry	.25	.28	.33	.25	.26	.29
Fertilizer	.22	.24	.28	.25	.23	.25

1.5). There is need for more information on the long-term effects of waste application on the composition of soils and crops.

Other Uses

Agricultural wastes can have benefits and uses in addition to the fertilizer nutrients present. For example, cattle slurry has been used as a medium for seeding

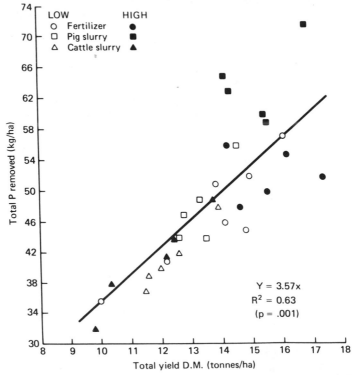

Fig. 1.10. Relationship between grass yield and phosphorus removed in crop with fertilizer and slurry treatments.

roadside banks. The grass seed is mixed with the slurry and then sprayed on the roadside banks. This method has the advantage that the seeds can be spread on areas that are inaccessible to machinery. In addition, the slurry provides a micro-environment that can facilitate seedling growth in otherwise inhospitable conditions. It is important to ensure that the ammonia content of the slurry is not so high that it can inhibit seed germination. Wastes can also be used to provide a micro-environment for seedling development in reclamation of land; it can be particularly valuable in the rehabilitation of mine heaps and other industrially disturbed soils. In these situations, the wastes are often superior to inorganic fertilizers as they have the dual benefits of improving the physical environment of the plant root and also supplying essential nutrients.

POLLUTION

The pollution risks of agricultural wastes and wastes in general have received considerable attention in recent years. The effects of nutrient enrichment from wastes on eutrophication of lakes has been given widespread attention. Phosphorus level is usually the limiting factor for aquatic plant growth (Kolenbrander, 1971), and nitrogen is probably the most important nutrient, after phosphorus, contributing to eutrophication.

In addition to nutrient enrichment, wastes can cause water pollution by depleting the oxygen in water. Animal manures have a high biochemical oxygen demand (B.O.D.), and relatively small quantities can deoxygenate large quantites of water (Loehr, 1974). Fish kills attributable to wastes usually result from oxygen depletion of water. If the ammonia in wastes enters the water, it makes fish life more susceptible to low oxygen levels.

There are risks of high nitrate levels in ground water if excessive levels of waste are applied to land (Azevedo and Stout, 1974). The risk of nitrate toxicity in animals is greater on land receiving high rates of animal manures (Hartmans, 1975). On a wider scale, it has been stated by Lande Cremer (1976) that excessive nitrogen, due to high rates of manure application, may be an important factor in limiting the yield and quality of crops. Water pollution by surface runoff from land treated with wastes is likely to be more of a problem in humid than in arid or semi-arid regions (Viets, 1971).

There is some danger of soil pollution from copper and other heavy metals in agricultural wastes. Copper is often added to pig feed and there may be several hundred parts per million copper in the manure (Haan *et al*, 1976). Sheep are particularly susceptible to toxicity on land treated with high rates of pig slurry rich in copper.

The odor and general nuisance problems of agricultural wastes may present many problems (Azevedo and Stout, 1974). These problems can be particularly acute where intensive livestock rearing is practiced in areas of high population density. There are chemicals on the market for controlling smells from manure but they are not very effective. Injecting slurry into the soil is perhaps the most effective way of reducing smell.

Slurry gases, particularly hydrogen sufide, are toxic and many fatal accidents involving both man and animal have been recorded. The danger is greatest when slurry is being agitated in enclosed buildings. Flies and dust are problems often associated with animal wastes. The dust problem is greatest in arid and semi-arid regions.

The disease hazards associated with animal wastes are well documented (Strauch, 1976). Spreading manures on to pasture land poses definite risks of spreading animal diseases. Microbial diseases and pathogenic worms can be spread in animal manures.

Dead earth worms are sometimes seen on the soil surface after spreading slurry. However, only a small proportion of earth worms are normally killed, and agricultural wastes usually increase the earthworm population of a soil by increasing their food supply.

Rainwater and washwater contaminated with milk, manure, and other wastes is a serious potential source of pollution if not properly handled. Dirty water of this type has a low fertilizer value and is normally of relatively low concentration and high volume. These wastes should be reduced by ensuring that only the minimum quantities of rainwater and washwater are contaminated by manure and other wastes. The dirty water can be safely disposed of by collecting and irrigating on land. One hectare of land should be adequate to receive the dirty water from up to 50 dairy cows.

Despite the potential pollution risks, there is considerable evidence (Azevedo and Stout, 1974) to suggest that if livestock wastes are properly managed, the danger of environmental pollution is minimal. If spread on soil at times and rates to supply the nutrient requirements of crops, these wastes can be a valuable asset. On the other hand, if managed carelessly and spread at heavy rates, without-regard for crop nutrient needs, they can be responsible for serious environmental pollution.

CONCLUSIONS

Agricultural wastes, if properly managed and used, represent a valuable natural world resource for maintaining soil fertility.

The 1976 world population for cattle, pigs, and poultry was estimated at 1,200 million, 655 million, and 6,116 million, respectively (F.A.O., 1977b). The annual manure production from all these animals would contain in the region of 54, 12, and 52 million tonnes of nitrogen, phosphorus, and potassium, respectively. This compares with the 1976 estimated world fertilizer production of 44, 11, and 20 million tonnes of nitrogen, phosphorus, and potassium, respectively

(F.A.O., 1977a). If the nutrients in other agricultural wastes are added, it can be seen that the overall potential fertilizer value is very large. Much of this potential value is not being realized at present.

There is little doubt that efficient utilization of the nutrients in agricultural wastes could make a valuable contribution to increasing world food production, particularly in developing countries. But there is a need for more information and advice on the use of wastes in agriculture to ensure that maximum benefit is obtained from the nutrients present.

REFERENCES

Anon. 1965. *Standard Methods for Examination of Water and Waste Water.* 12th ed., Amer. Pub. Health Ass., New York.

Anon. 1972. *Ceres-FAO Rev.* 5:13.

Anon. 1976. *Studies on Farm Livestock Wastes.* Agricultural Research Council, London.

Azevedo, J. and P. R. Stout. 1974. *Farm Animal Manures.* Univ. Calif. Agric. Ext. Serv., Manual **14.**

Barreveld, W. H. 1977. "World demand for animal products for human food, 1970–2000." *Animal Wastes.* E. P. Taiganides (Ed.). Applied Science Publishers, London, pp. 23–40.

Berryman, C. 1970. *Composition of Organic Manures and Waste Products Used in Agriculture.* N.A.A.S. Advisory Paper No. 2, Ministry of Agriculture, Fisheries and Food, London.

Bremner, J. M. 1977. "Role of organic matter in volatilisation of sulphur and nitrogen from soils." *Soil Organic Matter Studies.* 2:229–240. Int. Atomic Energy Agency, Vienna.

Broadbent, F. E. 1974. "Nitrogen release and carbon loss from soil organic matter during decomposition of added plant residues." *Proc. Soil Soc. Amer.* 12:246–249.

Broadbent, F. E. 1968. "Turnover of nitrogen in soil organic matter." *Pontificial Acad. Scient. Scripta Varia* 32:61–84.

Christensen, J. 1976. "Waste management systems in animal husbandry." *Utilisation of Manure by Land Spreading.* E.E.C. Luxembourg, pp. 507–520.

Chumley, C. 1975. *Profitable Utilisation of Livestock Manures.* Advisory Leaflet **171.** Ministry of Agriculture, Fisheries and Food, England.

Cothenie, A. and F. Van de Maeele. 1976. "Soil-water-plant relationships as influenced by intensive use of effluents from livestock." *Utilisation of Manure by Land Spreading.* E.E.C. Luxembourg, pp. 225–246.

F.A.O. 1977a. *Annual Fertiliser Review 1976*. Food and Agricultural Organization of the United Nations, Rome.

F.A.O. 1977b. *Production Year Book 1976*. Food and Agricultural Organization of the United Nations, Rome.

Flaig, W., B. Nagar, H. Sochtig, and C. Tietjen. 1978. *Organic Materials and Soil Productivity*. F.A.O. Soils Bulletin No. **35**.

Gisinger, L. 1950. "Organic manuring of grassland." *J. Br. Grassld. Soc.* **5**:63–79.

Gisinger, L. 1960. "New concepts on the value and use of gulle." *Schweiz. landw. mh.* **38**:433–450.

Haan, F.A.M. de, T. M. Lexmond, and F. Dijman. 1976. "Aspects of copper accumulation in soil following hog manure application." *Utilisation of Manure by Land Spreading*. E.E.C. Luxembourg, pp. 289–298.

Harrison, P. 1973. "Guard against silage run-off." *Farm Building Digest* **8**(1):6–7.

Hartmans, J. 1975. "The effect of intensification on crop composition and on health and production of livestock." *Stikstof* **18**:12–21.

Hendrick, J. 1915. *The Composition and Value of Liquid Manures*. Bulletin No. **19**, North of Scotland Agricultural College, Aberdeen, Publ. Milne and Hutchinson.

Hoibs, H. 1960. "Concrete slats for cattle pens." *Institute of Building Construction. Proc.* **40**. The Agricultural University of Norway, Vollebekk.

Iversen, K. 1934. "Loss of nitrogen by evaporation under application of liquid manure." Fordampningstabet Ved Ajlens Undbringning 1928–33. *Tidsskr. Pl. Avl.* **40**:199–202.

Johnston, A. E. 1969. *The Value of Residues from Long Period Manuring at Rothamsted*. Rothamsted Exp. Sta. Report, Part 2.

Kofoed, A. Dam. 1976. "Farmyard manure and crop production in Denmark." *Utilisation of Manure by Land Spreading*. E.E.C. Luxembourg, pp. 29–44.

Kofoed, A. Dam and O. Nemming. 1976. (Askov, 1894.) "Fertilisers and manures on sandy and loamy soils." *International Conference on Very Long-Term Fertiliser Experiments*. Paris, Grignon.

Kolenbrander, G. J. and L. C. N. de la Lande Cremer. 1967. *Stalmest en Gier Waarde en Mogelijkheden*, H. Veenman and Son, Wageningen, Netherlands.

Kolenbrander, G. J. 1971. *Contribution of Agriculture to Eutrophication of Surface Waters with Nitrogen and Phosphorus in the Netherlands*. Institute of Soil Fertility, Haren-Gr., Netherlands, Report No. **10**.

Kolenbrander, G. J. 1973. "Fertiliser practice and water quality." *Proc. Fertil. Soc. London* **135**:1–36.

Lande Cremer, L. C. N. de la. 1976. "Effect of rate of application of organic and inorganic nitrogen on crop production and quality." *Utilisation of Manure by Land Spreading*. E.E.C. Luxembourg, pp. 73–86.

Loehr, R. C. 1974. *Agricultural Waste Management*. Academic Press, New York and London.

May, D. M. and W. E. Martin. 1966. "Manures are a good source of phosphorus." *Calif. Agric*. **20(7)**:11–12.

McAllister, J. S. V. 1976. "Efficient recycling of nutrients." *Utilisation of Manure by Land Spreading*. E.E.C. Luxembourg, pp. 87–104.

McGrath, D. 1977. Personal communication. Agricultural Institute, Johnstown Castle Research Centre, Wexford, Ireland.

Mechi, A. 1859. *How to Farm Profitably*. Routledge, Werne and Routledge, London and New York.

Miner, J. R. and J. Smith. 1975. *Livestock Waste Management with Pollution Control*. MWPS, 10 North Central Regional Research Publication **222**, Iowa State University, Ames, U.S.A.

Naess, O. and K. Myhr. 1975. *Slurry Manuring for Ley in West Norway*. State Agric. Exp. St. Fureneset, Report No. **30**:145–159.

O'Callaghan, J. R., V. A. Dodd, J. O'Donoghue, and K. A. Pollock. 1971. Characterisation of waste treatment properties of pig manure." *J. Agric. Eng. Res.* **16**:399–419.

Porter, K. S. 1975. *Nitrogen and Phosphorus*. Ann Arbor Science, Michigan.

Pratt, P. F., F. E. Broadbent, and P. J. Martin. 1973. *Using Organic Wastes as Nitrogen Fertilizers*. Univ. Calif. Agric. Ext. Serv., Manual **44**.

Robertson, A. M. 1977. *Farm Wastes Handbook*. Scottish Farm Buildings Investigation Unit, Aberdeen.

Russell, E. W. 1961. *Soil Conditions and Plant Growth*. 9th ed., Longmans.

Sailsbury, F. B. and C. Ross. 1969. *Plant Physiology*. Wadsworth Publ. Co., Belmont, Calif.

Scarsbrook, C. E. 1965. "Nitrogen availability." *Soil Nitrogen*. Agron Monograph No. **10**, Amer. Soc. Agron, Madison, Wisconsin, pp. 486–507.

S.F.B.I.U. 1969. *Farm Building Progress* **18**. Scottish Farm Buildings Investigation Unit, Aberdeen.

Sherwood, M. 1968. Personal communication. Agricultural Institute, Johnstown Castle Research Centre, Wexford, Ireland.

Sluijsmans, C. M. J. and G. J. Kolenbrander. 1976. "The nitrogen efficiency of manure in short and long term." *Stikstof* **7(83/84)**:349–354.

Stevens, R. J. and I. S. Cornforth. 1974. "The effect of pig slurry applied to a soil surface on the composition of the soil atmosphere." *J. Sci. Fd. Agric.* **25**:1263–1272.

Stewart, T. A. and I. L. McCullough. 1974. "Silage effluent quantities produced composition and disposal." *Agric. N. Irl.*, Feb.: 368–372.

Stout, P. R. 1972. "Fertiliser, food production and environmental compromise." *Proc. Western Hemisphere Nutr. Congr.*, pp. 293–299.

Strauch, D. 1976. "Veterinary-hygienic aspects of land spreading and transport of animal manure." *Utilisation of Manure by Land Spreading*. E.E.C. Luxembourg, pp. 351–362.

Strauch, D., W. Baader, and C. Tietjen. 1977. *Abfalle aus der Tierhaltung*. Verlag Ulmer, Stuttgart.

Taiganides, E. P. 1977. *Animal Wastes*. Applied Science Publishers, London.

Thunegard, E. 1975. "On the persistence of bacteria in manure." *Act. Vet. Scand. Suppl.* **56**:1–86.

Tietjen, C. 1966. "Plant response to manure nutrients and processing of organic wastes." *Proc. Nat. Symp. Animal Waste Management.*, Publ. SP-**0366**:27–32, Amer. Soc. Agric. Eng.

Tietjen, C. 1976. "The yield efficient nitrogen portion in treated and untreated manure." *Utilisation of Manure by Land Spreading*. E.E.C. Luxembourg, pp. 129–142.

Tunney, H. 1975. "Fertiliser value of livestock wastes." *Managing Livestock Wastes, Proc. 3rd Int. Symp. Livestock Wastes*. ASAE, PROC. **275**:594–597.

Tunney, H. and S. M. Molloy. 1975. "Field test for estimating dry matter and fertiliser value of slurry." *Ir. J. Agric. Res.* **14**(1):84–86.

Tunney, H. 1976a. *Animal manures for grass production*. Dairy Herd Management. Handbook Series No. 4:98–101. An Foras Taluntais, Dublin, Ireland.

Tunney, H. 1976b. "Fertiliser value of animal manure." *Utilisation of Manure by Land Spreading*. E.E.C. Luxembourg, pp. 7–28.

Tunney, H. 1977a. *Fertiliser Value of Livestock Wastes*. Ph.D. thesis, plant physiology, The University of California, Berkeley.

Tunney, H. 1977b. *Grass Response to Slurry and Fertiliser Nitrogen*. Soils Research Report, 1976 (pp. 31–32) An Foras Taluntais, Dublin.

Tyler, K. B., A. F. Maren, O. A. Lorenz, and F. H. Takatori. 1964. "Sweet corn fertility experiments in the Coachella Valley." *Univ. Calif. Agric. Exp. Stat. Bul.* **808**:16 pp.

U.S.D.A., 1970. Agricultural Statistics–1970. U.S. Govt. Printing Office Washington, D.C.

Vanderholm, D. H. 1975. "Nutrient losses from livestock wastes during storage, treatment, and handling." *Managing Livestock Wastes, Proc. 3rd Int. Symp. Livestock Wastes.* ASAE, PROC. **275**:282–285.

Vetter, H. 1973. *Mist and Gulle.* DLG Verlag Frankfurt (Main).

Viets, F. G. 1971. "The mounting problem of cattle feed lot pollution." *Agric. Sci. Rev.* **9(1)**:1–8.

Viets, F. G., 1974. "Nitrogen transformation in organic wastes applied to land." *Factors Involved in Land Application of Agricultural and Municipal Wastes.* Publ. Agric. Res. Serv. U.S.D.A., pp. 51–65.

Virgil, P. M. c 79 A.D. *The Georgies.* Book 2, lines 346–348. *Great Books of the Western World*, No. **13**. Encyclopedia Brittanica (1954).

Voorhoeve, J. J. C. van. 1974. *Organic Fertiliser Problems and Potential for Developing Countries.* World Bank Fertiliser Study, Background Paper **4**, Washington, D.C.

2

Agricultural wastes as feedstuffs

E. Owen, Ph.D.
Department of Agriculture and Horticulture, University of Reading, Earley Gate, Reading, U.K.

WHAT CONSTITUTES WASTE AND FEEDSTUFF

It is difficult to provide comprehensive definitions for these terms. The word waste implies a useless material, and thus one which has to be disposed of, but this is not necessarily the case. A waste in one situation may not be one (or may not be one to the same extent) in another. For example, while cereal straw is a major agricultural waste by-product in the arable areas of eastern England, it is sought after as a winter feedstuff for ruminants in the western livestock producing areas, especially in years of poor climate for summer grass growth (and thus low conservation of winter feedstuff). Straw is a low density material and therefore is expensive to transport. This, together with its low nutritive value, limits its use. Thus, site of production, especially in relation to site of utilization, may influence the extent to which a material is a waste. Ease of storage is another factor; this is influenced considerably by moisture content. Country of production also has a bearing, especially the extent of development and intensification of agricultural practice.

In discussing feedstuff, it is necessary to differentiate between food for man directly and food for man indirectly (i.e., via animals). In the latter case, animal type must be specified: monogastric or ruminant. Use of agricultural wastes as food for man directly, even after suitable processing, can be largely dismissed at present, but this may not be the case in the future. Utilization of wastes as animal feedstuffs depends primarily on their content of available nutrients, especially in relation to alternative conventional feedstuffs and their cost. However, there are other factors, such as the cost of processing to increase nutrient availability, or—as in the case of recycling animal excreta as feedstuff—the cost of processing to eliminate the risk of disease transmission. Additional factors which may have a considerable effect on enhancing utilization in the future will be the need to minimize pollution, conserve energy, and maximize food production.

Returning to definitions, agricultural wastes will therefore be defined as agricultural by-products which are currently not used, little-used, or under-used as animal feedstuffs. This chapter will identify the wastes involved, indicate the magnitude of production, and outline current methods of disposal. The main objectives will be to consider their use or potential use as animal feedstuffs and to discuss methods of processing them to make them suitable feedstuffs.

AGRICULTURAL WASTES

Owen (1976) identified the following categories, based upon site of production and origin.

1. Crop wastes produced on the farm, generally in the field. These can be further sub-divided into high dry matter content materials, such as straws (cereal, legume, and grass) and shrub stems (e.g., cotton and soybean) and high moisture content materials, such as sugar beet and sugar cane leaves (tops) and horticultural by-products (carrot tops, pea haulms, and brassica stalks). It is debatable whether horticultural by-products and sugar beet leaves should be designated as wastes, since they are nutritious feedstuffs for ruminants and are fed. However, unless fed *in situ*, they are often not used as feedstuffs, being difficult to harvest and conserve.

2. Crop wastes produced in centralized processing plants. The principle examples are sugar cane bagasse and sisal leaf waste (pulp). There are other centrally produced by-products which cannot be considered wastes, as they are generally fed to livestock. These are materials such as sugar cane and sugar beet molasses, brewers grain, coffee and citrus pulp, and banana wastes. In certain circumstances, however, these materials are "wasted" because animal production enterprises are

either not yet developed or are too far away to allow their economic transport. Under-utilization of molasses in Mauritius is an example.

3. Animal wastes produced in large, intensive animal production enterprises. These are excreta and bedding from cattle, pigs, and poultry.

A further category (not a by-product, but representing a waste) is the vast amount of natural grassland, especially under tropical conditions, which is under-utilized annually.

PRODUCTION AND CURRENT DISPOSAL METHODS

It should be appreciated that the estimated production of agricultural waste by-products calculated in Table 2.1 is necessarily approximate. This is especially the case for animal excreta, where production within species is affected by size of animal, level of production, and type of diet. However, assuming 1977 to be typical, Table 2.1 shows that the annual production is enormous. Duckham *et al*, (1976) estimated that the annual production of fibrous agricultural waste dry matter (DM) approximated the annual quantity of photosynthate DM recovered as potential human food. Of the wastes shown in Table 2.1, cereal straw is by far the most voluminous.

Methods of waste utilization and disposal vary with type of waste and with site and country of production.

In most developing countries, cereal straws are utilized (albeit inefficiently) by feeding them to ruminant livestock. Indeed, they provide an important source of feedstuff during tropical dry seasons. In certain developing countries, the straws are under-utilized and burnt (Jackson, 1978); e.g., in Malaysia, rice straw (due to lack of development of ruminant animal production) and in Egypt, rice, maize, and cotton straws (due to abundance of wheat straw to satisfy ruminant requirements).

In developed countries, a substantial proportion of the cereal straw produced is unwanted and is disposed of by burning (Table 2.2). The amount burnt depends, among other things, on the availability of other feedstuffs and the climate. England suffered a major drought in 1976, and consequently, demand for feeding straw increased and the amount burnt decreased. Of all the straw produced in Denmark, 25 to 33% is burnt each year (Jackson, 1978). Burning is also practised in North and Central America (Miles, 1977). Burning represents the cheapest way of disposal and can be valuable in reducing disease carry-over and killing weed seeds and encouraging their germination (Wood, 1977). It is particularly valuable in preparing the field for the next crop, especially if this involves direct drilling and minimum cultivation (Cannell *et al*, 1977). However, burning has attracted criticism on account of its polluting effects. Legislation is already being developed—e.g., in Oregon, U.S.A. (Miles, 1977)—to prevent the practice of open field burning. Animal bedding and feeding have represented the main uses

TABLE 2.1. Estimated Production of Agricultural Waste By-Products in 1977.

Dry Matter (million tonnes)

	United Kingdom	Africa	North and Central America	South America	Asia	Europe	Oceania	U.S.S.R.	Developed	Developing	Centrally Planned	World
Crop wastes produced on the farm												
High dry matter materials[1]												
Cereal straws												
Wheat	4.4	7.0	65.7	7.7	91.8	70.0	8.3	78.2	119.2	72.1	137.3	328.6
Rye	–	–	0.7	0.2	0.7	11.5	–	7.2	4.7	0.8	14.7	20.2
Barley	9.2	2.6	17.8	1.0	23.0	55.5	2.4	44.8	63.8	13.3	70.1	147.1
Oats	0.7	0.1	12.9	0.6	2.1	12.3	0.9	15.6	22.7	1.1	20.8	44.6
Maize	–	44.5	300.9	52.8	92.7	84.3	0.7	18.7	349.5	131.6	113.4	594.5
Millet and sorghum	–	31.9	41.0	14.4	73.4	1.2	1.7	3.7	19.9	90.7	38.7	130.7
Rice, paddy	–	6.7	5.3	11.1	284.8	1.3	0.5	1.9	–	162.9	128.8	311.5
Cereal straws total	14.3	92.8	444.3	87.8	568.5	236.1	14.5	170.1	579.8	472.5	523.8	1577.2
Soybean straw	–	0.1	40.6	11.9	12.4	0.4	0.1	0.4	40.4	13.4	12.1	65.9
Cotton wood	–	3.7	10.6	3.6	17.0	0.7	0.1	9.7	9.7	18.0	17.7	45.4
Sunflower seed stover	–	0.5	1.1	0.8	0.5	2.0	–	5.0	2.5	1.4	6.2	10.0
Rapeseed straw	0.1	–	1.5	0.1	2.7	2.1	–	–	–	–	2.0	7.7
Low dry matter materials												
Sugar cane tops[2]	–	4.5	11.9	14.0	22.9	–	2.0	–	5.3	46.4	3.6	55.3
Sugar beet tops[3]	0.8	0.2	2.5	0.3	2.6	14.9	–	9.7	13.3	1.9	14.9	30.2
Crop wastes produced in centralized processing plants												
High dry matter materials[4]												
Oat hulls	0.1	–	2.2	0.1	0.4	2.1	0.2	2.7	3.1	0.2	3.5	7.5
Maize cobs	–	5.6	37.6	6.6	11.6	10.5	0.1	2.3	43.7	16.4	24.1	74.3
Rice husks	–	1.3	1.1	2.2	57.0	0.3	0.1	0.4	3.2	32.6	25.8	62.3
Groundnut hulls	–	0.8	0.3	0.2	1.7	–	–	–	0.3	2.1	0.5	3.0

TABLE 2.1. (Continued).

	United Kingdom	Africa	North and Central America	South America	Asia	Europe	Oceania	U.S.S.R.	Developed	Developing	Centrally Planned	World
Dry Matter (million tonnes)												
Sunflower seed husks	—	0.1	0.2	0.2	0.1	0.4	—	1.0	0.5	0.3	1.2	2.0
Cotton seed hulls	—	0.4	1.1	0.4	1.7	—	—	1.0	1.0	1.8	1.8	4.5
Low dry matter materials												
Sugar cane bagasse[2]	—	6.0	15.8	18.7	30.6	—	2.6	—	7.1	61.9	4.8	73.7
Sisal pulp[5]	—	1.0	—	1.1	—	—	—	—	—	2.2	—	2.2
Livestock feces[6]												
Ruminants and herbivores												
Horses	0.1	2.7	13.5	9.6	9.9	4.3	0.5	4.4	9.6	22.0	13.3	44.9
Mules	—	1.5	2.7	2.1	1.6	0.5	—	—	0.5	6.9	1.2	8.6
Asses	—	8.2	2.7	3.2	15.0	1.1	—	0.3	0.9	20.6	9.1	30.6
Cattle	12.7	147.5	170.1	201.3	327.2	122.5	38.0	100.7	269.6	641.7	196.1	1107.3
Buffalo	—	1.7	—	0.2	93.1	0.3	—	0.3	0.1	70.7	24.7	95.6
Camels	—	6.3	—	—	3.6	—	—	0.2	—	8.7	1.4	10.1
Sheep	4.7	27.0	3.8	17.3	47.4	21.1	32.6	23.5	55.5	72.8	44.4	172.7
Goats	—	22.0	2.1	3.1	38.9	1.9	—	0.9	2.8	53.7	12.5	68.9
Monogastric												
Pigs	1.6	1.8	17.1	11.3	62.8	34.0	0.9	13.4	36.9	24.3	80.1	141.3
Chickens (urine and feces)	0.5	2.1	2.8	2.1	9.8	5.3	0.2	3.0	6.7	8.1	10.6	25.3

All figures based on production statistics given by F.A.O., 1978.
[1] Straw: grain ratio 1:1 for wheat, rye, barley, oats, rice, soybean, sunflower, and rapeseed; 2:1 for maize, millet, sorghum, and cotton; and straw DM 850 g/kg.
[2] Sugar cane tops: cane ratio 1:5; sugar cane tops DM 300 g/kg; and bagasse DM 500 g/kg.
[3] Total DM production sugar beet top: root ratio 1:2.3; root DM 240 g/kg; and top DM 150 g/kg.
[4] Hull: husk grain ratio 1:5 for oats, rice, groundnut, sunflower, and cotton seed; cob: grain ratio 1:4 for maize; and hull/husk/cob DM 850 g/kg.
[5] Sisal pulp: sisal ratio 32.33:1; and sisal pulp DM 140 g/kg.
[6] Feces DM production/animal/day 2.0 kg for horses, mules, asses, buffalo, and camels; 2.5 kg for cattle; 0.58 kg for pigs; 0.46 kg for sheep; and 0.46 kg for goats; feces and urine DM production/bird/day 0.011 kg.

44

TABLE 2.2.   Straw Utilization and Disposal in England and Wales in
1972, and in England in 1976.

	England and Wales, 1972[1]	England, 1976[2]	
	All Straws (%)	Wheat Straw (%)	Barley Straw (%)
Burnt in fields	37	41	6
Animal bedding and crop storage	36	35	26
Sold between farms	9	8	14
Ploughed-in	2	9	3
Non-agricultural sales	1	1	1
Fed to animals	15	2	39

[1]N.F.U. (1973).
[2]Hughes (1977).

for cereal straw. Straw as a feedstuff will be discussed later. Use as a bedding has declined somewhat in recent years due to the development of slatted floor buildings for cattle and pigs and the handling of animal excreta as a slurry rather than admixed with straw in the form of farmyard manure.

In the United Kingdom, little straw is used for non-agricultural purposes (Table 2.2) and this appears to be the case elsewhere. During and immediately after World War II, about 0.25 million tonnes/yr was used in the United Kingdom for paper-making. In recent years, there has been discussion of developing industrial uses for straw (Radley, 1978). In addition to paper-making (Cobbett, 1977; Riddell, 1977; Triolo, 1977; Dean, 1977), chemical production (e.g., furfural) (Rijkens, 1977; Sachetto, 1977), fuel generation (Miles, 1977), and building board manufacture (Craven, 1973) are possibilities. As yet there is little practical application for these but this may well change in the future as fossil fuel supplies become scarcer. The attraction of straw as an industrial raw material is that it is a renewable resource which will inevitably be produced. A major factor limiting its present use is the high cost of harvesting and transporting it to centralized processing plants. Current research on densification of straw (Klinner and Chaplin, 1977) may remove this for the future.

Burning is the means of disposal of other unwanted high DM wastes, such as soybean straw, cotton stalks, sunflower seed stover and rapeseed straw.

Sugar cane tops are partially destroyed by the practice of trash-burning before cutting the cane. What remains is generally ploughed-in. Tops are little-used as animal feedstuff on large sugar estates because livestock are generally absent. On small farms, tops are cut and fed fresh or occasionally conserved by drying for livestock feeding in the dry season.

Sugar beet tops, after wilting, are fed *in situ* to sheep and cattle or are ploughed-in as green manure. They are therefore not strictly wastes as defined earlier, but they could be more extensively used as a ruminant feedstuff. Limiting factors are difficulties of removal from the field due to wet conditions, difficulties of conservation by ensiling, and the fact that many farms growing sugar beet do not have livestock. High transport costs also prohibit inter-farm sales.

Centrally produced dry crop wastes are sometimes included in compound animal feedstuffs, as they are often produced on the same site. They are of low nutritive value and, unless processed, their role is largely as an inert filler or a source of roughage in highly concentrated diets for cattle. Rice and sunflower seed husks are of very low nutritive value and are more often used as animal bedding. Small amounts of maize cobs, rice husks, and cotton seed hulls are used for the manufacture of furfural. The facts that they are centrally produced and easily conserved (being dry) are distinct advantages if they could be industrially utilized.

Modern methods of combine harvesting result in the loss of husks from barley and wheat, since they are disgorged in the field along with straw. Under temperate conditions, some grain loss occurs from shedding due to harvesting in over-ripe condition. There is therefore current research into whole-crop harvesting which would minimize grain loss and also harvest the husk portion.

Sugar cane bagasse is mainly (60% of production; Gohl, 1975) used for fueling sugar factories, especially in developing countries. Other uses involve paper manufacture, acoustical wall-board and other building material manufacture, and agriculture (Srinivasan and Han, 1969). Recently, bagasse has also been used for furfural manufacture (Radley, 1978). Agricultural purposes include use as animal bedding, composting into soil conditioner, and use as a filler in high concentrate diets for cattle. The high initial moisture content (50%) of bagasse requires that conservation techniques be applied if storage is to be undertaken. Disposal of bagasse would appear to be more of a problem in developed countries.

Sisal pulp represents 97% of the fresh leaf weight. After composting, it is occasionally used as a mulch and fertilizer for young sisal plants (Acland, 1975). Waxes, sodium pectates, alcohol, methane, and hecogenin (containing cortisone) can be extracted from sisal waste, but extraction is generally uneconomic (Acland, 1975). Sisal pulp can be fed to cattle, either fresh or after ensiling, but, as in the case of sugar estates, sisal estates rarely keep livestock, and the cost of transporting wet pulp is prohibitive.

The estimated total production of animal excreta is enormous (Table 2.1). However, the waste disposal problem implied is by no means as large since all except pigs, poultry, and a proportion of cattle spend much of their time outdoors grazing under fairly extensive conditions, and their excreta is returned to the soil as plant food nutrient.

The problem of animal waste disposal has developed during the past 20 years with the advent of large intensive livestock units, resulting in the production of waste in excess of the absorptive capacity of immediately available lands. The problem has attracted much attention and research, and literature on the subject is volumnious (e.g., Loehr, 1974 and 1977; Porter, 1975; Agricultural Research Council, 1976; Hobson and Robertson, 1977; and Weller and Willetts, 1977). It appears that most farmers will prefer traditional land spreading to the more sophisticated and expensive reclamation processes as methods of disposal. In certain circumstances, land spreading may not be possible. Reclamation to produce water, fertilizer, fuel, or animal feed will therefore be necessary.

LIMITATIONS OF AGRICULTURAL WASTES AS FEEDSTUFFS

The principal factors to be considered concerning wastes as feedstuffs are whether animals will eat them, their content of available nutrients for a given species, and whether they contain compounds toxic to animals eating them or to humans eventually consuming the animal products.

Table 2.3 indicates the composition and nutritive value of different wastes, but the data are not comprehensive and wide ranges in composition within given products are known to exist—e.g., in cereal straws (Owen *et al*, 1969) and in livestock excreta (Ward and Muscato 1976; Bhattacharya and Taylor, 1975). However, one common feature is their high content of fibrous constituents, which have been traditionally analyzed as crude fibre and more recently (Van Soest, 1976) analyzed as cell wall constituents: cellulose, hemicellulose, and lignin. This means they are potentially only suitable as major ingredients in diets for ruminants. However, with the exception of sugar cane and sugar beet tops, sisal pulp, and poultry excreta, their digestibility, even by ruminants, is low and generally less than 50% for organic matter (the energy yielding component). Low digestibility results in a low concentration (3.0 to 8.0 MJ/kg DM) of metabolizable energy (ME) in the DM. The ME values of barley grain and grass silage, both widely used feedstuffs, are also shown in Table 2.3, for comparison. In the case of poultry excreta, sugar beet tops, and sisal pulp, low ME concentration is caused by high ash content. All the agricultural wastes shown in Table 2.3 contain relatively low amounts of available (metabolisable) energy.

A second common limitation, except for animal excreta, is the low content of crude protein (CP) or nitrogen (N): less than 63 g CP/kg DM (<1% N). To sustain adequate activity of the cellulose and hemicellulose digesting micro-organisms in the rumen, the diet DM needs to contain at least 1% N, even if dietary energy digestibility is less than 50% (Pigden and Heaney, 1969). In more digestible fibrous diets, N requirements may increase to 1.5%, and if readily available carbohydrate (e.g., cereal grains) is included in the diet at levels greater than 20%, N requirements for efficient microbial activity are about 2% of the DM. The N content of many of the agricultural wastes shown in Table 2.3 is therefore too

TABLE 2.3. Chemical Composition, Digestibility, and Metabolizable Wastes for Ruminants.

	Dry Matter (g/kg)	g/kg DM								Organic Matter Digestibility (%)	Metabolizable Energy (MJ/kg DM)
		Ash	Crude Protein	Crude Fiber	Acid Detergent Fiber	Cell Wall Constituents	Cellulose	Hemicellulose	Lignin		
Crop wastes produced on the farm											
High dry matter materials											
Cereal straws											
Wheat	860[1]	62[1]	24[1]	426[1]	540[2]	850[2]	403[3]	292[3]	87[3]	41.5[1]	5.7[1]
Rye	860[1]	45[1]	36[1]	465[1]	–	–	–	–	–	45.0[1]	6.3[1]
Barley	860[1]	66[1]	37[1]	488[1]	501[3]	804[3]	432[3]	303[3]	69[3]	41.8[1]	5.8[1]
Oats	860[1]	57[1]	22[1]	402[1]	539[3]	834[3]	446[3]	282[3]	93[3]	48.8[1]	6.8[1]
Maize	850[1]	56[1]	59[1]	461[1]	390[2]	670[2]	–	–	–	54.0[1]	7.3[1]
Millet	950[4]	79[4]	32[4]	342[4]	–	–	–	–	–	–	–
Sorghum	937[4]	94[4]	37[4]	418[4]	–	–	–	–	–	–	–
Rice, paddy	850[3]	124[3]	58[3]	306[3]	460[3]	653[3]	362[3]	145[3]	98[3]	56.1[3]	[7.4]
Soybean straw	840[1]	121[1]	88[1]	311[1]	440[2]	580[2]	–	–	–	–	7.5[1]
Rapeseed straw	840[1]	45[1]	30[1]	450[1]	–	–	–	–	–	46.1[1]	6.5[1]
Low dry matter materials											
Sugar cane tops	256[5]	62[5]	63[5]	350[5]	–	–	–	–	–	59.7[14]	[8.4]
Sugar beet tops	230[1]	322[1]	104[1]	148[1]	–	–	–	–	–	73.7[1]	7.9[1]
Crop wastes produced in centralized processing plants											
High dry matter materials											
Oat hulls	900[1]	42[1]	21[1]	351[1]	–	–	–	–	–	34.4[1]	4.9[1]
Maize cobs	–	[50]	20[4]	350[4]	390[4]	890[4]	–	–	–	49.5[2]	[7.1]
Rice husks	941[6]	224[6]	30[4]	450[4]	720[4]	820[4]	–	–	–	14.1[4]	[1.6]
Groundnut hulls	823[7]	74[7]	49[7]	684[7]	–	–	–	–	–	–	–
Sunflower seed husks	885[8]	32[8]	55[8]	556[8]	–	–	–	–	–	20.7[8]	[3.0]
Cotton seed hulls	–	40	40[4]	500[4]	710[4]	900[4]	–	–	–	34.4[4]	[5.0]

Low dry matter materials											
Sugar cane bagasse	500[9]	24[9]	20[4]	500[4]	–	905[9]	460[9]	245[9]	200[9]	46.1[4]	[6.7]
Sisal pulp	149[10]	339[10]	61[10]	160[10]	–	–	–	–	–	63.2[10]	[6.3]
Livestock excreta											
Beef cattle feces (finishing)	–	70[11]	188[11]	–	–	580[11]	–	–	–	50.0[11]	[7.0]
Dairy cattle feces (lactating)	–	90[11]	163[11]	–	–	560[11]	–	–	–	30.0[11]	[4.1]
Cattle feces, diet of maize: lucerne hay 17:83	176[12]	297[12]	129[12]	413[12]	–	–	–	–	–	–	–
Cattle feces, diet of maize: lucerne hay 75:25	229[12]	74[12]	187[12]	261[12]	–	–	–	–	–	–	[6.3]
Pig feces (growing)	–	160[11]	200[11]	440[11]	–	–	–	–	–	50.0[11]	–
Poultry excreta (caged layers)	–	260[11]	388[11]	–	–	400[11]	–	–	–	67.0[11]	[7.4]
Poultry excreta (caged broilers)	–	220[11]	406[11]	–	–	340[11]	–	–	–	61.0[11]	[7.1]
Poultry excreta, dried (caged layers)	897[13]	280[13]	280[13]	127[13]	–	–	–	–	–	–	6.4[13]
Poultry litter, dried (broilers)	847[13]	150[13]	313[13]	168[13]	–	–	–	–	–	–	8.3[13]
Barley grain	860[1]	26[1]	108[1]	53[1]	–	–	–	–	–	88.3[1]	13.7[1]
High digestibility ensiled grass (British conditions)	200[1]	100[1]	116[1]	300[1]	–	–	–	–	–	74.0[1]	10.2[1]

NOTE: Metabolisable energy in brackets derived from $0.15 \times$ (% digestible organic matter in dry matter). Other values in brackets, assumed.

[1] MAFF, 1975.
[2] Van Soest, 1973.
[3] Jayasuriya, 1974.
[4] Gohl, 1975.
[5] Butterworth, 1963.
[6] ARC, 1976.
[7] Oyenuga, 1968.
[8] Owen, 1967.
[9] Srinivasan and Han, 1969.
[10] Franck, Meissner, and Hofmeyr, 1973.
[11] Smith, 1977.
[12] Ward, and Muscato, 1976.
[13] Bhattacharya and Taylor, 1975.
[14] McDowell, et al., 1972.

low for optimum microbial activity in the rumen. The dietary CP requirements of the host animal range from 100 to 120 g/kg DM for maintenance and moderate growth to 150 to 170 g/kg DM for rapid growth in young ruminants or high milk yield in lactating animals. Some N or CP supplementation of many of the agricultural wastes is therefore required. The extent to which this can be in the form of low cost non-protein nitrogen (e.g., urea), as opposed to more expensive preformed true protein, depends on the extent to which rumen synthesized microbial protein can meet the host animal's requirements and the amount of dietary protein escaping degradation in the rumen. Considerable information concerning these has been collected in recent years (Roy *et al*, 1977) and the future assessment of protein feedstuffs for ruminants will include indicating the content of rumen degradable and non-degradable protein. Although animal excreta contain substantial quantities of N, much of it, at least in poultry excreta (50 to 60%; Bhattacharya and Taylor, 1975) is in the form of non-protein nitrogen (e.g, uric acid in poultry excreta) and there is little information concerning the rumen degradability of the true protein fraction.

The mineral and soluble carbohydrate contents of many fibrous wastes may also be too low for efficient digestion of cellulose and hemicellulose by rumen micro-organisms. Supplementation of the diet with 5 to 10% of readily available carbohydrate is therefore suggested (Pigden and Heaney, 1969). However, because of the low appetite for straw diets concentrate, supplementation may need to be more than 20% (Andrews *et al*, 1972) to achieve even moderate levels of animal productivity.

Negligible vitamin A content is another limitation of straws and seed hulls.

An important limitation and second only to low digestibility is the fact that ruminants will consume only limited quantities of the wastes listed in Table 2.3. In the case of cereal straws, daily intakes are generally less than 15 g DM/kg liveweight, even when nitrogen and minerals are not limiting, whereas consumption of good quality feedstuffs ranges from 25 to 30 g. This means that the low intake of ME provided by wastes such as straw is generally inadequate to provide even for maintenance in adult ruminants. Low digestibility is the prime cause of the low intake (Campling, 1970). Low palatability may also play a part; e.g., in the case of straw (Greenhalgh and Reid, 1967) and also poultry excreta (Bhattacharya and Taylor, 1975).

A limitation concerning the use of animal excreta as feedstuff is the possible risk of transmitting diseases and pollutants to both animals and man (Bhattacharya and Taylor, 1975). Future legislation in many countries will probably dictate adequate sterilization before recycling as feedstuff or indeed may even prohibit the feeding of certain types of excreta.

Other factors, noted earlier, limiting the use of many agricultural wastes as ingredients for the manufacture of compound animal feedstuffs are site of production, low density, and high moisture content. Unless processed, these materials are only suitable as feedstuffs at the site of production. Even then, their basic nutritional limitations are such that they must be supplemented and also processed if they are to be anything more than relatively inert diet fillers. Processing may also be necessary to allow storage and reduce the risk of disease transmission.

PROCESSING METHODS TO REDUCE LIMITATIONS

Particle Size Reduction

Optimum particle size reduction of long roughages such as straw, by grinding, can increase their intake. This occurs because of a faster passage of digesta out of the reticulo rumen. Mechanical processing of roughages has been extensively researched during the past twenty years and has been the subject of several reviews (Minson, 1963; Moore, 1964; Beardsley, 1964; Campling and Milne, 1972; Greenhalgh and Wainman, 1972; and Swan and Clarke, 1974). Despite this, it is difficult to be precise about effects, but the poorer the material the higher the responses appear to be. Greenhalgh and Wainman (1972), reviewing the effects of processing roughages for beef cattle and sheep, concluded that, in the more reliable long-term trials involving roughages alone, appropriate processing increased intake by about 30%. With concentrate-containing diets, the effect of processing was reduced, or even negative, with diets containing only 20 to 30% roughages. This was probably due to the diet digestibility being sufficiently high so that intake was not limited by the physical capacity of the reticulo rumen. The point at which changeover from physical to physiological control of intake occurs (Fig. 2.1) is not yet precisely defined, but is likely to differ between

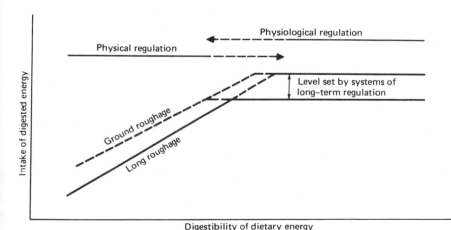

Fig. 2.1.   To illustrate that the improved intake due to grinding roughages varies with digestibility. *Source:* Baumgardt. 1970; Lonsdale. 1976.

classes of ruminants, being dependent on their energy requirements. Intake improvement due to grinding roughage is greater in sheep than in cattle (Greenhalgh and Reid, 1973). The definition of optimum processing is also unclear, largely because researchers have not adequately defined particle size (e.g., modulus of fineness (MF) or modulus of uniformity (MU); ASAE, 1967) in their investigations. For roughages generally, MF ranges from 1 to 2 for ground and pelleted material to up to 5 to 6 for simply chopped feeds (Greenhalgh and Wainman, 1972). Wilkins (1973), feeding high digestibility dried grass to sheep, found an increase in intake of 22% with decreasing particle size over the MF range 3.5 to 1.0. Swan and Clarke (1974), on the other hand, compared the effect of grinding barley straw through different screen sizes (3.18 mm, giving MF 2.38; 6.35 mm, MF 2.93; 9.26 mm, MF 3.39; 12.70 mm, MF 3.68). A screen size of 6.35 mm was the optimum in terms of effect on diet digestibility, intake, and growth rate, when the straw was given at 30% in a diet containing 70% concentrates fed *ad libitum* to beef cattle. It seems likely, therefore, that the optimum particle size reduction will differ with the type of animal to be fed and possibly also with the quality of the other dietary constituents.

In practice (the production of dried grass and lucerne), grinding is frequently followed by pelleting. This facilitates handling of the product. In the case of ground straw, pelleting does not improve nutritive value unless the process causes a further reduction in particle size (Swan and Clarke, 1974). However, it reduces dustiness and increases the rate of eating.

Increased intake is the major nutritional effect of particle size reduction. Although digestibility may be reduced (due to greater rate of passage), the effect on ME is small because loss of energy as methane is reduced. In growing and fattening ruminants there is some evidence (Wainman and Blaxter, 1972) that ME is utilized more efficiently following grinding of roughages.

Potential nutritional disadvantages, especially with fine grinding and when the diet contains no long roughages, are increased risks of bloat and hyperkeratosis (clumping of rumen papillae) and in the case of dairy cows, production of low fat milk. Despite the foregoing, straw grinding has not been extensively practiced —largely because of the absence of suitable machinery and lack of economic incentive. However, in recent years, tractor-driven and mobile machines for chopping and coarse grinding (e.g., tub grinders) straw have become increasingly available and utilized. This has occurred partly because of increasing awareness of the need to utilize agricultural wastes and partly because of chopped/coarsely ground straw having a role as a source of roughage and diluent (at 10 to 20% levels) in 'complete' diets for intensively fed beef and dairy cattle (Preston and Willis, 1970; Owen, 1978). In these circumstances, though, it is unlikely that the particle size reduction applied is sufficient to increase intake; in any case, response would not occur because of the high digestibility of the total diet (cf. Fig. 2.1). Inclusion of straw, or any other long roughage, in a "complete" diet necessitates its chopping or coarse grinding beforehand. The current cost of coarse grinding in the U.K. is about £7.00/tonne. Fine grinding to promote improved intake of straw would cost more. A consideration of increasing importance in the future will be the energetic efficiency of processing. The energy required to grind straw to different degrees of fineness is not well documented but would appear to be low (0.14 MJ/kg; Owen, 1976) in relation to its ME value (Table 2.3) and the improved efficiency of ME utilization and intake accruing.

The preceding discussion has concerned the processing of mainly barley and wheat straw, for which there is documented evidence, but it is likely that similar effects would occur on processing other long straws. Apart from maize cobs, other centrally produced high DM materials (seed hulls) are inherently coarsely ground and are frequently used as diluents in compound feed manufacture. There is little information on the extent to which their nutritive value could be improved by further reduction in particle size. Maize cobs are not consumed and cannot be incorporated in a complete diet unless ground. There is need for research to define optimum particle size reduction and the energy input required.

Low DM materials, such as sugar cane and sugar beet tops, if they are to be ensiled, require some particle size reduction (i.e., chopping) to facilitate adequate consolidation and generate anaerobic conditions. Suitable machinery for chopping sugar cane tops is rarely available in developing country situations. Other low DM wastes, bagasse and animal excreta, are already finely comminuted; but further comminution would probably not improve their nutritive value.

Treatment with Alkali

The objective is increased digestibility, resulting from cellulose and hemicellulose in lignified materials such as straw (Table 2.3) becoming more susceptible to microbial digestion in the reticulo rumen. The chemical reactions during treatment with alkali are not well understood, but bonds between lignin and hemicellulose and, to a lesser extent, lignin and cellulose are known to be alkali-labile. A further benefit is that the increased digestibility is accompanied by improved intake, roughage intake in ruminants being highly correlated with digestibility (Campling, 1970). To date, most of the research and application has centered on improving cereal straw in developed countries, mainly Europe, using sodium hydroxide.

The Traditional Method (Beckmann). The Beckmann (1922) system was applied on German farms during World War I. It involves immersing loosely baled straw in a weak (1.0 to 1.5%) sodium hydroxide solution for 4 to 24 hours, using about 10 parts solution/part straw and then draining away the liquor, which is later reused after recharging with alkali. This is followed by an extensive washing of the straw (using about 40 liters water/kg straw) to remove residual alkali before feeding. The method has the advantages of improving digestibility of organic matter to 65% or more, with consequent improvements in intake. Other advantages are that it requires no sophisticated machinery and can be applied relatively safely under farm conditions. The system was used on British farms during World War II and is still used in some Scandinavian countries, notably Norway.

The system also has several disadvantages. It is very extravagant in the use of alkali, water, and labor. The product is wet (200 to 300 g DM/kg) and furthermore, the extensive washing results in about 20% of the organic matter being lost by leaching. The possible pollution of water courses by effluent from treatment plants is an added disadvantage, and is currently under scrutiny in Norway. Apart from Scandinavia, other countries have not used the system extensively since World War II.

Developments Since 1964. Following a Canadian investigation by Wilson and Pigden in 1964, there has been an international revival of research interest in alkali treatment of poor quality roughages with over 200 research papers being published (Jackson, 1977). Investigators have aimed to overcome the disadvantages of the traditional method by using less alkali and eliminating the washing stage. The effect of factors such as volume and concentration of sodium hydroxide, temperature, pressure, time, physical form of straw, and neutralization prior to feeding have been investigated (F.A.O., 1977; Jackson, 1977; Owen, 1978). Commercial application of this research, both on an industrial scale and on farms, has developed only since 1975.

Industrial Processing of Straw. This was developed both in the U.K. by Unilever Limited (Robb, 1976; Wilson and Brigstocke, 1977) and Denmark (Rexen *et al*, 1976). Both Unilever and British Petroleum are now operating a total of 10 plants in the U.K., each plant having an annual capacity to process about 25,000 tonnes of straw.

The method is sophisticated, involving grinding straw and mixing it with a low volume (0.10 to 0.17 l/kg straw) of highly concentrated (27 to 47%) sodium hydroxide solution. The mixture is then pelleted, which generates heat and pressure and speeds up the chemical reaction so that it is over in minutes. The process can therefore be made a continuous one. Because the product is dry (900 g DM/kg), there are no storage problems and it can be used as an ingredient in the compound animal feed industry. This is further facilitated by the ease with which the material can be handled; it is about five times as dense as unprocessed straw and can be transported economically. A further advantage is that, because it is an industrial process, skilled labor can be employed and adequate safety precautions adopted to handle the potentially hazardous concentrated chemical involved. This, and the advantages of "turn-key" principles, suggest that industrial processing could have considerable application in developing countries. The major disadvantage is that high transport costs are inevitably incurred; this would be particularly so in developing countries.

Treatment of Straw on Farms. This is less developed than industrial processing and involves various approaches: mobile straw processing machines, modified Beckmann-type plants, and utilization and modification of existing farm machinery. Compared with industrial processing, a less complex range of machinery is involved, with no pelleting and less emphasis on producing a dry product.

Purpose-built machines (Farmhand Feed Processor, JF SP2000, Taarup 801, Taarup 805) have been marketed since 1977. Their common feature involves chopping and shredding baled straw and application of very low volumes of concentrated sodium hydroxide solution, as in the industrial process. Their technical efficiency has not yet been assessed in controlled experiments but their throughput rate is 2 to 3 tonnes/hr. One disadvantage of using such low water volumes is fire risk, since considerable heat is liberated during the exothermic reaction between alkali and straw.

The Norwegian Torgrimsby method (Jackson, 1978) is an ingenious modification to the Beckmann system involving the use of three immersion tanks and reuse of the steeping liquor and washwater, and producing no liquid effluents or loss of straw nutrients. Additions to the system merely involve replacing the water (3.0 l/kg straw) and sodium hydroxide (40 g/kg straw) removed in each batch of straw treated. The method is currently under test in Norway and Sweden.

The Swedish Boliden method is another modification. Batches of baled straw (c 500 kg) are sprayed with alkali solution and the effluent is recirculated over a period of $3\frac{1}{2}$ hours. The straw is then stood for $17\frac{1}{2}$ hours, after which acid is applied to neutralize before feeding. Although neutralization is probably not essential (Raine and Owen, 1978) it does provide a means of fortifying treated straw with minerals. Both the Torgrimsby and Boliden methods are attractively simple, but they do suffer the disadvantages of being batch processes.

Widespread processing on farms is unlikely unless low cost, simple methods are developed. Recent research in the U.K. has investigated the adaptation of existing farm machinery such as mixer trailers (Greenhalgh and Pirie, 1977) and forage harvesters (Wilkinson, 1978). Wilkinson's method involves picking up the straw from the swath with a precision chop forage harvester and applying solid sodium hydroxide and water, separately, into the cutting chamber. Treated straw is then ensiled. Investigations with barley straw (Gonzalez Santillana, 1977) treated with 50 to 75 g NaOH/kg DM and wate to produce 50 to 60% DM silage have shown the material to store satisfactorily under anaerobic conditions, with the pH ranging from 8.0 to 10.0, after 90 days. A pH in excess of 9.0

would appear necessary to prevent mold growth, but further investigations are needed. Ensiled alkali treated straw has been satisfactorily fed to pregnant sheep at up to 58% of the diet DM, as a replacer for hay (Owen *et al*, 1978). The principle of forage harvesting straw from the swath, treating with alkali, and ensiling the material for subsequent use is attractive. The operation could be independent of climatic conditions, and therefore less competitive with grain harvesting and would eliminate the baling and bale carting operation. The method may also have considerable application for harvesting and improving standing hays, especially under tropical conditions. Its apparent major disadvantage stems from the need to handle solid sodium hydroxide.

Nutritive Value of Alkali Treated Straw. Direct comparisons of straw treated with sodium hydroxide using industrial and on-farm processing methods have not yet been made, although improvements in digestibility, at least with barley straw, appear to be comparable (Owen, 1978). Thus, organic matter digestibility in sheep fed at maintenance increases from 40 to 50% for untreated straw, to 60 to 65% after treatment with c 60 g NaOH/kg DM. Estimated ME content is increased: from 6.0 to 7.0 MJ/kg DM to 8.0 to 9.0 MJ/kg DM. The improvement in ME content is less than that implied from the digestibility increases because of the diluting effect of ash added during treatment (ash in treated straw c 100 to 120 g/kg DM). A potential disadvantage of the modern methods, where treated straw is not washed before feeding, is the physiological effect of the high sodium content (c 30 g/kg DM) upon the animal. This results in greater water consumption, especially where treated straw forms a high proportion of the diet; the problem will also assume greater significance with housed animals and under tropical conditions. The high sodium content in unwashed treated straw is also considered to limit the intake improvement associated with increased digestibility (Jackson, 1977). Nevertheless, treatment clearly improves intake (Owen, 1978), and it is likely that industrial processing, involving greater particle size reduction than on-farm treatment, will promote better intake improvements. A possible benefit of the alkaline nature of treated straw is that its inclusion in high concentrate diets would assist in reducing the problem of acidosis. Alkali treated straw and acidic grass silage would also appear to be complementary feedstuffs (Terry *et al*, 1975).

The economics of treating straw with sodium hydroxide are difficult to assess. A major determinant of whether or not to process is the availability and cost of comparable alternative feedstuffs. Treated straw can be equated to a high digestibility grass hay containing little protein. Clearly, it would be uneconomic to purchase and process straw at a cost greater than the cost at which comparable grass hay could be purchased, if it were to be used as a hay replacer. Nevertheless, industrially processed straw has the additional advantage of being suitable for inclusion in complete diets or compound feedstuffs. In the U.K., it is currently included at levels of 10 to 15% in some compounds for dairy cows and up to 30% for beef cattle. Higher levels of inclusion would be possible, particularly for beef cattle, if other components of the diet were concentrates. For example, Pirie and Greenhalgh (1977) found a diet of 50% farm treated straw and 50% grain to promote good growth rates (1.18 kg/day) in beef cattle. For moderate to low producing stock (store cattle, non-lactating cows) even higher inclusion rates of treated straw would be feasible, provided its cost compared favorably with alternative feedstuff such as hay.

Besides economics, the energetic efficiency of processing straw with alkali merits consideration. It is notable that the improvement in straw ME achieved by treatment with, say, 40 g NaOH/kg DM, is matched by a similar input of energy during the manufacture of sodium hydroxide (51.3 MJ/kg NaOH; Leach and Slesser, 1973).

Alternatives To Sodium Hydroxide. A number of chemicals, known to have an effect on lignin, have been investigated (Jackson, 1977) but none appears as effective, economic, and non-toxic to animals as sodium hydroxide. However, assessments (e.g., Chandra and Jackson, 1971) have generally involved short reaction periods, thus penalizing slow-reacting chemicals such as ammonia and calcium hydroxide.

Investigations in Norway (Arnason and Mo, 1977) showed that anhydrous ammonia applied to baled straw enclosed in polythene sheets increased organic matter digestibility by sheep from 48.3% (untreated) to 61.6% after 3 weeks and 65.7% after 8 weeks. The optimum amount of ammonia was 35 g/kg DM. An obvious attraction of ammonia is that it also adds nitrogen. A further attraction is the simplicity of the treatment operation.

Calcium hydroxide is cheaper and less dangerous to handle than sodium hydroxide and would therefore be more suitable for use on farms. A potential disadvantage of high-calcium diets would be the need for increased supplementation with phosphorus. Gonzalez Santillana (1977) found calcium hydroxide (50 g/kg DM) ineffective in improving the digestibility *in vitro* of barley straw organic matter (49.0%) after 1 day of treatment, but after 90 days' ensilage (silage contained 430 g DM/kg) digestibility was increased to 60.7%. Comparable treatment with sodium hydroxide gave a digestibility of 65.9%. Recent research at Reading University (Table 2.4) showed ensiled calcium hydroxide-treated straw to be only marginally better digested by sheep than untreated straw, yet intake was improved to the same extent as after treatment with sodium hydroxide. Ammonia and sodium carbonate treated straws were also well consumed. Further research is needed on the use of these and other chemicals. Sodium carbonate is of special interest because of its non-hazardous nature and the fact that impure deposits are available in several tropical countries.

Treatment with superheated and pressurized steam also improves digestibility.

TABLE 2.4.   Digestibility and Intake by Sheep of Chopped Barley Straw Ensiled[1] with Different Chemicals.

Chemical	Rate of Application (g/kg DM)	pH at Feeding	Derived Digestibility of Straw Organic Matter		Ad Libitum Intake of Straw Organic Matter Live Weight/Day (g/kg)
			Restricted Feeding[2] (%)	Ad Libitum Feeding[3] (%)	
Sodium hydroxide	60.0	9.6	67.4	60.4	16.7
Ammonia (anhydrous)	35.0	8.0	67.3	59.6	16.7
Sodium carbonate	79.5	9.6	61.1	54.2	15.9
Sodium carbonate plus calcium hydroxide	{ 39.8 27.8	9.0	58.1	54.5	16.1
Calcium hydroxide	55.5	7.7	56.9	53.8	17.1
Control (untreated)	0	–	54.8	49.6	12.5

Source: B. S. Nwadukwe and E. Owen, 1978 (unpublished).
[1] Apart from ammonia, chemicals were dissolved in water so that treated straw contained 500 g DM/kg. Treated straws were ensiled for 90 days.
[2] Diet DM content 72% straw, 28% concentrates. Concentrate organic matter digestibility assumed to be 87%.
[3] Diet DM content 68% straw, 32% concentrates. Concentrate organic matter digestibility assumed to be 87%.

Bender *et al* (1970) found wheat straw digestibility *in vitro* to increase from 36.4% to 50.3% following steam treatment (1 hour at $160°C$ and 7.04 kg/cm^2). Though interesting, the method has not been developed because of difficulties of application, especially under farm conditions.

Alkali Treatment of Other Wastes.   Although most of the wastes shown in Table 2.3 have low digestibility as a common feature, alkali treatment appears to be less effective in improving the digestibility of materials other than cereal straws (Jackson, 1977). Wastes of leguminous plants show particularly poor response, presumably as a reflection of the lower solubility in alkali of legume lignin compared with grass lignin (van Soest, 1964). For temperate grass hays (*Lolium perenne*), at least, the lower the initial digestibility, the greater the digestibility improvement following alkali treatment (Mwakatundu and Owen, 1974). This is also the case for treated cereal straws (Jayasuriya, 1974; Raine, 1978). The response of tropical grasses to alkali treatment appears less than for temperate species (Owen, 1977), but further research is needed.

Rice husks seem particularly immune to improvement by alkali treatment (Guggolz *et al*, 1971; Choung and McManus, 1976). Ground maize cobs, on the other hand, show a considerable response (Klopfenstein *et al*, 1972; Kategile, 1978).

Initial investigations by Smith *et al* (1971) showed the DM digestibility of dried dairy cow feces treated with sodium hydroxide (3.0 g/100 g dried feces) to increase from 27.4% to 42.0%. Although considerably improved, the treated feces remained a low digestibility material. Subsequent feeding trials with sheep showed less promise, with no increase in digestible DM intake following treatment (Smith, 1977). From the earlier discussion, it is conceivable that the feces of legume fed animals will be less responsive to alkali treatment than will grass fed animals. As well as improving digestibility, alkali treatment of feces may be beneficial in reducing its pathogen content (Wilkinson, et al, 1978). These aspects require further research.

Irradiation

Gamma and electron irradiation, at optimum levels, has been shown to increase the *in vitro* digestibility of straw and cattle feces (Pigden et al, 1966), but animal feeding trials have apparently not been conducted. In any case, the quantity of irradiation needed is considered far too expensive for commercial application at the present time.

Biological Treatment To Increase Nutritive Value

Use of bacteria, yeasts, and fungi to improve the nutritive value of agricultural wastes for ruminants has received relatively little research and, as yet, no practical application. Investigations have centered more on the possible use of wastes for the production of single-cell protein for use as a feedstuff for non-ruminants (Bellamy, 1976; Birch et al, 1976).

In reviewing treatment of lignocellulose wastes, Linko (1977) categorized the possible biological methods as follows:

1. Enzymatic hydrolysis of cellulose and hemicellulose to fermentable sugars followed by cultivation of organisms suitable for single-cell protein production.
2. Direct cultivation of cellulolytic organisms on cellulosic materials.
3. Production of single-cell protein from processed wastes already in hydrolyzed form.
4. Partial hydrolysis of wastes to improve their digestibility by animals.

With the possible exception of methods 3 and 4, the above would involve industrial rather than on-farm processing, and thus would most likely make use of centrally produced wastes such as bagasse.

The difficulties with enzymatic hydrolysis are the high costs of cellulases and hemicellulases and the fact that lignification of wastes (Table 2.3) restricts their activity so that some pretreatment is needed. To date, cellulase production has been largely based on the fungus *Trichoderma viride*. Linko (1977) draws attention to the possibility of acid rather than enzymatic hydrolysis. Although this would be more rapid, the high temperature and pressure required would involve a more expensive plant. However, acid hydrolysis of cellulosic materials is industrially applied in the Soviet Union. Linko (1977) emphasizes that application of method 2 also necessitates pretreatment, such as with alkali. For example, the Louisiana State University method for single-cell protein production from bagasse (Rockwell, 1976) involves treatment with 2 to 4% sodium hydroxide solution, then heating at 110 to 130°C for 30 to 60 minutes. This effectively sterilizes the material and solubilizes the lignin and hemicellulose. The remaining cellulose is then fermented with a mixed culture of *Cellulomonas* bacteria and yeast such as *Candida guillermondii* or *Trichosporon cutaneum*. In this process, 100 kg bagasse DM yields about 13 kg protein and 13 kg non-protein cell constituents.

Method 3 is already applied to sulfite waste liquor produced during wood pulping. Harmon (1976) and Day (1977), at Illinois University, have developed the principle for recycling pig waste. Pig slurry (liquid mixture of feces, urine, hair, and waste food) is aerobically fermented to single-cell protein. The oxidation ditch mixed liquor (containing 45 g DM/kg) is fed as a mineral and protein nutrient solution to supplement pig diets containing 15% below normal protein levels. Two possible disadvantages are nitrate toxicity due to excessive aeration and the build-up of intestinal worms. The Cereco system (Day, 1977) developed by Ceres Ecology Corporation, Colorado, for processing cattle slurry, involves separating the liquid and solid portions. The liquid is fermented to encourage single-cell protein production and then pasteurized and dried for inclusion as a 30% crude protein feed ingredient. The solid portion is ensiled for subsequent feeding to ruminants.

As noted earlier, on-farm biological treatment is most likely to be of the method 4 type. However, Autrey *et al* (1975) recently reported only modest increases in digestibility following treatment of poor hay with cellulase. Cellulose digestibility of coastal Bermuda grass hay increased from 41.7% to 51.0% following a five day treatment with 7.2 g *Trichoderma viride* cellulase in 7:1 water/kg DM. A pretreatment soaking in 2.0% sodium hydroxide increased digestibility to 58.7%. Clearly, further research is needed to determine optimum treatment conditions, but the current high cost of cellulase and the improvement following pretreatment with sodium hydroxide suggests that this method has little to offer over conventional alkali treatment, discussed earlier.

Recent investigations at the University of Aston, in Birmingham (Seal and Eggins, 1976) suggest that the protein value of mixtures such as straw and excreta could be enhanced using thermophilic fungi (optimum activities at about 50°C) which are capable of solid substrate fermentation. Further attributes are their natural occurrence in plant and animal wastes, their highly cellulolytic property, their lack of obnoxious odor, and their pH (slightly alkaline) and temperature optima, which tend to discourage pathogenic bacteria and inactivate animal parasites. The latter are especially important in regard to treatment of animal excreta for use as feedstuff. The essentials of a pilot plant constructed by Seal and Eggins (1976) involved the capacity to heat and maintain a temperature of 50°C. Mixtures of straw and pig slurry placed in the plant were found to be much reduced in bulk over a 14 day period and the final product had no obnoxious smell. Amino acid content of the DM had also increased during the 14 days. Although rabbits were found to consume the product readily and with no ill effects, assessments of digestibility were not undertaken. The usefulness of the method as a means of improving digestibility therefore remains to be established. Information is also required on the loss of DM occurring during treatment. The potential usefulness of aerobic thermophilic micro-organisms has also been investigated in the United States by Bellamy (1976), who has made a plea for much more research into the development of labor-intensive, low-technology, low-capital techniques for single-cell protein production from lignocellulose wastes in the less developed countries. In this regard, the lack of research attention given to harnessing the cellulolytic capacity of termites is surprising. Other potential biological systems for treating animal wastes (Calvert, 1977) include protein production from algae, coprophagous insects, earth worms, and other miscellaneous invertebrates.

Ensilage

Anaerobic fermentation of good quality forage crops to produce silage is extensively practiced as a means of conserving high moisture animal feedstuffs. It is also the method of conserving high moisture wastes such as sugar beet tops and sisal pulp. Sugar cane tops, bagasse, and animal excreta do not ensile as readily due to their low content of fermentable carbohydrate, but this can be overcome by adding a carbohydrate source such as molasses, cereal, or—in the case of wet excreta—hay.

In the case of animal excreta, ensiling has the additional role of tending to reduce the number of potentially pathogenic micro-organisms (*Salmonella* and coliforms, McCaskey and Anthony, 1975) and Nematode larvae, thus reducing the risk of disease transmission when recycling as feedstuff. A further benefit is that ensiling appears to increase the palatability compared with feeding fresh excreta.

Most of the research emphasis and application to date has been in the United States. It has concerned ensiling beef feedlot excreta with other feedstuffs and refeeding, the objectives being to reduce feed costs and waste disposal problems. The best known system "wastelage" (Anthony, 1971) involves ensiling a mixture of about 60 parts wet excreta and 40 parts ground hay. This undergoes a typical lactic acid fermentation, provided the total moisture content does not exceed

50%. One operating difficulty is that silage-making facilities must be continuously available (Smith, 1977). The system is therefore only readily feasible with top-loading, bottom-unloading tower silos. Caged poultry excreta and poultry litter have also been successfully ensiled when mixed with other feedstuffs (Smith, 1977); e.g., 60% grass hay and 40% poultry excreta, 15 to 30% of DM as poultry litter and 70 to 85% as whole-crop maize.

Ensiling is attractive as an on-farm method of processing animal excreta, because of its simplicity, low cost, and already widespread application for conserving conventional feedstuffs. It is also undoubtedly useful as a means of preventing nitrogen losses from animal excreta. However, it is important to remember that the process does nothing to overcome the low ME content of wastes (cf Table 2.3). The scope for inclusion of ensiled wastes such as cattle feces in diets is bound to be limited, unless digestibility is improved (e.g., by alkali treatment). There is more scope for poultry excreta because of its higher nitrogen content.

The long-term prospect for recycling ensiled (or any other form of) animal excreta as a feedstuff will depend, among other things, on the development of legislation regulating their use. This, in turn, will depend on an assessment of the human and animal health hazards likely to be involved. Current regulations vary from country to country (Ward and Muscato, 1976). Bhattacharya and Taylor (1975) listed the many potential hazards, but Hojovec (1977) considers there is need for further information concerning these. On the other hand, de Moor (1977) feels, from experience to date, that risks are small especially if proper control is exercised.

Chemical Sterilization

Within the context of recycling as animal feed, chemical treatment of animal excreta to kill pathogenic organisms has, surprisingly, received only little research. Work by Hatfield at Illinois University (Bullock, 1978) suggests that formaldehyde might be suitable. In his system, the waste is simply placed in a mixer wagon and mixed with formalin (4.2:1 of 40% formaldehyde/tonne). One commercial chemical solution available in the United States for treating cattle manure is Grazolin (Grazon, 1975). The product is designed to improve palatability and control smell as well as kill pathogens (Day, 1977).

Dehydration

High temperature drying of poultry excreta and poultry litter has been extensively practiced to facilitate their conservation and use as raw materials for manufacture of compound ruminant feedstuffs. Drying is also considered to reduce the risk of disease transmission. There is a volumnious literature on the subject. Of the 230 or so research publications during the period 1951 to 1977

(C.A.B., 1974, 1978), 180 concerned poultry wastes and nearly all involved dehydrated products. There is little doubt that dried poultry wastes can be useful as sources of nitrogen and minerals in the diets of less productive ruminants (Bhattacharya and Taylor, 1975; Smith, 1977). However, since the dramatic rise in fuel oil costs in the early 1970's, the economics of drying have become doubtful and there seems little prospect of it being economical in the future. Research (and practice) are now moving toward cheaper and less energy-intensive methods such as ensiling. In countries with suitable climates, sun drying has been and will continue to be used. High temperature drying of agricultural wastes other than poultry has never been economic.

CONCLUSIONS

The annual DM production of agricultural waste by-products is enormous, but the question remains as to how much is truly wasted. Even accepting the definition of agricultural wastes given at the outset, it has not been possible to be precise about the proportions used as animal feedstuff. It *is* clear that animal wastes are the least used. Furthermore, over two-thirds of the wastes produced are cereal straws.

Because of their fibrous nature or their non-protein nitrogen content (poultry excreta), animal wastes are only suitable as feedstuffs for ruminants, and even for these animals they are unsuitable as sole feeds. Their role is limited to being a component of the diet and, unless processed, to being only a very small component, especially in the case of highly productive ruminants.

From the review of processing methods to improve intake and ME content, treatment with alkali—and in particular with sodium hydroxide—offers the greatest improvement. This method is currently gaining popularity at both the farm and factory levels, but in the future, treatment with alternative chemicals (e.g., ammonia or naturally occurring minerals) may assume a greater role. Biological treatment to increase ME and protein content requires much further research and development. It is conceivable that this approach will assume major importance in the future, both as a method for upgrading wastes for ruminant feeding and for converting wastes into single-cell protein for non-ruminants. The evidence reviewed suggests a role for animal waste, especially poultry excreta, as a source of nitrogen and minerals in ruminant diets. The risk of disease transmission is not yet clearly defined and future legislation may well demand control measures which will make animal waste recycling uneconomic.

Attention was drawn earlier to the enormity of the annual DM production of agricultural wastes. This is reinforced in Table 2.5 by considering the number of cattle which could theoretically be maintained on crop wastes produced on farms. In practice, this would not be possible because of nutritional limitations,

TABLE 2.5.   Proportion of Cattle in Different Regions Which Could Have Been Supplied With Annual Maintenance Metabolisable Energy from Crop Wastes Produced on the Farm in 1977.

	ME Yielded by Crop Wastes Produced on the Farm[1] (MJ × 10^{12})	Cattle Population[2] (Head × 10^6)	Cattle Which Could be Maintained[3] (%)	
			On Crop Wastes Produced on the Farm	On Alkali Treated Crop Wastes Produced on the Farm[4]
Africa	0.722	161.6	22.7	29.5
North and Central America	3.598	186.3	98.0	127.4
South America	0.860	220.5	19.8	25.7
Asia	4.406	358.4	62.4	81.1
Europe	1.659	134.1	62.7	81.6
Oceania	0.107	41.6	13.0	16.9
U.S.S.R.	1.188	110.4	54.6	71.0
Developed countries	4.446	295.3	76.4	99.3
Developing countries	3.931	702.8	28.4	36.9

[1] Assuming yields and ME contents shown in Table 2.1 and 2.3, respectively.
[2] Based on production statistics given by F.A.O., 1978.
[3] Assuming maintenance ME requirement 54 MJ/head/day (MAFF, 1975).
[4] Treatment assumed to increase ME content by 30%.

and it should be remembered that the ME requirements of growing and lactating cattle would be two to four times their maintenance needs. Nevertheless, the data is impressive, especially for alkali treated wastes. Ironically, the table also demonstrates, at least for crop wastes, that production in relation to potential utilization (by cattle) is greater in the developed regions. It is also the case that intensive animal production—and hence the problem of animal waste disposal—is greater in the developed regions.

REFERENCES

Acland, J. D. 1971. *East African Crops*. F.A.O. and Longman Group, London.

Agricultural Research Council. 1976. *Studies on Farm Livestock Wastes*. Agricultural Research Council, London.

Andrews, R. P., J. Escuder-Volonte, M. K. Curran, and W. Holmes. 1972, "The influence of supplements of energy and protein on the intake and performance of cattle fed on cereal straws." *Anim. Prod.* **15**:167–176.

Anthony, W. B. 1971. "Cattle manure as feeds for cattle." *Livestock Waste Management and Pollution Abatement. Proc. Int. Symp.* (Ohio State University), Am. Soc. Agric. Eng. Pub. Proc. 271, pp. 293–296.

ARC. 1976. *The Nutrient Requirements of Farm Livestock. No. 4. Composition of British Feedingstuffs.* Agricultural Research Council, London.

Arnason, J. and M. Mo. 1977. "Ammonia treatment of straw." *Report on Straw Utilization Conference.* Min. Ag. Fish. and Food, Oxford, Feb. 24–25, 1977, pp. 25–31.

ASAE. 1967. "Method of determining modulus of uniformity and modulus of fineness of ground feed." *Am. Soc. Agric. Eng. Yearbook.* Recommendation ASAE **2461**:269.

Autrey, K. M., T. A. McCaskey, and J. A. Little. 1975. "Cellulose digestibility of fiberous materials treated with *Trichoderma viride* cellulase." *J. Dairy Sci.* **58**:67–71.

Baumgardt, B. R. 1970. "Regulation of feed intake and energy balance." *Physiology of Digestion and Metabolism in the Ruminant.* A. T. Phillipson (Ed.). Oriel Press, Newcastle-upon-Tyne, pp. 235–253.

Beardsley, D. W. 1964. "Nutritive value of forage as affected by physical form. Part II, beef cattle and sheep studies." *J. Anim. Sci.* **23**:239–245.

Beckmann, E. 1922. "Conversion of grain straw and lupins into feeds of high nutrient value." *Chem. Abstr.* **16**:765.

Bellamy, W. D. 1976. "Production of single cell protein for animal feed from lignocellulose wastes." *Wild Animal Review* **18**:39–42.

Bender, F., D. P. Heaney, and A. Bowden. 1970. "Potential of steamed wood as a feed for ruminants." *For. Prod. J.* **20**:36–41.

Bhattacharya, A. N. and J. C. Taylor. 1975. "Recycling animal waste as a feedstuff: A review." *J. Anim. Sci.* **41**:1438–1457.

Birch, G. C., K. J. Parker, and J. T. Worgan. 1975. *Food From Waste.* Applied Science Publishers, London.

Bullock, S. 1978. *Report on a Tour of the United States to Study the Recycling of Animal Wastes as Feedstuffs.* Nuffield Farming Scholarships.

Butterworth, M. H. 1963. "Digestibility trials on forages in Trinidad and their use in the prediction of nutritive value." *J. Agric. Sci.* **60**:341–346.

C.A.B. 1974. *Animal Excreta in Feeds for Livestock.* A.B. No. **38**. Commonwealth Agricultural Bureaux, Farnham Royal, Slough.

C.A.B. 1978. *Animal Excreta in Feeds for Livestock.* A.B. No. **38A**. Commonwealth Agricultural Bureaux, Farnham Royal, Slough.

Calvert, C. C. 1970. "Systems for the indirect recycling of animal and municipal wastes as a substrate for protein production." *New Feed Resources*. F.A.O. animal production and health paper **4**. Food and Agricultural Organization of the United Nations, Rome, pp. 245–264.

Campling, R. C. 1970. "Physical regulation of voluntary intake." *Physiology of Digestion and Metabolism in the Ruminant*. A. T. Phillipson (Ed.). Oriel Press, Newcastle-upon-Tyne, pp. 226–234.

Campling, R. C. and A. A. Milne, 1972. "Nutritive value of processed roughages for milking cattle." *Proc. Br. Soc. Anim-Prod*. N.S. **1**:53–60.

Cannell, R. Q., F. B. Ellis, and J. M. Linch. 1977. "Toxicity of straw to germinating seeds and seedlings of cereals." *Report on Straw Utilization Conference*. Min. Ag. Fish. and Food, Oxford, Feb. 24–25, 1977, pp. 11–13.

Chandra, S. and M. G. Jackson. 1971. "A study of various chemical treatments to remove lignin from coarse roughages and increase their digestibility." *J. Agric. Sci.* **77**:11–17.

Choung, C. C. and W. R. McManus. 1976. "Studies on forage cell walls. 3. effects of feeding alkali-treated rice hulls to sheep." *J. Agric. Sci.* **86**:517–530.

Cobbett, W. G. 1977. "Technical problems associated with the pulping of straw." *Report on Straw Utilization Conference*. Min. Ag. Fish. and Food, Oxford, Feb. 24–25, 1977, pp. 49–50.

Craven, T. A. 1973. "The use of straw for packaging and for the manufacture of compressed straw slabs." Paper given at the R. Agric. Soc. Engl. Symp.: *Straw Utilization and Disposal*.

Day, D. L. 1977. "Utilization of livestock wastes as feed and other dietary products." *Animal Wastes*. E. P. Taiganides (Ed.). Applied Science Publishers, London, pp. 295–314.

Dean, T. W. R. 1977. "The background to proposals for action." *Report on Straw Utilization Conference*. Min. Ag. Fish. and Food, Oxford, Feb. 24–25, 1977, pp. 56–58.

de Moor, A. G. 1977. "Potential health hazards and legal implications of waste recycling." *New Feed Resources*. F.A.O. animal production and health paper **4**. Food and Agricultural Organization of the United Nations, Rome, pp. 295–300.

Duckham, A. N., J. G. W. Jones, and E. H. Roberts. 1976. *Food Production and Consumption: The Efficiency of Human Food Chains and Nutrient Cycles*. Amsterdam, North-Holland.

F.A.O. 1977. *New Feed Resources*. F.A.O. animal production and health paper **4**. Food and Agricultural Organization of the United Nations, Rome.

F.A.O. 1978. *Production Year Book*. Volume **31**. Food and Agricultural Organization of the United Nations, Rome.

Franck, F., H. H. Meissner, and H. S. Hofmeyer. 1973. "Feeding value of sisal pulp for sheep." *South African J. Anim. Sci.* **3**:63–64.

Gohl, B. 1975. *Tropical Feeds*. Food and Agricultural Organization of the United Nations, Rome.

Gonzalez Santillana, R. 1977. *The Ensiling of Alkali Treated Straw*. M. Phil. Thesis, University of Reading.

Grazon, 1975. *Cattle Diets*. Grazon, Champaign, Illinois.

Greenhalgh, J. F. D. and G. W. Reid. 1967. "Separating the effects of digestibility and palatability on food intake in ruminant animals." *Nature* **214**:744.

Greenhalgh, J. F. D. and G. W. Reid. 1973. "The effects of pelleting various diets on intake and digestibility in sheep and cattle." *Anim. Prod.* **16**:223-233.

Greenhalgh, J. F. D. and F. W. Wainman. 1972. "The nutrative value of processed roughages for fattening cattle and sheep." *Proc. Brit. Soc. Anim. Prod.*, N.S. **1**:61-72.

Greenhalgh, J. F. D. and R. Pirie. 1977. "Alkali treatment of straw: A summary of experiments made at the Rowett Research Institute." *Report on Straw Utilization Conference*. Min. Ag. Fish. and Food, Oxford, Feb. 24–25, 1977, pp. 23-24.

Guggolz, J., G. M. McDonald, H. G. Walker, W. N. Garrett, and G. O. Kohler. 1971. "Treatment of farm wastes for livestock feed." *J. Anim. Sci.* **33**:284.

Harmon, B. G. 1976. "Recycling of swine waste by aerobic fermentation." *Wild Animal Rev.* **18**:34-38.

Hobson, P. N. and A. N. Robertson. 1977. *Waste Treatment in Agriculture*. Applied Science Publishers, London.

Hojovec, J. 1977. "Health effects from waste utilization." *Animal Wastes*. E. P. Taiganides (Ed.), Applied Science Publishers, London, pp. 105–109.

Hughes, R. G. 1977. "Fate of cereal straw 1976." *Report on Straw Utilization Conference*. Min. Ag. Fish. and Food, Oxford, Feb. 24–25, 1977, pp. 2-5.

Jackson, M. G. 1977. "Review article: the alkali treatment of straws." *Anim. Feed Sci. and Tech.* **2**:105-130.

Jackson, M. G. 1978. *Treating Straws for Animal Feeding*. F.A.O. animal production and health paper **10**. Food and Agricultural Organization of the United Nations, Rome.

Jayasuriya, M. C. N. 1974. *Treatment of Cereal Straws with Sodium Hydroxide to Improve their Value for Ruminants*. Ph.D. thesis, University of Reading.

Kategile, J. A. 1978. *Improvement of the Nutritive Value of Maize Cob using Sodium Hydroxide*. Ph.D. Thesis, University of Dar es Salaam.

Klinner, W. E. and R. U. Chaplin. 1977. "Developments in straw densification." *Report on Straw Utilization Conference*. Min. Ag. Fish. and Food, Oxford, Feb. 24–25, 1977, pp. 14–20.

Klopfenstein, T. J., V. E. Krause, M. J. Jones, and W. Woods. 1972. "Chemical treatment of low quality roughages." *J. Anim. Sci.* **35**:418–422.

Leach, G. and M. Slesser. 1973. *Energy Equivalents of Network Inputs to Food-Producing Processes*. Report Strathclyde University, Glasgow.

Linko, M. 1977. "Biological treatment of lignocellulose materials." *New Feed Resources*. F.A.O. Animal production and health paper **4**. Food and Agricultural Organization of the United Nations, Rome, pp. 39–50.

Loehr, R. C. 1974. *Agricultural Waste Management*. Academic Press, New York.

Loehr, R. C. 1977. *Pollution Control for Agriculture*. Academic Press, New York.

Lonsdale, C. R. 1976. *The Effect of Season of Harvest on the Utilization by Young Cattle of Dried Grass Given Alone or as a Supplement to Grass Silage*. Ph.D. Thesis, University of Reading.

MAFF, 1975. *Energy Allowances and Feeding Systems for Ruminants*. Tech. Bull. **33**., H.M.S.O., London.

McCaskey, T. A. and W. B. Anthony. 1975. "Health aspects of feeding animal waste conserved in silage." *Managing Livestock Wastes. Proc. 3rd. Int. Symp.* (University of Illinois), Amer. Soc. Agric. Eng., pp. 230–233.

Miles, T. R. 1977. "Straw uses as feed, fibre and fuel in Oregon." *Report on Straw Utilization Conference*. Min. Ag. Fish. and Food, Oxford, Feb. 24–25, 1977, pp. 75–81.

Minson, D. J. 1963, "The effect of pelleting and wafering on the feeding value of roughage: A review." *J. Brit. Grassl. Soc.* **18**:39–44.

Moore, L. A., 1964, "The nutritive value of forage as affected by physical form. Part 1. General principles involved with ruminants and effect of feeding pelleted or wafered forage to dairy cattle." *J. Anim. Sci.* **23**:230–238.

Mwakatundu, A. G. K. and E. Owen. 1974. "*In vitro* digestibility of sodium hydroxide-treated grass harvested at different stages of growth." *E. Afr. Agric. For. J.* **40**:1–10.

NFU. 1973. *Report on the Use and Disposal of Straw*. National Farmers' Union, London.

Owen, E. 1967. Unpublished data.

Owen, E., F. G. Perry, A. W. A. Burt, and M. C. Pearson. 1969. "Variation in the digestibility of barley straws." *Anim. Prod.* **11**:272.

Owen, E. 1976. "Farm wastes: Straw and other fibrous materials." *Food Production and Consumption: The Efficiency of Human Food Chains and Nutrient Cycles*. A. N. Duckham, J. G. W. Jones, and E. H. Roberts (Eds.). Amsterdam, North-Holland, pp. 299–318.

Owen, E. 1978. "Processing of roughages." *Recent Advances in Animal Nutrition 1978*. W. Haresign (Ed.). Butterworths, London, pp. 127–148.

Owen, E., M. Herrod-Taylor, M. Tetlow and J. M. Wilkinson. 1978. "Ensiled alkali-treated straw for sheep." *Anim. Prod.* 26:401.

Oyenuga, U. A. 1968. *Nigeria's Food and Feedingstuffs*. Ibadan University Press.

Pigden, W. J., G. I. Prichard, and D. P. Heaney. 1966. "Physical and chemical methods for increasing the available energy content of forages." *Proc. 10th Int. Grassl. Congr.* (Helsinki), pp. 397–401.

Pigden, W. J. and D. P. Heaney. 1969. "Lignocellulose in ruminant nutrition." *Cellulases and Their Applications*. R. F. Gould (Ed.), Am. Chem. Soc., Washington, D.C., pp. 245–261.

Pirie, R. and J. F. D. Greenhalgh. 1977. "Alkali-treated barley straw in complete diets for beef cattle." *Anim. Prod.* 24:147.

Porter, K. S. 1975. *Nitrogen and Phosphorus. Food Production, Waste and the Environment*. Ann Arbor Science, Michigan.

Preston, T. R. and M. B. Willis. 1970. *Intensive Beef Production*. Pergamon Press, Oxford.

Radley, R. W. 1978. "Straw for industry." *Are Today's Cereals Tomorrow's Crops?* M. G. Hughes (Ed.). Reading University Agricultural Club, pp. 51–56.

Raine, H. D. 1978. *On Farm Treatment of Cereal Straws with Sodium Hydroxide—Effect of Alkali Level, Post Treatment Neutralization and Concentrate Supplementation Upon Nutritive Value for Sheep*. Ph.D. Thesis, University of Reading.

Raine, H. D. and E. Owen. 1978. "On farm treatment of barley straws. Effect of sodium hydroxide, post treatment neutralisation and concentrate level on digestibility and intake by sheep." *Anim. Prod.* 26:399.

Rexen, F., P. Stigsen, and V. F. Kristensen. 1976. "The effect of a new alkali technique on the nutritive value of straws." *Feed Energy Sources for Livestock*. H. Swan and D. Lewis (Eds.). Butterworths, London, pp. 65–82.

Riddell, M. C. 1977. "Feasibility studies by the paper industry of pulp mill producing straw pulp." *Report on Straw Utilization Conference*. Min. Ag. Fish. and Food, Oxford, Feb. 24–25, 1977, pp. 51–53.

Rijkens, B. A. 1977. "Some possibilities for multiple uses of straw." *Report on Straw Utilization Conference*. Min. Ag. Fish. and Food, Oxford, Feb. 24–25, 1977, pp. 64–68.

Robb, J. 1976. "Alternatives to conventional cereals." *Feed Energy Sources for Livestock*. H. Swan and D. Lewis (Eds.). Butterworths, London, pp. 13–27.

Rockwell, P. J. 1976. *Single Cell Proteins from Cellulose and Hydrocarbons*. Noyes Data Corporation, Park Ridge, New Jersey.

Roy, J. H. B., C. C. Balch, E. L. Miller, E. R. Orskov, and R. H. Smith. 1977. "Calculation of the nitrogen requirement for ruminants from nitrogen metabolism studies." *Protein Metabolism and Nutrition*. Centre for Agricultural Publishing and Documentation, Wageningen, pp. 126–129.

Sachetto, J. P. 1977. "A project for producing chemicals from straw." *Report on Straw Utilization Conference*. Min. Ag. Fish. and Food, Oxford, Feb. 24–25, 1977, pp. 71–72.

Seal, K. J. and H. O. W. Eggins. 1976. "The upgrading of agricultural wastes by thermophilic fungi." *Food From Waste*. G. C. Birch, K. J. Parker, and J. T. Worgan (Eds.). Applied Science Publishers, London, pp. 58–78.

Smith, L. W. 1977. "The nutritional potential of recycled wastes." *New Feed Resources*. F.A.O. animal production and health paper 4. Food and Agricultural Organization of the United States, Rome, pp. 227–244.

Smith, L. W., H. K. Goering, and C. H. Gordon. 1971. "Nutritive evaluations of untreated and chemically treated dairy cattle wastes." *Livestock Waste Management and Pollution Abatement. Proc. Int. Symp.* (Ohio State University), Amer. Soc. Agric. Eng. Pub. Proc. **271**, pp. 314–318.

Srinivasan, V. R. and Y. W. Han. 1969. "Utilization of bagasse." *Cellulases and their Applications*. R. F. Gould (Ed.). Am. Chem. Soc., Washington, D.C., pp. 447–460.

Swan, H. and V. J. Clarke. 1974. "The use of processed straw in rations for ruminants." *University of Nottingham Nutrition Conf. for Feed Manufacturers*: H. Swan and D. Lewis (Eds.). Butterworths, London, **8**:205–233.

Terry, R. A., M. C. Spooner, and D. F. Osbourn, 1975. "The feeding value of mixtures of alkali-treated straw and grass silage." *J. Agric. Sci.* **84**:373–376.

Triolo, L. 1977. "Straw pulping for paper in Italy." *Report on Straw Utilization Conference*. Min. Ag. Fish. and Food, Oxford, Feb. 24–25, 1977, pp. 54–55.

van Soest, P. J. 1964. "New chemical procedures for evaluating forages." *J. Anim. Sci.* **23**:838–845.

van Soest, P. J. 1973. "Revised estimates of the net energy value." *Proc. Cornell Nutr. Conf.*, p. 11.

van Soest, P. J. 1976. "Laboratory methods for evaluating the energy value of feedstuffs." *Feed Energy Sources for Livestock*. H. Swan and D. Lewis (Eds.). Butterworths, London, pp. 83–84.

Wainman, F. W. and K. L. Blaxter. 1972. "The effect of grinding and pelleting on the nutritive value of poor quality roughages for sheep." *J. Agric. Sci.* **79**:735–745.

Ward, G. M. and T. Muscato. 1976. "Processing cattle waste for recycling as animal feed." *Wild Animal Rev.* **20**:31–35.

Weller, J. B. and S. L. Willetts. 1977. *Farm Wastes Management.* Crosby, Lockwood, Staples, London.

Wilkins, R. J. 1973. "The effect of processing on the nutritive value of dehydrated forages." *Proc. 1st Green Crop Drying Cong.*, Oxford, pp. 119–134.

Wilkinson, J. M., J. E. Cook, and B. F. Pain. 1978. "The nutritive value of wastelage—ensiled mixtures of cattle manure and forage. *Anim. Prod.* **6**:377.

Wilkinson, J. M. 1978. Personal communication.

Wilson, P.N. and T. Brigstocke. 1977. "Commercial straw process." *Process Biochem.* **12**:17–21.

Wilson, R. K. and W. J. Pigden. 1964. "Effect of a sodium hydroxide treatment on the utilization of wheat straw and poplar wood by rumen micro-organisms." *Can. J. Anim. Sci.* **44**:122–123.

Wood, R. S. 1977. "Straw burning—purposes, methods and costs." *Report on Straw Utilization Conference.* Min. Ag. Fish. and Food, Oxford, Feb. 24–25, 1977, pp. 6–8.

3

The use of sewage sludge as a fertilizer

Professor B. R. Sabey

Department of Agronomy, Colorado State University, Fort Collins, Colorado

INTRODUCTION

The use of sewage sludge and other organic materials as fertilizers for plant growth improvement is not new. The practice is almost as old as man's history on earth. Many nations or societies have used human and animal excreta for soil fertility enhancement throughout their history. These materials have been a source of nitrogen, phosphorus, potassium, and sulfur for plants, and have contributed to the maintenance of desirable physical properties of soils, thus allowing for more abundant food and fiber production than would have been possible without them. Yet, in recent times, man has come to recognize that application of excessive quantities of these materials to land poses potential pollution problems, and a rising concern has become evident in many localities.

Why did these localities become so concerned all of a sudden (historically speaking)? It was probably out of necessity. First, this pollution potential has resulted from the tendency for large human and animal populations to become concentrated in small areas, thereby concentrating organic residues to be utilized on limited land areas near production points.

Second, many of these societies have become more environmentally oriented. Concern over environmental pollution has risen greatly, causing us to evaluate more closely the effect that many of our waste disposal practices are having on our natural resources: the air, the water, and the soil. What would be the eventual effect on our waters of continued sewage disposal in streams, lakes, and oceans? What eventual influence will land filling have on our soils and ground water? What will happen to our air resource with continued incineration of sewage sludge?

These and other questions, designed to evaluate more closely the quality of our environment, have caused societies to restrict some or all of these disposal practices. Such restrictions have caused us to re-evaluate land application as a "waste treatment" or "resource recycling" mechanism, even though the economics of such a practice may not be optimum as presently conceived.

The terms waste and resource are used advisedly because the same product can be a waste or a resource, depending on the economics at the time of consideration. If the product has a negative economic value, it is considered a waste; if it has a positive value, it is considered a resource.

There are some contemporary aspects that tend to move sewage sludge away from a waste connotation and toward acceptance as an economic resource. World energy needs have increased due to apparent shortages of inexpensive energy sources (e.g., fossil fuels). This and economic inflation have caused the cost of inorganic fertilizers to increase. As the price of inorganic fertilizers increased, the comparative value of the lower nutrient level sludge also increased, making it more competitive as a plant nutrient supplier. In addition, with a tendency toward the development of greater environmental and resource conservation conscience, society feels that it is somewhat "immoral" to waste valuable soil amendments by disposal of sewage sludge; thus, social pressures dictate beneficial recycling of sludge. This newly acquired social conscience with respect to the environment has urged legislatures to pass laws prohibiting disposal of sewage sludge by ocean dumping, land filling, or incineration. These legislative restrictions rapidly changed the economic consideration, since they left fewer alternatives for sludge disposal (even though there were apparent economic advantages of such disposal methods). These and other factors have forced us into considering sewage sludge for its value as a fertilizer that supplies nitrogen, phosphorus, potassium, sulfur, and several other essential plant nutrients.

It takes an average of about 10,000 people in a community to produce the equivalent of 1 tonne of dried, anaerobically digested sewage sludge per day. In a year, this would be 365 tonnes or 365,000 kg. If 3.3% of that sludge were nitrogen, that would be about 11,000 kg of nitrogen per year. If all that were available, it could fertilize from 40 to 100 ha, depending on the crop and management system.

Actually, all the nitrogen will not be available. Very likely, even with excellent management, only 40 to 50% of the total nitrogen will be available during the first year, depending on the proportion of the total nitrogen that is in the inorganic form. Therefore, possibly 20 to 40 ha of productive land could be adequately fertilized from the sludge of a community with a population of 10,000. If we were to assume that 75% of a country with a population of 10 million had modern toilet facilities and treatment plants with anaerobic digestion systems, 15,200 to 30,300 ha of that nation's land could be adequately fertilized by the optimal use of the sludge. In the United States, no more than 0.4 to 0.8 million ha could be fertilized with the sludge produced by the present population. This is a very small proportion of the total agricultural land requiring fertilizer addition, amounting to a strip of land only about 160 km long and 26 to 52 km wide. In fact, if all the nitrogen contained in the sludge produced in the U.S.A. were available and usable, it would represent only about 4% of the total nitrogen fertilizer sold in that country.

It is apparent, then, that although sewage sludge can be used in localized areas as a valuable resource, it cannot play a very significant role in providing fertilizer for a very large segment of the United States.

There are some negative aspects that are aptly pointed out by those who oppose sludge application to the land. These include the potential problems of pathogenic populations in sewage sludges and distribution of disease (Hoadley and Goyal, 1976; Rudolfs *et al*, 1950); the application of toxic levels of heavy metals that may inhibit plant growth or may enter food chains and cause lethal or sub-lethal effects to humans and/or animals (Chaney and Giordano, 1976; Page, 1974); possible nitrate accumulation and leaching into ground water, eventually ending up in drinking water wells (Viets and Hageman, 1971; Committee on Nitrate Accumulation, 1972); the accumulation of polychlorinated biphenyls —PCB's (Parr, 1974; Tucker *et al*, 1975) and the other toxic organic compounds that may enter food chains; and finally, excessive salt accumulation in the soil and the resultant effect on germination, emergence, and soil fertility (Sabey and Hart, 1975).

The objective of studying the effect on sewage sludge application on the land would be to gain knowledge and understanding so as to allow the development of proper management systems. A management system that would allow the economic utilization of sewage sludge as a fertilizer and soil conditioner without causing pollution to our environment would be ideal.

SEWAGE TREATMENT PROCESS FOR LAND APPLICATION

Methods of treatment of raw sewage vary greatly. They consist of everything from essentially no treatment—where the raw material is used directly on the land, to well digested and irradiated sludge that has been dried and bagged. Applying raw sewage to land is practiced extensively in many areas of the world. In other areas, this practice is considered unsanitary and some kind of primary and even secondary treatment is recommended or required. In Fig. 3.1, several possibilities and pathways for treatment systems are illustrated. This is not meant to show all the potential ramifications, and only an elementary explanation will be attempted.

Generally, the raw sludge contains a multiplicity of materials consisting of various solids and liquids. The large solids are separated by a bar screen or rack, whereas the finer materials or grit are removed by a grit removal system. From here the flow stream takes the sewage to a primary settling tank or clarifier. In this tank, the solids with a density greater than that of the water settle on the bottom, while the solids that float are skimmed off the top. Both of these groups of solid materials are then sent to either an aerobic digester or an anaerobic digester.

The water that is separated can be piped to a trickle filter or to an activated sludge tank. As it passes through the trickle filter or activated sludge tank, many of the soluble and/or biodegradable organic compounds are converted into biomass and CO_2. Following either of these treatments, the solid biomass is separated from the water in another clarifier, whereupon the water is chlorinated and released to a stream.

The solids or sludge from either the top or the bottom of the primary settling clarifier can be piped to anaerobic or aerobic digesters for further processing. The solids from the clarifier separation following the trickle filter may be sent either to an activated sludge tank or directly to the digesters. Solids from the activated sludge tank clarifier are sent directly to the anaerobic or aerobic digesters.

The products produced with secondary digestion will be as variable as the content of the raw sewage supplied to the treatment plant. The chemical constituents found in typical aerobic and anaerobic sludges are shown in Table 3.1. The solids contents typically vary from less than 2% up to about 6%. The relatively stable digested sludge can be handled in several possible ways when land application and recycling are considered as a means of utilization. Without further treatment, the sludge can be applied to a field soil—on the surface, using flood or furrow irrigation, or by sprinkler or dribbled distribution systems. It also can be incorporated underground with subsoil injection equipment. If injected, several potential problems, such as nitrogen loss, pathogen exposure, and odor generation, are partially if not completely eliminated (which is obviously impossible in surface application).

One of the biggest problems associated with using the products of the secondary digester directly for land application is the high cost of transporting large quantities of water by tank truck or pipe pumping to use small quantities of solids on

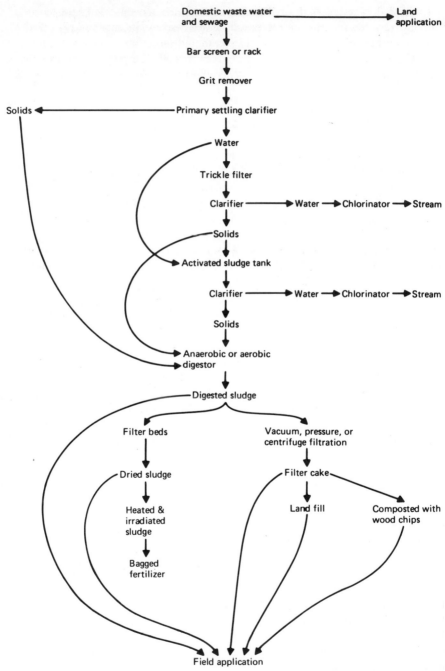

Fig. 3.1.   Generalized biological sewage treatment for land utilization of sludge.

TABLE 3.1. Typical Analysis of Anaerobically and Aerobically Digested
Sewage Sludge Produced by Metropolitan Denver Sewage
Disposal District No. 1.

	Elemental Analysis			
	Anaerobically Digested Primary Sludge (Total Solids−5.7%)		Aerobically Digested Waste Activated Sludge (Total Solids−4.4%)	
Element	Dry Weight Basis (%)	Wet (mg/l)	Dry Weight Basis (%)	Wet (mg/l)
N (Organic)	2.800	1,600.00	6.600	2,900.00
P	1.300	741.00	3.100	1,360.00
K	1.100	627.00	1.300	572.00
Ti	0.180	103.00	0.180	79.00
Cr	0.021	12.00	0.065	29.00
Mn	0.035	20.00	0.035	15.00
Fe	1.700	970.00	1.000	440.00
Co	0.001	0.60	0.001	0.44
Ni	0.024	14.00	0.025	11.00
Cu	0.130	74.00	0.230	101.00
Zn	0.400	228.00	0.480	211.00
Br	0.002	1.10	0.003	1.30
Rb	0.004	2.30	0.008	3.50
Sr	0.037	22.00	0.023	10.10
Y	0.010	5.70	0.004	1.80
Zr	0.041	23.00	0.014	6.20
Mo	0.006	3.40	0.001	0.44
Ag	0.007	4.00	0.011	4.80
Cd	0.003	1.70	0.009	4.00
Sn	0.020	11.00	0.016	7.00
Ba	0.170	97.00	0.084	37.00
Pb	0.170	97.00	0.093	41.00
U	0.002	1.10	—	—
As	—	—	0.002	0.88
Se	—	—	0.006	2.60

the land. Therefore, some treatment plants de-water or dry their sludges. This can be done in several ways (Fig. 3.1), including sand bed filtration, vacuum or pressure filtration, and centrifugation. Sand bed filtration and drying is a slow process. It may take six weeks to two or three months, but the product is relatively dry, ranging from 40 to 60% solids, is easily handled with a pitch fork or shovel, and can be stored in piles or hauled to the field by open bed truck and spread on the land. Some communities are heating or irradiating and bagging this material for use as fertilizers and soil conditioners.

Filter cake, a material that is about 15 to 20% solids and has the appearance of a thick mud, is a result of vacuum filtration, pressure filtration, or centrifugation. Flocculation of the solids is required for successful vacuum or pressure filtration. This can be accomplished by adding iron chloride and lime or by adding a number of suitable polymers presently available. The filter cake may be spread on the land surface or it can be injected underground using screw-type pumps which can handle mud-like materials. Filter cake is also disposed of via landfill. Further treatment of filter cake by mixing with wood chips and composting has been studied extensively in Maryland recently by USDA scientists. The composted product is separated from the wood chips and used as a fertilizer and soil conditioner. The filter cake can be spread on the land surface and incorporated into the soil with tillage equipment or it can be injected underground using the screw-type pumps.

SLUDGE CHARACTERISTICS

Like most products of biological systems, sewage sludges vary greatly in chemical, biological, and physical properties (Sommers *et al*, 1976). These characteristics will depend on the sources of sewage, the treatment system, and the management of the treatment system. Sludges from industrialized areas with laboratories and chemical processing facilities will generally have greater heavy metal contents than sludges from so-called "bedroom communities," even though zinc and copper originating in the plumbing pipes of homes are added to sewage systems of all modern communities. It is of interest to note that the sewage from homes varies in chemical composition with time of day. The waters in sewers in the morning are higher in zinc and copper than at other times of the day, resulting from water being stationary in the plumbing pipes at night, causing zinc and copper to dissolve slightly. Sludge composition changes with season of the year, especially when there are seasonal industries in the community. For example, pickle processing plants use large quantities of sodium chloride brine. As the used brine is discarded into the sewer system, the content of sodium and chloride is greatly increased. This often happens during the "pickle run" in the fall of the year. Other industries use other ingredients that may be undesirable in sewage sludge when land application is being considered. Table 3.2 shows typical ranges of chemical elements contained in sewage sludges (Dowdy *et al*, 1976). The differences between the minimum and maximum concentrations of most elements is from less than one to many orders of magnitude. Similar conclusions were reached by Sommers *et al* (1976) after analyzing sludges from eight Indiana communities. These data indicated the necessity of analyzing each sludge for land application, if appropriate application rates are to be recommended.

Not only do sludges vary greatly in chemical composition, but they vary in physical properties, too, depending on the method of treatment and treatment

TABLE 3.2. Total Elemental Composition of Sewage Sludge.[1]

Component	Minimum	Maximum	Median
	Concentration[2] (%)		
Organic C	6.5	48.0	30.4
Inorganic C	0.3	43.0	1.4
Total N	<0.1	17.6	3.3
NH_4^+-N	<0.1	6.7	1.0
NO_3^--N	<0.1	0.5	<0.1
Total P	<0.1	14.3	2.3
Inorganic P	<0.1	2.4	1.6
Total S	0.6	1.5	1.1
Ca	0.10	25.0	3.9
Fe	<0.10	15.3	1.1
Al	0.10	13.5	0.4
Na	0.01	3.1	0.2
K	0.02	2.6	0.3
Mg	0.03	2.0	0.4
	Concentration[2] (ppm)		
Zn	101	27,800	1,740
Cu	84	10,400	850
Ni	2	3,515	82
Cr	10	99,000	890
Mn	18	7,100	260
Cd	3	3,410	16
Pb	13	19,730	500
Hg	<1	10,600	5
Co	1	18	4
Mo	5	39	30
Ba	21	8,980	162
As	6	230	10
B	4	757	33

[1] Data compiled from over 200 samples from 8 states. (Dowdy et al, 1976.)
[2] Values expressed on 110°C weight basis.

management. Some sludges have as little as 1 or 2% solids carried in a water suspension. Typical of anaerobically digested sludge is a solids content of 4 to 5%. This material can be applied satisfactorily to soils by open furrow methods, sprinklers, and dribble systems, or by underground injection. Transportation costs become prohibitive if this type of sludge has to be transported great distances from the treatment facility, because of the large quantities of water involved. In order to overcome this problem, many treatment plants vacuum filter their sludge to de-water from 95 to 96% water down to 80 to 85% or less. This product is a mud-like material that can be transported in an open box truck rather than a tank truck. The de-watered material will have high concentrations of iron chloride and lime, a polymer, or a mixture of both since these are needed to flocculate the sludge before vacuum filtering. The addition of these chemicals and the filtration process not only change the physical condition but also change the chemical and biological nature of the resulting so-called filter cake. Whereas the unfiltered sludge has a considerable population of organisms (pathogenic and non-pathogenic), the lime and iron chloride treated filter cake has a very low population due to the extremely high pH of the sludge.

A comparison between anaerobically digested and aerobic waste activated sewage sludge indicates considerable differences in chemical properties. Table 3.1 illustrates some of the differences between the two types of sludges handled at the Metropolitan Denver Sewage Disposal District No. 2 plant in Denver, Colorado (Sabey and Hart, 1975). Note the higher nitrogen, phosphorus, and potassium content of the "waste activated sludge." There were some considerable differences between the two types of sludges in the content of some of the other elements as well. These may not be typical comparisons between the anaerobically digested and the waste activated sludges, but they do illustrate that the differences can be considerable.

NUTRIENT SUPPLY

Sewage sludge must be considered a low analysis fertilizer and soil conditioner. As indicated in Table 3.1, sewage sludges range from less than 0.1 to 17.6% nitrogen (with a median value of 3.3), less than 0.1 to 14.3% phosphorus (with a median value of 2.3) and less than 0.02 to 2.6% of potassium (with a median value of 0.3). Other nutrient elements are contained in greater or lesser amounts but are generally of less concern fertility-wise than the above mentioned elements. Only a portion of each of the essential nutrients contained in sewage sludge is available for plant growth during the first plant growing season after application. Those elements that are an integral part of the organic molecule in sludge are not immediately available to plant growth until mineralization (decomposition of the organic molecules to release inorganic ions) occurs. However, most of the nutrients that are found in organic molecules—such as nitrogen, phosphorus, and sulfur—are also found in inorganic forms in the sludge. Some of these inorganic forms are usable by the plants. For example, the median total nitrogen figure is 3.3%, whereas the median inorganic nitrogen value is about 1.0%. Thus, potentially 30% of the total nitrogen is immediately available for plant growth. According to the data for phosphorus in Table 3.1, about 70% is in the organic form, some of which will be available for plant use during the first growing season after sludge application. Most of the potassium will be in the inorganic form, and will be available to the plants immediately. Although some other elements contained in sewage sludge are essential for plant growth, adding sludge as a fertilizer to supply adequate amounts of most of them is not considered important except in areas where those specific nutrient deficiencies are encountered.

Of greater concern in the management of sludge application as a fertilizer to supply nitrogen, phosphorus, and potassium, is the concomitant addition of excessive amounts of elements such as zinc, copper, lead, nickel, and cadmium that can accumulate in the soil to toxic levels—toxic to either the plants that grow in the soil or to the animals (including man) that eat the plants. The possibility of toxic levels of nitrate, heavy metals, or toxic organic compounds entering into food chains represents one of the greatest deterrents to acceptance of land application of sewage sludge as an effective recycling practice.

SEWAGE SLUDGE DECOMPOSITION IN SOILS

Unlike fresh plant and animal residues that are incorporated into the soil, most sewage sludges have been through a biological treatment, wherein partial decomposition and stabilization have occurred. This results in a decrease in total amount of organic material, due to loss of carbon, either as carbon dioxide in an aerobic environment, or as methane and carbon dioxide in an anaerobic digestion. When the sludge is added to soil, the rate of decomposition will be somewhat slower than that of most fresh organic residues since the available energy material (soluble organic carbon compounds) and nutrients for microbial growth are less. The result is an increase in the level of soil organic matter, sustained over a longer period of time. In a given environment, this may have its advantages, including an increased source of plant nutrients and improved air and water relationships, resulting in greater plant growth. However, the possible increased mobility of some of the heavy metals due to the addition of naturally occurring chelating compounds in the sludge (Parsa and Lindsay, 1972) may cause leaching of the chelated metals into the ground water.

The specific decomposition rates of many plant and animal residues in soils have been determined, but little data are available on the rates of decomposition of sewage sludges added to soils. Estimates have ranged from 2% to greater than 60%. Rothwell and Hortenstine (1969) and Miller (1973a) determined that 17 to 20% of the organic carbon was mineralized in a six month laboratory incubation study, but Larson et al (1972) estimated only about 6% of the sludge was decomposed in a year. Agbim et al (1977) determined that 20 to 28% of the sewage sludge added to a soil in a laboratory incubation study decomposed in a 365 day period at typical room temperature ($21°C \pm 2$), based on CO_2 production. Generally, the higher the proportion of the sludge in the soil (higher application rate), the lower the decomposition percentage was in a given time period. In a laboratory column study, Miller (1973b) estimated that 3.3 to 3.4% of the sludge organic nitrogen was mineralized and leached through the columns in a six month period. At the other extreme is the work of King (1973), indicating that about 60% of the applied nitrogen in the sludge was mineralized in a laboratory study in 18 weeks of incubation. Ryan et al (1973) reported that 4 to 48% of the organic nitrogen in sludge was apparently mineralized in a 16 week incubation study, while a delay series of 35%, 10%, and 5% of the organic nitrogen in sewage sludge becoming mineralized for the first three years after application to soils is suggested by Pratt et al (1973). This series is based on the best estimates of the authors rather than from data produced in a specific study.

Nitrogen

No one of whom the author is aware has yet done a definitive study which gives primary evidence showing the rate of organic nitrogen mineralization from sludge. Most studies have not been able to distinguish between the NH_4^+ and NO_3^- that originated from the soil organic matter and the sludge. One method of approach to this difficulty would be to feed humans foods isotopically labelled with ^{15}N, then digest the fecal wastes in an anaerobic digester, after which the resulting sludge could be added to soil and the rate of mineralization of the ^{15}N labelled organic nitrogen compounds measured. This would allow one to distinguish between the inorganic nitrogen coming from the soil organic matter and that coming from the sludge. Other alternatives would include adding highly enriched ^{15}N to an anaerobic digester to which raw sewage sludge is gradually added, thus allowing the production of ^{15}N enriched sludge. This has been carried out in the author's laboratory. The enrichment, however, has only been about 2% ^{15}N, which is further diluted as the sludge is mixed with the soil, thus making precise measurements somewhat difficult.

Data for such a study should be analyzed and summarized soon, and this, it is hoped, will shed further light on the problem. Another alternative approach is that of Sommers (1977), wherein a population of yeasts was produced that was fed on ^{15}N enriched nutrients. These ^{15}N enriched yeast cells were then added to an anaerobic digester from which an artificial sewage sludge was produced. The sludge was then added to soil and measurements of the ^{15}N enriched NO_3^- and NH_4^+ production gave the net mineralization rate and provided a reasonably adequate data base for making recommendations of sludge application rates to meet specific nitrogen requirements for optimum plant growth.

An interesting phenomenon relating to this problem is the so-called "priming" effect that the addition of sewage sludge has on the native soil organic matter decomposition rates. A decomposition rate of 0.5 to 2% of the native soil organic matter may be increased considerably by addition of sewage sludge, thus releasing extra nitrogen for plant growth beyond what is normally anticipated from the soil organic matter. This further complicates the problem of recommending an accurate and sound application rate for sewage sludge.

An additional approach to determining the rate of mineralization of the organic nitrogen to sludge could be to choose a very sandy soil with little or no organic matter and then thoroughly leach the sludge and the soil to be used for the incubation study with KCl or $CaCl_2$ to exchange and remove the NH_4^+ and NO_3^- prior to the study. Any NH_4^+ and NO_3^- produced would then presumably have originated from the organic nitrogen of the sludge.

Neither of the last two approaches would be entirely satisfactory since both are artificial to a considerable degree, but they may add useful information.

Several aspects of the nitrogen cycle are of significance to environmentally sound and successful management of a land sludge recycling system. If we do not want to pollute the ground water by sludge or effluent applications to soil, we must either add only enough to supply the nitrogen for plant growth or we must learn to manipulate the cycle to decrease nitrogen in the soil-sludge mixture. This latter alternative is possible at two places in the cycle. We can enhance ammonia volatilization or enhance denitrification. Ammonia volatilization can be promoted by raising the pH, spraying, or drying the material before incorporating it into the soil. Denitrification can be increased in soils by adding readily available sources of energy material, such as sugars, alcohols, or other easily oxidizable carbon compounds, and by depleting the oxygen by saturating the soil with water after nitrite or nitrate has been produced.

When evaluating use of sewage sludge as a supplier of nutrients, the addition of excessive sewage sludge is not often considered, so mechanisms to promote denitrification or volatilization are ignored, although these mechanisms will occur naturally in the presence of excess sludge, so an understanding of them is pertinent.

In an attempt to better evaluate sewage sludge as a nitrogen supplying fertilizer, a greenhouse study was set up in the author's laboratory in which sewage sludge was added to a Nunn clay loam at rates ranging from 0 to 40 tonnes/ha (Sabey, 1977). Also, inorganic fertilizer was added to the same soil at rates of 0, 50, 100, and 200 kg of nitrogen/hectare as ammonium nitrogen. These soil and sludge or fertilizer mixtures were placed in greenhouse pots and wheat was grown for seven weeks after germination, emergence, and seedling establishment. After harvest of the wheat plants, nitrogen analyses of the plants and of the soil allowed an evaluation of the proportion of the total nitrogen added by the sludge that was taken up by the wheat growth in soils to which sludge had been added. The wheat growth due to inorganic nitrogen additions help us to evaluate the equivalent fertilizer value of sewage sludge. Table 3.3 shows the plant growth and nitrogen uptake in relation to sludge and inorganic nitrogen additions. Based on the data in Table 3.3 and the nitrogen analyses of the sludge, the plants, and the

TABLE 3.3. Plant Growth and Nitrogen Uptake in Relation to Sludge and Inorganic Nitrogen Additions.

Sewage Sludge Application Rate (tonnes/ha)	Plant Growth (g/pot)	N Uptake by Plants (mg/pot)
0	1.8	45
22.4	4.6	212
56	8.9	341
112	8.3	335
224	6.0	265
Inorganic Fertilizer Only (kg/ha)		
0	1.8	45
67	3.0	65
134	3.7	100
269	5.3	170

soil, the information in Table 3.4 was generated. It appears that the equivalent nitrogen fertilizer values for 0, 4, 8, 20, and over 40 tonnes of sludge per hectare are 0, 167, 458, 410, and 263 kg/ha, respectively, or 287, 502, 490, and 375 kg/ha, respectively, depending on whether the comparison between sludge and inorganic nitrogen is made on the basis of plant growth or on the basis of nitrogen uptake. These values may be quite valid for the soil, the plant used, and the other conditions of this particular study, but care must be exercised in extrapolating to other soils, crops, and conditions.

Although there has probably been less total effort expended on elucidating similar problems related to phosphorus and sulfur, some of the same considerations would be pertinent to a better understanding of mineralization of the organic forms of these elements.

Phosphorus

Phosphorus may be the factor governing the rate of application of sewage sludge to land once the patterns of nitrogen release are determined. Even though many sludges contain less than half as much phosphorus as nitrogen, there are not as many ways to promote phosphorus losses from the soil as there are for nitrogen. Short of incineration, little or no phosphorus volatilization occurs in soils, except possibly phosphine (PH_3) in paddy soils (Tsubota, 1959), but even this possibility is questioned by Burford and Bremner (1972). However, because of the large amounts of aluminium, iron, and calcium they contain, many medium to fine textured soils have an almost unlimited capacity to convert phosphorus into

TABLE 3.4. Parameters Measured and Calculated in the Greenhouse Study.

Parameter Measured	Sewage Sludge Application Rate (tonnes/ha)				
	0	22.4	56	112	224
Total N concentration in soil after harvest (mg/pot)	1886	2544	3326	5098	7028
Total N in pots initially (mg/pot) soil N plus sludge N	2100	2834	3934	5720	9440
Total N recovered in plant tissue and in the soil after harvest (mg/pot)	1931	2756	3667	5433	7293
Total N recovered after harvest (%)	91.9	97.3	93.2	94.1	77.2
Equivalent fertilizer N value based on plant growth (kg/ha)	–	210	575	515	330
Apparent N availability (%)	–	26	28	13	4
Equivalent fertilizer N value based on N uptake (kg/ha)	0	360	630	615	470
Apparent N availability (%)	–	44	31	15	6
N use efficiency (%)	–	29	19	9	4
N uptake due to sludge addition only (mg/pot)	0	167	296	290	220
N in the original pots minus N in the original soil (mg/pot)	0	734	1835	3670	7340
N use efficiency from treatment only (%)	0	22.7	16.1	7.9	5.0
Overall N use efficiency (%)	2.1	7.5	8.7	5.8	2.8

nearly insoluble compounds with these elements. It is therefore doubtful that on these soils phosphorus will limit sewage sludge applications.

Potassium

The quantities of potassium in most sewage sludges are quite low, as noted in Table 3.1; therefore, when using sludge as a fertilizer for soils that are deficient in potassium, it may be necessary to add supplemental potassium. However, in arid areas, many of which have soils with ample available potassium, some sewage sludges make a balanced fertilizer, supplying ample nitrogen and phosphorus as well as some of the micro-nutrients.

Essentially, all of the potassium contained in sewage sludge will be in the inorganic form; therefore, when added to the soil, it does not have to undergo mineralization to become available to plants, as is true of much of the nitrogen and phosphorus. Some potassium fixation can occur, however, especially in the presence of soils containing clays such as smectites and hydrous micas with a $2:1$ crystal structure.

Micro-nutrients

In the past, concern for micro-nutrients, including the essential heavy metals such as copper and zinc, has been in the area of deficiencies, and questions have not arisen concerning excessive amounts that could be toxic to either plants or the animals that ingest the plants. With respect to sludge application, concern has turned to toxicities and how much of the heavy metals, including cadmium, nickel, lead, and chromium, as well as zinc and copper, can be added before the plants are detrimentally affected and toxic levels accumulate in food chains. This problem will be discussed in a subsequent section. Most municipal sludges contain appreciable quantities of copper, zinc, iron, and manganese, with lesser quantities of boron, molybdenum, and chlorides (see Table 3.1). Generally, when sludges are added to most soils at rates supplying adequate nitrogen, they supply ample quantities of these micro-nutrients for plant growth. Additionally, the naturally occurring chelating compounds contained in—or produced during the decomposition of—the sludge tend to render these micro-nutrients more available to plants. The solubility and mobility of the metals added to the soil with sludge also vary with soil conditions and organism activity.

Soil micro-organisms not only affect the oxidation-reduction potential that influences solubility of the metals but also produce organic chelates that can result in leaching of the metals into the ground water, especially in acid soils. Metals are generally more soluble and mobile in low pH than in high pH soils and this fact makes it possible to either decrease or increase solubility, plant availability, and leachability of metal ions as the situation warrants.

POTENTIAL PROBLEMS OF SEWAGE SLUDGE

Nitrate Accumulation in Waters

The accumulation of nitrate in ground water, well water, lakes, and streams is considered by many to be undesirable. The United States federal government regulations state that 10 parts per million of nitrate nitrogen is the maximum permissible concentration in water used for human consumption. One of the reasons for such a guideline is the fear of methemoglobinemia (Blue Baby Disease) when water containing a high concentration of nitrate is used for baby formulas. Additionally, high nitrate levels have been cited as one of the causes of algal bloom in streams and autrophication of lakes (Commoner, 1968; Hoadley and Goyal, 1976). However, there is no universal agreement on this point. Ryther and Dunstan (1971), Frink (1971), and Commoner (1968) have implicated fertilization of agricultural lands as a major cause of higher nitrate levels in surrounding lakes and streams. Others have indicated that there is enough nitrogen in rainwater and in most lakes to fully support algae growth. Unless the nitrate level is too low for algae growth, the addition of more nitrate will not trigger algal bloom (King, 1970).

A mechanism that has considerable potential for triggering algal bloom is the addition of organic debris to water bodies. From the decomposing organic matter, carbon dioxide is produced. This carbon dioxide, along with water and sunlight, allows for rapid algae growth if sufficient levels of other essential elements are present. Most elements, including nitrogen and phosphorus, are contained in most water bodies and streams in high enough concentrations (0.01 parts per million phosphorus, 0.3 parts per million nitrogen) for algal growth. The extent to which sewage sludge applications will contribute to nitrate pollution will largely be dependent on management. If only sufficient sludge is added to supply adequate nitrogen for plant growth, there is unlikely to be a problem of nitrate pollution. Even when amounts of sludge are excessive of that needed for plant nutrition, nitrate pollution may not occur if denitrification can be promoted. This would convert much of the leachable nitrate to nitrogen oxide gases or to nitrogen gas, both of which would diffuse back into the atmosphere. Many sewage treatment plant operations are interested in disposing of the sewage sludge produced by their plants. There is, therefore, a tendency to apply as much sludge per unit area of land as possible. With this concept in mind, excessive amounts of sludge will be added to the field. This can be illustrated by the following example. Assume that 47 tonnes/ha of dry, anaerobically digested sludge is added to a field. The total nitrogen contained in the sludge could be about 1 tonne; of the 1,000 kg of nitrogen about 300 to 400 kg could be ammonium and nitrate nitrogen. This inorganic nitrogen is immediately available for plant utilization unless it is lost by ammonia volatilization during drying at the soil surface or by denitrification once nitrate is formed. Most plants will use no more than about 225 to 340 kg/ha of nitrogen. If no losses occurred, this would leave 75 to 175 kg of ammonium nitrogen that would eventually be converted to nitrate and have a high potential of being leached into ground water, wells, lakes, or streams. Fortunately, there will most likely be some losses occurring during or soon after surface application and during the decomposition process, especially if the system is managed with this objective in mind. In addition to the available inorganic nitrogen, some of the organic nitrogen will be mineralized and supply potentially leachable nitrate. At 67 tonnes/ha, the inorganic nitrogen contained in the sludge or produced during decomposition is excessive compared to that needed for optimum plant growth. Unless management practices are designed for removal of the excess inorganic nitrogen, ground and well-water pollution or runoff pollution are feasible.

Phosphorus Accumulation

Phosphorus differs markedly from nitrogen in several ways that relate to its pollution potential. First, phosphorus in its various forms is much less mobile than nitrate. The tendency to leach through the soil is much less. Second, phosphorus is not as easily lost from the soil by volatilization as is nitrogen. Third, phosphorus is used by plants to a lesser extent than nitrogen. All these facts lead to the conclusion that phosphorus could accumulate in the soil if sewage sludge were added in excessive quantities. The major removal mechanisms are by plant growth and erosion caused by surface runoff. When sludge application rates are governed by nitrogen requirements and when the phosphorus content of sludge is typically about one-half that of nitrogen, excessive phosphorus will be added to the soil. Fortunately, much of the phosphorus is converted to insoluble forms in many soils so it may not present a problem in those soils. In other soils, however, an overabundance of phosphorus can cause zinc deficiencies in plants, as well as other problems. After some period of time of sludge applications, phosphorus may have to be used as the element controlling sludge application rate instead of nitrogen.

Heavy Metal Accumulation

Many sewage sludges, especially from highly industrialized communities, are laden with high concentrations of heavy metals, each of which can be potentially toxic to plants, animals, or humans. Although there have been no documented cases of human deaths due to heavy metals accumulating in the food chain from a sludge source added to soils, the threat of sub-lethal effects concern some people. The possibility of lethal effects to plants growing on soils which have been treated with sludge for many years is more likely, but even this has not been fully documented, so irrefutible proof of potential serious problems of heavy metal accumulation and uptake on sludge applied land has yet to be established. Excellent reviews of heavy metal concerns have been prepared by Page (1974) and Chaney and Giordano (1976). They point out that plants vary greatly in their tendency to concentrate metals. The green leafy vegetables such as chard, beet, and lettuce are excellent concentrators of the metals, whereas most grains are less so. Additionally, various plant parts differ greatly in their concentration of the metals, with the leafy portions concentrating more than the stem and seed portions. An interesting aspect of the problem is that some metals which can be accumulated by plants can be toxic to the plant before the concentrations in the plant become toxic to humans eating the plant. Such is the case with zinc and copper, but it has been shown that cadmium can be concentrated in the plant sufficiently to become toxic to humans without being lethal to the plant (Menzies and Chaney, 1974). This makes cadmium a metal of great concern in considering sludge application to land for food production; thus, application rate based on cadmium content is becoming a primary or at least secondary guideline along with nitrogen and phosphorus contents.

Other metals, though of some concern, are considered much less of a problem than cadmium. Chromium is not readily taken up by plants because it is generally found in sludge in the Cr^{+++} form, which is not generally available to plants. Lead can be toxic in soils with very low phosphate levels, but in the presence of phosphate, lead is generally rendered unavailable to plants. Mercury and other metals are not typically found in sludges at high concentrations or are not soluble enough to be a serious problem when the sludges are added to the soil. This is especially true in soils containing high levels of organic matter. The organic matter increases the cation exchange of the soil, thereby increasing the tendency for metals to be adsorbed. The increase in organic matter in sludge also adds chelating compounds, some of which form stable complexes with the metals and withhold the metals from the plants. The extent of this mechanism depends on the stability constants of the chelate-metal complexes, the concentration of the chelate and the metals, and the pH of the amended soil mixture.

Chaney (1973) pointed out that the following factors control the metal toxicity to plants, especially that of zinc, copper, and nickel:

1. The amount and the combination of species of toxic metals present in the soil. Some metals are more toxic than others. Chembley (1971) developed the zinc equivalent concept based on studies that indicated that copper was two times more toxic and nickel was eight times more toxic than zinc. He recommended that a zinc equivalent (parts per million zinc + 2x parts per million copper + 8x parts per million nickel) of no more than 250 parts per million be added to soils with a pH of 6.5 or higher. Others have questioned the validity and/or the accuracy of the concept, but it does give a guideline for sludge application rates that some planners have used. Cadmium is considered the heavy metal of greatest concern when considering land application of sewage sludge. Chaney (1973) states that "we can only conclude that if we control the cadmium to zinc ratio so that when the zinc is injuring the crop enough that the farmers will add lime or change the crop to prevent injury from zinc, that the cadmium content will not be a hazard by Federal Drug Administration regulatory standards; hence the 0.5% cadmium of the zinc content."

2. The pH of the soil. This may be the most important factor of all, since as the pH is raised up to 7.0, the solubility of most of the metals decreases. Zinc, copper, and nickel concentrations that are not toxic to plants at a soil pH of 7.5 may be very toxic if the soil pH were decreased to 5.5. Chaney (1973) reported the data in Table 3.5 showing the effect of the amount of zinc added and soil pH on zinc content and yield reduction of chard. Additions of 262 parts per million of zinc were necessary before toxicity was evident at pH of 7.2, whereas at a pH of 5.3, toxicity was manifest at a zinc addition of 65.4 parts per million.

3. The amount of organic matter. This factor was discussed previously.

TABLE 3.5.   Effect of Zinc Added and Soil pH on Zinc Content and Yield Reduction of Chard Leaves.

Added Zn	Soil pH		
	5.3	6.4	7.2
ppm	μg Zn/g dry weight		
1.31	210	116	12
32.7	754	237	74
65.4	1058(7)[1]	337(5)	100
131.0	2763(41)	765(9)	177
262.0	2692(95)	1678(22)	406(27)

Source: Chaney (1973).
[1] Yield reduction in parentheses (%).

4. The phosphate content of the amended soil. The role of phosphate in reducing zinc availability to plants has been well documented. A high phosphate content in soil generally decreases the availability and toxicity of many of the metals to plants. However, phosphate enhances iron deficiency chlorosis caused by excessive copper and nickel (Spencer, 1966). Although the chemical reactions and interactions are complicated and sometimes counteracting, the addition of rather large amounts of phosphate in the sludge generally has a tendency to decrease metal toxicity.

5. The cation exchange capacity of the soil. A higher soil cation exchange capacity tends to cause greater binding of all cations, including the toxic metals. Chaney (1973) suggested that soils with a higher cation exchange capacity are inherently safer for disposal of sludge than are soils of low cation exchange capacity.

6. Reversion to unavailable forms. Generally, unless the pH of the soil is lowered, most of the toxic metals revert to a form that is insoluble and unavailable to plants. The reversion process is more rapid in high pH soils than in lower pH soils.

7. The plant grown on amended soil. Plant species and even plant varieties can vary as much as three- to ten-fold in tolerance to metal toxicity. Generally, vegetables are more sensitive to metal toxicity than are the cereal crops. The beet family is especially sensitive. Corn, small grains, and soya beans are moderately tolerant, whereas most grasses are very tolerant of high concentrations of metals by comparison.

Pathogens

Pathogenic organisms added to soil when sewage sludges are applied to land present a potential hazard to human and animal health. Most sludges are not pathogen-free, even though the numbers of most species are much lower in digested sludges than in raw or primary sludges. Some methods of treating sewage sludges prior to application to land help to decrease the hazards. These include composting, liming, irradiation, chlorination, pasteurization, and heat-drying. Although these treatments decrease the population of pathogens, it is questionable that even these, on a practical on-going basis, will render sludges 100% free of disease organisms. Fortunately, most of the pathogens "die-off" quite rapidly upon being incorporated into the soil, as noted in Table 3.6. However, some species persist for rather long time periods by various protective mechanisms. Many factors affect the persistence of pathogens in soils, including available energy supply, moisture, temperature, aeration, soil pH, and competition with the native population. The degree of seriousness and hazard associated with applying well digested sewage sludge is not fully known. Some would contend that since workers associated with sewage treatment plants do not have a higher incidence of disease than does the normal population, the hazard is not very great and the practice should not be prohibited for fear of epidemics. Others would say that this is only *prima-facie* evidence of the safety of the practice, and that there may be sub-epidemic illness prevalent among treatment plant workers or among those exposed to land spreading sites. They would advocate a very detailed epidemiological study over a long time period before they could endorse sludge application to land.

The fact is that most, if not all, serious disease outbreaks that have been traced to sewage wastes have been due to raw or poorly digested sewage rather than to digested, well treated sludge. Burge and Marsh (1978) stated that salmonellosis in the United States has infrequently been associated with sewage sludge even though salmonella organisms are found in digested sewage. They further state cases of enteric viral disease arising from well treated sewage are relatively few

TABLE 3.6. Pathogen Survival in Soils. Adapted from Parsons *et al* (1975).

Organisms	Media	Survival Time
Coliforms	Soil Surface	38 days
Streptococci	Soil	35–63 days
Fecal streptococci	Soil	26–77 days
Salmonellae	Soil	15–>280 days
Salmonella typhi	Soil	1–120 days
Tubercle bacilli	Soil	>180 days
Leptospira	Soil	15–43 days
Entamoeba histolytica	Soil	6–8 days
Enteroviruses	Soil	8 days
Ascaris ova	Soil	Up to 7 years
Hookworm larvae	Soil	42 days

considering the estimates of these organisms in sludge. He concludes that there is little evidence to show that disease is disseminated by well managed and controlled land spreading systems.

Pathogen Survival and Movement.   One of the greatest concerns of scientists and of the general populace in well developed nations associated with sludge and effluent application to soils is the potential pathogenic contamination of the water supply and food chains. As a result, some societies prohibit the application of raw sewage to soils. We are aware that in the Far East, raw sewage has been applied to soils, keeping them fertile and productive for hundreds of years. In that area, those who survive childhood have an apparent immunity to the illnesses caused by eating fresh, raw vegetables grown in those soils. A foreigner travelling in the area needs only to experience a severe case of dysentery, however, to realize the pathogen problem caused by frequent application of raw human wastes to soils.

Well digested sewage sludges have far fewer pathogens than do raw sewage sludges, although they are not pathogen-free. Some pathogens can and do survive the digestion process in the sewage treatment plant. Generally, during the time after sludge addition to soil, the pathogenic population decreases quite rapidly. Disappearance of these organisms is caused by antagonistic effects of saprophytic soil organisms, predation, nutritional factors, or other adverse environmental conditions. If the soil has a healthy, active native population, the rate of pathogen "die-off" is rapid. Data from Van Donsel *et al* (1967), as adapted by Miller (1973c), on survival of fecal coliform and fecal streptococcus in soils are given in Fig. 3.2. These data indicate a rapid decrease in the numbers of fecal coliform and fecal streptococci. Other investigators have not shown population decreases as dramatic as this figure shows (Rudolfs *et al*, 1950; Dunlop, 1968).

The data in Table 3.7 show that the numbers of some groups of organisms found in plots to which Denver sewage sludge was added were greater after five months of crop growth and summer weather than the numbers of organisms in the control plots, where no sludge was added. This was true of total numbers of aerobic bacteria and generally of total numbers of coliform and fecal streptococci, but was not true of fecal coliform.

Klein and Casida (1967) reported a reduction of 90% of *E. coli* numbers within five days in unamended soils, but if glucose was added, the lives of these organisms were prolonged in the soils. Evans and Owens, as reported by Robinson (1972), found that it took 57 days to reach 90% reduction of *E. coli* (applied to soils) in pig manure. Their studies showed that *E. coli* appeared in subsurface drainage tiles within 1.5 hours after application to an English soil and that the number of these organisms carried through the soil was related to the flow rate of water to the tile system. If one assumes that salmonella and related organisms act similarly, there is a possibility that pathogenic organisms found in sewage

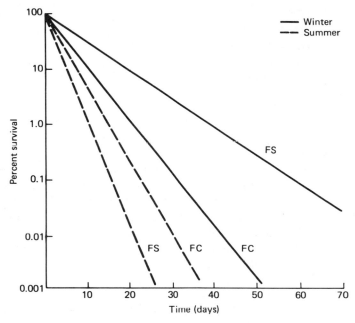

Fig. 3.2. Survival of fecal streptococcus (FS) and fecal coliform (FC) organisms in soils. *Source:* Van Donsel *et al.* 1967.

TABLE 3.7. Effect of Metropolitan Denver Sewage Sludge Application Rate of Survival of Four Bacterial Groupings* (Five Months After Sludge Application).

Sludge Applied (tonnes/ha) (dry basis)	Soil Depth (cm)	Total Aerobic Bacteria	Total Coliform	Fecal Coliform	Fecal Strept.
		Fallow Plot			
0	0–6.3	3.3×10^6	2.2×10^3	<1000	100
55	0–6.3	24×10^6	6.9×10^3	<1000	320
		Sorghum Plot			
0	0–3.1	4×10^6	1.4×10^3	<100	<100
11	0–3.1	13×10^6	0.3×10^3	<100	<1300
55	0–3.1	14×10^6	30×10^3	<100	<100

*Appreciation is expressed to Dr. Kenneth G. Doxtader and N. N. Agbim for making the microbial counts on these soils. (No./g of dry soil.) Unpublished data.

effluent and sludge and in animal manures may escape from soil in drainage water before being inactivated by soil. Work done by Peterson (1970) showed that there was a rapid decrease in fecal coliforms with lateral movement in ground water even in sandy soils. However, in test wells on the sludge application sites, fecal coliforms were detected, especially after a heavy rain. Organism movement into the well was due to gravitational water movement. In a test well, off the sludge application site, no test organisms were detected, indicating the filtering effect of soils as ground water moved laterally. Allen (1972) showed that a high percentage of mountain well waters are contaminated by fecal organisms from septic tank leach fields, especially where there is insufficient fine textured soil to provide adequate filtering between the well and the leach field. Bouwer (1968) reported that percolation through 5 to 10 feet of soil effectively removes fecal bacteria. Coliforms below 5 to 10 foot depths can sometimes be detected, but they are usually of soil origin rather than human origin (McMichael and McKee, 1965). Bouwer (1968) suggests that lateral ground water travel through 500 to 1000 feet of soil with a transit time of several months is sufficient to provide hygienically safe well water. Kardos (1970) found no bacterial contamination in well waters when sewage effluent was recycled in forested lands with the leachate being filtered through the soil to the well.

Although there is a rapid decrease with time in pathogens added to soil via effluents and sludge, some organisms have the capacity to survive in the soil by spore formation or other protective mechanisms. However, though the potential for disease is there, many persons work around sewage sludge and effluents daily with few documented cases of illness due to the pathogens contained in the wastes. With reasonably low rates and infrequent applications of wastes, the possibility of problems caused by pathogens will be small.

Viruses present a more difficult problem to handle, since culturing and isolating them is more complicated than with many other organisms. Most work on pathogens in the soil due to sewage effluents or manure application has been with bacteria. Merrill and Ward (1968) reported an absence of viruses after 1500 feet of soil percolation, but presence of viruses when the distance was only 400 feet.

Other work on pathogenic organism survival and movement through soils has been reported by Krone (1968) and Dunlop (1968), but much more research is needed on bacteria, viruses, and other pathogens added with effluents, sludge, and other organic waste materials.

Organic Compounds

Little work has been done on the effects of some of the toxic organic compounds that are contained in sewage sludges. These include insecticides, herbicides, and other pesticides, along with polychlorinated biphenyls (PCB's). Guenzi (1974) has treated the subject of the nature and stability of some of the pesticides that are added to the soil and the subsequent impact on the soil and food chains, and the possibility of ground water contamination. His discussion involves the addition of the pesticides to plants which eventually reach the soil, and he raises many questions about the long-range influence of persistent compounds leading to potentially toxic levels. It is assumed that although sewage sludges contain low concentrations of many of these pesticides, the effects would not approach those of direct spraying on the crops grown on the land. However, the long-term effects must be investigated and considered in designing land spreading systems.

Several workers have investigated the chemical and biological stability of PCB's in soils and found them generally to be quite persistent and toxic (Parr, 1974; Tucker *et al*, 1975). Tucker *et al* (1975) reported that the persistence and solubility of the PCB's are related to the degree of chlorination.

When sludge containing PCB's is added to soils, the PCB's are absorbed both by the clay minerals and the organic matter and generally have low solubilities. However, the potential for plant uptake and leaching into the ground water is ever-present. The information on absorption of the PCB's by the plant roots or the leaves after volatilization from the soil, the effective adsorption by the soil, and the amount of leaching in the ground water, is sketchy and fragmented. The amount of PCB uptake by plants cannot be estimated accurately from the amount of the PCB's in the soil because many variable factors influence plant uptake. Undoubtedly, these mechanisms are dependent on many environmental factors that need further research to elucidate the many ramifications associated with this problem.

Salts

Some sewage sludges contain considerable quantities of salts. If such sludges are applied to poorly drained lands where the water table is near the soil surface and the net water movement is upward, a salt problem could arise, particularly if concentrations of sodium salts (especially carbonate) are present. The difficulty associated with salt accumulation in the soil revolves around increased osmotic pressure in the soil solution, which may decrease plant uptake or seed germination and produce a degradation of the soil physical properties. A basic understanding of the mechanisms surrounding this problem exists, and if a knowledge of the initial salt content of the soil and the sludge, the moisture movement relationships of the soil, and the quantities of water to be added to the soil is available, the magnitude of the salt problem can be predicted (Martin *et al.*, 1976). In well drained irrigated soils, soil salinity is not likely to be a problem, although salts leached out of the zone of incorporation of the sludge may move through the soil profile and into the water table and degrade ground water quality. The difficulty in assessing the seriousness of this problem is in knowing the degree of dilution of the salts as they move into ground water.

SLUDGE MANAGEMENT

Application Techniques

Numerous methods of application of sewage sludge to land exist, each with advantages and disadvantages. The moisture condition of sludge to be spread controls, to some degree, the techniques used. Sludges that have less than 70% water are relatively dry and can be spread with a manure spreader or equipment similar to it. Sludges with moisture contents of 75 to 85% (so-called "filter cake") can be hauled in an open box truck and be dumped or spread with some success with a manure spreader, especially at the drier end of the range. There are screw-type pumps that can push this material through pipes behind sweeps or in furrows into the soil. High moisture sludges ranging from 90 to 99% moisture can be spread by pump and sprinkler or dribble systems, including adapted pivot sprinkler or other self-propelled systems. The traveling sprinkler gun has been used successfully for this kind of sludge. Special provisions are usually necessary to insure that larger chunks of solids do not clog the outlets in these distribution systems. These require pump power and pipe conveyance. Open ditch conveyance is possible with high moisture sludges. The ditch feeds a ridge and furrow or flood irrigation system in order to evenly distribute the sludge over the land. This type of sludge can also be distributed by tank truck spreading on the soil surface or by tank truck fed injection apparatus similar to that developed by Smith of Colorado and reported by Hall (1975), or into a plow furrow or disk plot slit in the soil. One of the problems associated with high moisture sludges is the tendency for solids to settle to the bottom of holding lagoons and transport and feed tanks, thereby resulting in heterogeneous distribution of the solids on the land.

Application Rates

One of the most serious problems associated with land spreading of sewage sludge is the lack of understanding of the physical, biological, and chemical reactions in the soil and thus the inability to make accurate recommendations for application rates. Several criteria have been suggested, including nitrogen and phosphorus supply for plant growth and heavy metal loadings. Nitrogen supply has been the most persistent criterion suggested by most experts. This is because of the possibility of nitrate accumulation over and above the nitrogen needs of the plant and thus the potential for nitrate leaching into the ground waters. This, it is felt, should be avoided if at all possible. However, since the amount of nitrate that accumulates in the soil is dependent on several biochemical reactions, each of which is dependent on a variety of environmental conditions, nitrate accumulation and leaching potential is extremely difficult to predict.

Nitrogen

We need to know the amount of total organic and inorganic nitrogen in the sludge as well as the nitrogen supplying capacity of the soil. The latter can be estimated by soil testing procedures and the former from standard chemical analyses. There are other factors that are not so easily determined or estimated. These include an estimate of the amount of ammonium that will be lost by volatilization when added in a high liquid sludge on the surface of the soil or plant tissue. This is highly dependent on the pH at which the sludge dries. At a high pH, more ammonia will be lost by volatilization than at low pH. It further requires an estimate of the amount of denitrification that occurs. Anaerobic conditions in the presence of ample available organic material are conducive to rapid loss of nitrogen into the atmosphere, resulting in less nitrate remaining in the soil. It also requires an estimate of the amount of organic nitrogen that will be mineralized in a growing season. This depends on the stability of the sludge and the environmental conditions that influence microbial activity in the soil. If all these estimates were available, it would be relatively simple to make accurate recommendations for the proper sludge application rates necessary to supply the nitrogen needed for plant growth and avoid applications of excess nitrogen that could be leached into the ground water. A formula for making application recommendations for optimum nitrogen levels was presented by the Colorado State Health Department (1978). These calculations were adapted from those of Keeney *et al* (1975). The following information and sample calculations were taken from the above policy statement.

Nutrient Availability Due to Methods of Application. Different methods of application result in different percentages of ammonia nitrogen in the sludge that will be available to the crop being grown. For design purposes the following percentages should be used:

1. Surface application of high liquid sludge without incorporation results in an estimated 80% loss of ammonia nitrogen, leaving about 20% of ammonia available for plant growth.

2. Surface application of high liquid sludge with immediate incorporation results in 30 to 70% loss of ammonia nitrogen, depending on how soon incorporation occurs.

3. Subsurface application of high liquid sludge results in negligible loss, so most of the ammonia nitrogen would be available.

4. Some losses of nitrate can occur when denitrification proceeds as soils are aerated and then flooded for a period of several days. The amount depends on the conditions in the soil.

5. There is a five year cycle of organic nitrogen availability. In the first year after the sludge application, 20% of the organic nitrogen is estimated to be

available to the crop. In the second, third, fourth, and fifth years, it is assumed that another 10%, 5%, 2.5%, and 1.25%, respectively, of the organic nitrogen added becomes available to the crop from the sludge applied the first year. After the fifth year it is thought that no appreciable organic nitrogen will become available for crop usage from the first year application. Calculations should be made to determine in what year equilibrium will be reached for any design scheme. Sample calculations for any given year are given below.

Calculation for Determination of Nitrogen Requirements for Sludge Application to Land. These procedures are recommended for calculating sewage sludge application rates for a field, with or without a soil test report.

1. With a soil test (this is most desirable):
   A. Obtain nitrogen recommendation in kg/ha = (A) from test results.
   B. Calculate the available N from NH_4^+ in sludge added for the current year, using the following formula:

   $$\% \ NH_4^+\text{-N in sludge} \times \frac{1000 \ \text{kg/tonne}}{100 \ (\text{conversion from } \%)}$$

   $$= \% \ NH_4^+\text{-N} \times 10 = (B) \ \text{kg } NH_4^+\text{-N/tonne of sludge.}$$

   If surface-applied and not incorporated until dry, reduce this value (B) to one-fifth ($\times$ 0.2).

   C. Calculate the available N from organic N in sludge added for the current year, using the following formula:

   $$\% \ \text{organic N} \times \frac{1000 \ \text{kg/tonne}}{100 \ (\text{from } \%)} \times 0.2 \ (\text{mineralization rate, 20\% during}$$

   the first year) = % organic N $\times$ 2 = (C) kg of available N from organic N/tonne of sludge.

   D. Residual sludge N is soil = (D) kg N/ha. If soil has received sludge in the past five years, calculate residual N from Table 3.8.
   E. Sludge application rate, tonnes/ha (E)

   $$= \frac{\text{Nitrogen recommendations kg/ha} - \text{residual N, kg/ha}}{\text{kg available N/tonne sludge}}$$

   $$= \frac{(A) - (D)}{(B) + (C)} \ \text{tonne/ha.}$$

   Example calculation for determining how much sludge to apply: corn for grain; Weld County, Colorado; yield potential—very high; surface application of high moisture sludge for the third year; 11 tonnes/ha were applied in the first and second years.

Soil Test Results

Texture: Clay loam
Organic matter: 1.5%
NO_3^-: 12 ppm

Fertilizer Recommendations

Corrective and maintenance
N: 180 kg/ha

Sludge N Analyses

NH_4^+ = 1.5%, organic N = 2.5%

a. Fertilizer N recommended = 180 kg/ha = (A).
b. Available N from NH_4^+ in sludge (current year) = 1.5 (% NH_4^+-N) × 10 × 0.2 (for surface application) = 3 kg/tonne = (B).
c. Available N from organic N in sludge (current year) = 2.5 (% organic N) × 2 = 5 kg/tonne = (C).
d. Residual N from Table 3.8 for 2.5% organic N sludge added one year previously 11 tonne/ha × 2.5 kg/tonne = 28 kg/ha. Sludge added two years previously 11 tonnes/ha × 1.25 = 14 kg/ha.
e. Sludge application rate = $(A - D)/(B + C)$ = $(180 - 42)/(3 + 5)$ = 17.25 tonnes/ha.

TABLE 3.8. Estimated Release of Available N (Per 1000 kg of Solids) During Sludge Decomposition. (This table can be used to calculate D in the formula.)

Years After Sludge Application	Mineralization Rate (%)	N in Sludge (%)						
		2.0	2.5	3.0	3.5	4.0	4.5	5.0
		kg of N available/tonne sludge						
1st	20	4	5	6	7	8	9	10
2nd	10	2	2.25	3	3.5	4	4.5	5
3rd	5	1	1.25	1.5	1.75	2	2.25	2.5
4th	2.5	0.5	0.62	0.75	0.87	1.0	1.12	1.25
5th	1.25	0.25	0.31	0.38	0.44	0.5	0.56	0.6

Example for use of Table 3.8: If you were calculating the amount of available N that would be released from the organic N contained in the sludge added in previous years (D in the formula) for the current year, this table could be used.

Assume that 5 tonnes of sludge had been added each year for four previous years and you want to know how much residual N will be supplied by the fifth year from the previous sludge additions. If the sludge had 2.5% organic N, then 0.31 kg, 0.62 kg, 1.25 kg, and 2.5 kg would be supplied from 1 tonne of sludge applied during the four previous years. That would be a total of 4.68 kg of N/tonne of sludge. Since 5 tonnes/ha of sludge were added each year, then 4.68 × 5 = 23.4 kg/ha of N would be available during the fifth year from the previously applied sludge.

2. Without a soil test, the N requirement can be obtained from Table 3.9 instead of from a soil test report (this is not as desirable as getting the N requirement from a soil test).

    A. Obtain N requirement from Table 3.9 = (*A*) kg/ha.

B. and *C*. Calculate available N in sludge as in Items *B* and *C* (with soil test). (*B*) and (*C*) kg N/tonne of sludge.

    D. Residual sludge N in soil = (*D*) kg/ha. If soil has received sludge the past four years, calculate residual N from Table 3.8 as in item *D* above.

    E. Sludge application rate, tonnes/ha

$$= \frac{\text{kg crop N requirement} - \text{kg residual N}}{\text{kg available N in sludge}}$$

TABLE 3.9. Example Nitrogen Requirement for Colorado Crops.

Crop	Typical Yields (tonnes/ha)	Available N Required (kg/tonne)	Estimated N Required (kg/ha)
Corn for grain	7.5–8.8	24	191
Corn for silage	60–70	3.5	224
Wheat[1]	3.8–5 (irrigated)	41	179
	1.3–1.9 (dry land)		67
Oats[1]	5–6.3 (irrigated)	20	112
	2.5–3.8 (dry land)		45
Barley[1]	5–6.3 (irrigated)	28	157
	2.5–3.8 (dry land)		56
Malting barley[1]	5–6.3 (irrigated	24	135
	2.5–3.6 (dry land)		50
Grain sorghum for grain	9 (irrigated)	20	179
Grain sorghum for silage	35–45 (irrigated)	3.7	146
Forage grasses	9–14 (irrigated)	21	224
Potatoes	28–35 (irrigated)	5	157
Sugar beets	45–60 (irrigated)	3	157
Alfalfa[2]	9–14 (irrigated)	19[2]	224
Clovers[2]	9–14 (irrigated)	19[2]	224
Dry beans[2]	2.2 (irrigated)	102[2]	224

[1] If the straw is removed.
[2] Legumes can obtain most of their N from the air and are normally not fertilized with N. However, if included in a crop rotation with N using crops, they will use the available N in the soil and will fix very little N from the air. Therefore, it can be assumed that they will remove as much N as corn for grain in the same rotation.

$$= \frac{(A) - (D)}{(B) + (C)} = \text{tonnes/ha.}$$

Example calculation:

From Table 3.9, N needed for 7.5 to 8.8 tonnes of corn = 191 kg N per acre = (A). The remainder of the calculations are as shown previously in (1), therefore:

$$E = \frac{191 - 42}{3 + 5} = 18.6 \text{ tonnes/ha.}$$

This estimated value is slightly higher than the previous calculation of 17.25 tonnes/ha, but for estimation purposes, it is quite close.

As can readily be surmised, the values used in Table 3.8 indicating the amount of available nitrogen from the organic nitrogen in the sludge during the year of application and during the subsequent four years are "best estimates" based on a composite of data available, and variations due to change in conditions can and will occur. Therefore, the values in the table are only rough guidelines to illustrate the use of the formula.

The other values that may be suspect are those estimating losses of ammonium due to volatilization of ammonia. Values that are higher or lower may be more appropriate under any given set of environmental conditions. Again, the values used are illustrative only.

Phosphorus

The phosphorus added to the soil in sewage sludge is not generally lost from the soil as is nitrogen, although much of it can become unavailable due to precipitation or adsorption. Typically, when sludge is added to the soil to supply sufficient nitrogen, adequate phosphorus will be added for optimum plant growth, since plants need only about $\frac{1}{10}$ to $\frac{1}{5}$ as much phosphorus as nitrogen, and sludges typically contain about $\frac{1}{2}$ as much phosphorus as nitrogen. The primary phosphorus removal mechanism from the soil is plant utilization and harvest from the land. Therefore, phosphorus could potentially accumulate in some soils to toxic levels and may eventually become the governing factor for sludge loading rates. After applying sludge to the land for five years or more, the phosphorus content of the soil should be determined. If the phosphorus content as determined by the Olsen bicarbonate test (Olsen, 1954) exceeds 450 kg/ha, then the sludge application rates should be reduced or ceased altogether.

Potassium

The potassium level of most sludges is quite low compared to the levels nitrogen and phosphorus. Since potassium is needed by plants in rather large amounts, a supplemental source of potassium may have to be supplied to provide a balanced nutrient level for optimum plant growth. This will be especially true in soils of low pH or soils of the more humid and highly leached parts of the world. In many of the arid areas, the soils have an ample supply of available potassium and supplemental potassium is not needed.

Trace Elements of Micro-nutrients

From a short-term perspective, the loading rates calculated to provide nitrogen for growing crops usually result in the application of trace elements to the soil in amounts that would not result in toxic levels in water and food chains. In the long run, the trace element content of the soil should be monitored to ensure that these trace elements do not build up in the soil. The frequency of monitoring of the soil will be dependent upon the quality of the sludge. Monitoring of the trace elements in the sludge soil mixture should be done at least once every five years starting with the fifth year of continuous application.

Most of the trace elements in sludge applied to soils are retained in surface soils; only small amounts may go into solution. Thus, pollution potential for surface waters is greater than that of ground water. Since trace elements applied to the soil are largely concentrated in the surface, runoff into surface waters may contribute to the pollution of water bodies. Concentration in water of trace elements considered to be toxic to aquatic organisms are, in many cases, lower than those considered toxic to higher plants and animals. Since these tolerances are so low, surface runoff of either sediment or solution into surface waters must be held to a minimum.

Optimum Conditions For Use of Sewage Sludges. It is difficult, if not impossible, to specify exactly the soil properties that are optimum for land application of sludge as a fertilizer. There are so many factors that influence so many aspects of the practice and some are incongruous with others. As a generalization, however, we could say that for optimum soil fertility, the extremes should be avoided. For example, if a soil is too sandy, the infiltration may be too rapid and the adsorptive capacity may be too low. Therefore, positively charged ions and particles, as well as pathogens, may move with percolating waters to a greater extent than is desirable. On the other hand, if a soil is too high in clay, it may be poorly drained and runoff and erosion could occur. Flach and Carlisle (1974) suggested that the information in Table 3.10 could be used as a guide for choosing the most desirable soils for sludge application. Certainly, texture, structure, organic matter content, and pH are important soil properties that influence infiltration, permeability, runoff, adsorption, and precipitation—and, ultimately, whether the soil is suitable for sludge application or not.

TABLE 3.10. Soil Limitations for Accepting Non-toxic Biodegradable Solids and Sludges.[1]

Item Affecting Use[2]	Degree of Soil Limitations		
	Slight	Moderate	Severe
Permeability of the most restricting layer above 150 cm.	Moderate, rapid, and moderate (1–5–15 cm/hr)	Rapid, and moderate, slow[2] (15–50 and 0.5–15 cm/hr)	Very rapid, slow, and very slow (50 and 0.5 cm/hr)
Soil drainage class	Well drained and moderately well drained	Somewhat excessively drained and somewhat poorly drained	Excessively drained, poorly drained, and very poorly drained
Runoff	None, very slow, and slow	Medium	Rapid and very rapid
Flooding	Soil not flooded during any part of the year		Soil flooded during some part of the year
Available water capacity from 0 to 150 cm or to a limiting layer[3]	19.5 cm	7.5–19.5 cm	7.5 cm

[1] Modified from a draft guide dated April 27, 1973, for use in the Soil Conservation Service, USDA. Solid wastes are those that cannot be moved by pumps.
[2] For definitions, see *Soil Survey Manual, USDA Handbook* 18, 1951.
[3] Available water capacity, as used here, is the difference between the amount of soil water at field capacity and the amount at wilting point.

REFERENCES

Agbim, N. N., B. R. Sabey, and D. C. Markstrom. 1977. "Land application of sewage sludge: V. Carbon dioxide production as influenced by sewage sludge and wood waste mixtures." *J. Environ. Qual.* 6:446–451.

Allen, M. 1972. "How safe is mountain well water." *CSU Research* 22(2):10–12. Jan.–March, Colorado State University, Fort Collins.

Bouwer, H. 1968. "Returning waste to the land. A new role for agriculture." *J. Soil Water Conserv.* 23:5.

Burford, J. R. and J. M. Bremner. 1972. "Is phosphate reduced to phosphine in waterlogged soils?" *Soil Biol. Biochem.* 4:489–495.

Burge, W. D. and P. B. Marsh. 1978. "Infectious disease hazard of landspreading sewage wastes." *J. Environ. Qual.* 7:1-9.

Chaney, R. L. 1973. "Crop and food chain effects of toxic elements in sludges and effluent." *Proc. Joint Conf. on Recycling Sewage Sludges and Effluents on Land, sponsored by USEPA, USDA, and NASULGC,* Champaign, Illinois, July 9-13, pp. 129-140.

Chaney, R. L. and P. M. Giordano. 1976. "Microelements as related to plant deficiencies and toxicities." *Soils for Management of Organic Wastes and Waste Waters.* L. F. Elliot and F. J. Stevenson (Eds.). American Society of Agronromists, pp. 235-279.

Chumbley, C. G. 1971. "Permissable levels of toxic metals in sewage used on agricultural land." *A.D.A.S. Advisory Paper No.* **10.**

Colorado State Health Department. 1978. "Policy on land treatment of municipal wastewater." Unpublished.

Committee on Nitrate Accumulation. 1972. *Accumulation of Nitrate.* Natural Academy of Sciences, Washington, D.C.

Commoner, B. 1968. "Balance of nature." *Providing Quality Environment in Our Communities.* W. W. Konkle (Ed.). Graduate School Press, USDA, Washington, D.C., pp. 37-62.

Dowdy, R. H., R. E. Larson, and E. Epstein. 1976. "Sewage sludge and effluent use in agriculture." *Land Application of Waste Materials.* Soil Conservation Society of America, Ankeny, Iowa, pp. 118-153.

Dunlop, S. G. 1968. "Survival of pathogens and related disease hazards." *Proc. Symp. on Municipal Sewage Effluent for Irrigation.* P. E. Wilson and F. E. Beckett (Eds.). Louisiana Polytechnic Institute, pp. 107-122.

Flach, K. W. and F. J. Carlisle. 1974. "Soils and site selection." *Factors Involved in Land Application of Agricultural and Municipal Wastes.* National Program Staff—Soil, Water and Air Science ARS, USDA, Beltsville, Maryland, p. 17.

Frink, C. R. 1971. "Plant nutrients and water quality." *Agric. Sci. Rev.* pp. 11-25.

Guenzi, W. D. 1974. "Introduction." *Pesticides in Soil and Water.* W. D. Guenzi (Ed.). Soil Science Society of America, Madison, Wisconsin.

Hall, D. 1975. "Subsurface injection of wastewater residuals at the City of Boulder," *Proc. Williamsburg Conf. of Wastewater Residuals.* J. L. Smith and E. H. Bryan (Eds.). Williamsburg, Virginia, pp. 12-13.

Hoadley, A. W. and S. M. Goyal. 1976. "Public health implications of the application of wastewaters to land." *Land Treatment and Disposal of Municipal and Industrial Wastewater.* R. L. Sanks and T. Asano (Eds.). Ann Arbor Science Publishers, Ann Arbor, Michigan, pp. 101-132.

Kardos, L. T. 1970. *A New Prospect: Preventing Eutrophication of Our Lakes and Streams.* Reprint Series No. 12, Institute for Research on Land and Water Resources, Pennsylvania State University, University Park.

Keeney, D. R., K. W. Lee, and L. M. Walsh. 1975. *Guideline for Application of Wastewater Sludge to Agricultural Land in Wisconsin.* Technical Bulletin No. 88, Department of Natural Resources, University of Wisconsin, Madison.

King, D. 1970. "The role of carbon in eutrophication." *J. Water Poll. Control Fed.* **42**:2035-2051.

King, L. D. 1973. "Mineralization and gaseous losses of nitrogen in soil-applied liquid sewage sludge." *J. Environ. Qual.* **2**:356-358.

Klein, D. A. and L. E. Casida. 1967. "*E. coli* die-out from normal soils as related to nutrient availability and indigenous microflora." *Can. J. Microbiol.* **13**:1461-1470.

Krone, R. B. 1968. "The movement of disease producing organisms through soils." *Proc. Symp. on Municipal Sewage Effluent for Irrigation.* P. E. Wilson and F. E. Beckett (Eds.). Louisiana Polytechnic Institute pp. 75-104.

Larson, W. E., C. E. Clapp, and R. H. Dowdy. 1972. *Interim Report on the Agricultural Value of Sewage Sludge.* USDA-ARS and Department of Soil Science, University of Minnesota, St. Paul.

Martin, W. P., R. G. Gast, and G. W. Meyer. 1976. "Land application of waste materials: unresolved problems and future outlook." *Land Application of Waste Materials.* Soil Conservation Society of America, Ankeny, Iowa, pp. 300-309.

McMichael, F. C. and J. F. McKee. 1965. *Research on Waste Water Reclamation at Whittier Narrows.* W. M. Keck, College of Environmental Health Engineering, California Institute of Technology, Pasadena.

Menzies, J. D. and R. L. Chaney. 1974. "Waste characteristics." *Factors Involved in Land Application of Agricultural and Municipal Wastes.* ARS-USDA, Natural Staff Program.

Merrill, J. C. and P. C. Ward. 1968. "Virus control at the Santer, Calif. Project." *J. Am. Water Works Assoc.* **60**:145-153.

Miller, R. H. 1973a. "The soil as a biological filter." *Recycling Treated Municipal Wastewater and Sludge through Forest and Crop Land.* W. E. Sapper and L. T. Kardos (Eds.). Pennsylvania State University Press, University Park, pp. 77-94.

Miller, R. H. 1973b. *The Microbiology of Sewage Sludge Decomposition in Soil.* Report to USEPA.

Miller, R. H. 1973c. "Soil microbiological aspects of recycling sewage sludges and waste effluents on land." *Proc. Joint Conf. on Recycling Sewage Sludges and Effluents on Land*, sponsored by USEPA, USDA, and NASULGC, Champaign, Illinois, July 9–13, pp. 79–88.

Olsen, S. R., C. V. Cole, F. S. Watanalie, and L. A. Dean. 1954. "Estimation of available phosphorus in soils by extraction with sodium bicarbonate." USDA circular 939.

Page, A. L. 1974. *Fate and Effects of Trace Elements in Sewage Sludge When Applied to Agricultural Lands*. Project No. EPA-670/74/005 USEPA, Washington, D.C.

Parr, J. F. 1974. "Effects of pesticides on microorganisms in soil and water." *Pesticides in Soil and Water*. W. D. Guenzi (Ed.). Soil Science Society of America, Madison, Wisconsin.

Parsa, A. A. and W. L. Lindsay. 1972. "Plant value in organic wastes." *Iranian J. Agr. Res.* 1:60–71.

Parsons, D., C. Brownlee, D. Welter, A. Maurer, E. Haughton, L. Kornder, and M. Selzar. 1975. *Health Aspects of Sewage Sludge Effluents Irrigation*. Pollution Control Branch, B.C. Water Resource Service, Department of Lands, Forests and Water Resources. Victoria, B.C.

Peterson, J. R. 1970. Personal communication.

Pratt, P. F., F. E. Broadbent, and J. P. Martin. 1973. "Using organic wastes as nitrogen fertilizer." *Calif. Agric.*, Jan., pp. 10–13.

Robinson, J. B. 1972. "Manure handling capacity of soil for a microbiological point of view." *CSAE Conf.*, June 1972, Charlottetown, P.E.I. Paper No. 72, p. 210.

Rothwell, D. F. and C. C. Hortenstine. 1969. "Composted municipal refuse: its effect on carbon dioxide, nitrate, fungi, and bacteria in Aredonda fine sand." *Agron. J.* 61:837–840.

Rudolfs, W., L. L. Falk, and R. A. Ragotzkie. 1950. "Literature review of the occurrence and survival of enteric pathogenic, and relative organisms in soil water, sewage sludges, and on vegetation." *Sewage Ind. Waste* 22:1261–1281.

Ryan, J. A., D. R. Keeney, and L. M. Walsh. 1973. "Nitrogen transformations and availability of an anaerobically digested sewage sludge in soil." *J. Environ. Qual.* 2:489–492.

Ryther, J. H. and W. M. Dunstan. 1971. "Nitrogen, phosphorus, and eutrophication in the costal marine environment." *Science* 171:1008–1013.

Sabey, B. R. 1977. "Availability and transformation of sewage sludge nitrogen." *Food Fertilizer and Agricultural Residues. Proc. 1977 Cornell Agric. Waste Mgnt. Conf.* R. C. Loehr (Ed.). Ann Arbor Science, Ann Arbor, Michigan, pp. 257–269.

Sabey, B. R. and W. E. Hart. 1975. "Land application of sewage sludge: effect on growth and chemical composition of plants." *J. Environ. Qual.* **4**:252–256.

Sommers, L. E. 1977. Personal communication.

Sommers, L. E., D. W. Nelson, and K. J. Yost. 1976. "Variable nature of chemical composition of sewage sludge." *J. Environ. Qual.* **5**:303–306.

Spencer, W. F. 1966. "Effects of copper on yield and uptake of phosphorus and iron by citrus seedlings grown on various phosphate levels." *Soil Sci.* **102**:296–299.

Tsubota, G. 1959. "Phosphate reduction in a paddy field." *Soil and Plant Food* **5**:10–15.

Tucker, E. S., W. J. Litscho, and W. M. Mees. 1975. "Migration of poly-chlorinated biphenyls in soil induced by percolating water." *Bul. Environ. Contamination and Tox.* **13**(*1*):86–93. *USDA Handbook* **18**, 1951; Soil Survey Manual.

Viets, G. and R. H. Hageman. 1971. *Factors Affecting the Accumulation of Nitrate in Soil Water and Plants. USDA–ARS Agricultural Handbook No.* **413**, Washington, D.C.

Van Donsel, D. J., E. E. Geldreick, and N. A. Clarke. 1967. "Seasonal variations in survival of indicator bacteria in soil and their contribution to storm water pollution." *Applied Micro.* **15**:1362–70.

4

The composting of municipal wastes

E. G. Hughes, M.I.PI.E., A.M.R.S.H.

Manager, Wanlip Composting Plant, Fillingate, Wanlip, Leicester, England

COMPOSTING OF ORGANIC WASTES

From time immemorial it has been known that decaying organic wastes improve the fertility of the soil. The civilizations that practiced good soil husbandry prospered for thousands of years while other civilizations that ignored this basic fact have disappeared. The Mayas of Central America were a people who were probably more advanced in the mathematical sciences than any other, as the fantastic cities that still stand show, but they were on the point of extinction long before Cortes and his Spaniards landed to plunder the Aztecs.

The Mayas civilization lived in forest countries and the people grew their basic crop of maize by burning off the trees and growing crop after crop on the cleared land until the fertility was exhausted. A great effort of reconstruction was then undertaken at a new location, as the date sequences of the cities show. The peasants from the pale villages that surround the cities brought their produce in but none of the wastes were returned to the land. As domestic animals and wheeled transport were not known to these people, the inevitable happened, and within a thousand years the civilization had burned itself out, with the constant reconstruction involved due to the poor agricultural policy followed.

The tragedy of the Mayas' story is in sharp contrast to the history of the Chinese. For fifty centuries the Chinese have followed a policy of the conservation of soil fertility, and their wastes—animal, human, and vegetable—have been utilized to build up the humus content of their soils. It is interesting to note that in Great Britain, two and a half acres of land are required to support one person, while in the Chinese delta countries, one acre will support two people.

An even older civilization was the Minoan. These people had a vast knowledge of sanitation and water engineering. Water was brought in pressure pipes to Knossos, the capital. Refuse was placed in large pits, called Kouloura, where it was layered with earth and sprayed with the used water and the sewage. The contents of the pits were then allowed to compost before being returned to the soil.

In the Western world of the eighteenth and nineteenth centuries, the farmers would bring their products into the growing cities and return to their farms with the wastes from the cities, which they used to manure their soil. A ring of fertility spread around the cities but as the Industrial Age gathered momentum, the cities expanded into the food producing areas and this fertile land, which had been nurtured for hundreds of years, was lost.

The Industrial Age brought fertilizers into farming and a temporary respite was granted. The use of fertilizers ensured good crops and the need for good husbandry became apparently less important in the short term. Composting fell into disuse except for isolated pockets and the need for compost and humus was lost sight of.

The wastes that had previously been returned to the soil were now becoming a vast problem to the cities that produced them, and open raw tipping became the order of the day. Every city had its foul tip and vast quantities of raw sewage were spread on the land of sewage farms. These farms were frequently used for cropping, but as the cities grew, the sewage application became heavier than the land could deal with and new methods of sewage disposal were looked for. Filtration was introduced, with the effluent being discharged into rivers and streams where it upset the ecology. Even now, in the Space Age, raw sewage is dumped at sea in the belief that the sea is a limitless sink which will accept anything that is put into it.

America has taken the major and far-seeing decision to ban the ocean dumping of sewage sludge in the near future, and this one act, by the wealthiest and most powerful nation on earth, is likely to lead to a major return to composting as a means of refuse disposal.

It must not be thought that composting is the be-all and end-all of waste disposal. In some situations, it will not be the answer, and it would also be wrong to assume that all wastes can be composted. Composting can in itself bring difficulties that other methods do not. However, every waste disposal authority should be fully aware of the advantages of composting, and it is the author's opinion that composting should always be examined as the primary choice of methods. If local conditions show that there are difficulties, in disposing of the compost for example, then by all means other methods should be examined before the decision is made. Other methods should *only* be considered after it has been proved that composting is not the answer—and not only the answer to the waste disposal problem, but also the answer to the agricultural needs of the locality.

A BRIEF HISTORY OF THE COMPOSTING PROCESS

The agricultural and horticultural composting of organic wastes has usually been accomplished by placing the wastes in a heap and allowing nature to decompose them. After a suitable time has elapsed, the compost is then used.

This conversion of the wastes is brought about by microbiological action. In modern composting systems this process is aerobic, i.e. carried out in the presence of oxygen.

In 1843, a patent was granted in America to George Bommer for his Bommer Method of Making Manure. In this, he placed the different types of farm wastes on a grid. The leachate that drained through the grid was collected in a tank and recirculated over the top of the heap by means of a wooden pump, for which a patent was also granted. The Bommer method was claimed to result in usable compost in 15 days and can probably be said to be the forerunner of scientific compost making (Rodale, 1962).

However, the first significant development in composting as a systemized process took place in India in 1925, when Sir Albert Howard developed a process involving the anaerobic degradation of leaves, refuse, animal manure, and sewage in pits. Again, the materials were placed in layers and the pit walls conserved some of the heat of degradation, resulting in higher temperatures than when composting was carried out in the open and was subjected to the cooling effects of the wind and the lower night temperatures. This was known as the Indore Process, and it took approximately six months to produce usable compost. As labor is plentiful in India, the process was modified to include more turning in an attempt to hasten the action (Howard, 1935).

Following this, the Indian Council of Agricultural Research improved the method by laying down alternate layers of wastes and sewage. This system is still used in India and is known as the Bangalore Process.

In India, the high humic degradation that occurs in the land requires that for soil fertility to be maintained, large amounts of humus are required. Coupling this with the fact that composting is an ideal method of recycling organic wastes, the Indians realised that this was an ideal system for them. Currently, approximately 2500 composting plants using the Bangalore method are in operation in India.

From 1922 onwards, patents for the mechanization of composting started to appear in Europe. In 1922, Baccari patented a process which composted material in a closed system using both anaerobic and aerobic decomposition.

The first full scale composting plant was established in Europe in The Netherlands in 1932 by a non-profit company, N.V. Vuilafvoer Mactschapij (VAM). This was a modification of the Indore process, where unground wastes were composted in large windrows. The system is known as the Van Maanen process.

In 1933, the Dano process was patented. The Dano process is currently the most successful system, with something like 240 units in operation. The history of the Dano is quite interesting. Kai Petersen, in association with Christopher Muller, founded the Dano Engineering Company in 1912 to manufacture mechanical under-carriage stokers for industrial furnaces. The stokers were excellent and raised the efficiency of combustion from 60 to 80%. This enabled the firm's customers to pay for the grates out of their fuel savings, and Dano prospered. However, Petersen, having a social conscience, was concerned about the situation his stokers were creating. One order he received meant that the customer's work force dropped from 50 to 17.

At the same time, a man called Folke Jacobsen had started a course of study, in Jutland, the basis of which put great emphasis on man's dependence on the soil. Kai Petersen interested himself in this and provided financial backing for the purchase of a house and land. The two hectares involved were subdivided into plots and the students came on weekends to stay in the small summer houses provided and to tend their plots as they wished.

George Muller, the brother of Kai Petersen's partner, also became interested. He was a deeply religious man who not only believed in the Mosaic laws of hygiene but put them into such strict practice that he would have nothing to do with water closets. He attended to his household dry privies personally and mixed the nightsoil with waste vegetable matter to make compost. Muller presented the Corporation of Copenhagen with the proposition that they should do away with their sewage system and put composting receptacles on street corners where the population could empty their night soil into them. Not unnaturally, the city fathers did not believe that sweet smelling compost would be produced instead of noxious odors.

To prove his point, Muller persuaded Petersen to let him stand some 45 gallon oil drums within the grounds of the Dano factory. After the drums had been filled and left to stand for three weeks, they were emptied. The resulting odor from the foul contents emptied the factory. Petersen realized that the difference between foul smelling degradation and sweet smelling composting was the difference between anaerobic and aerobic methods. To introduce oxygen to the process, he mounted one of Muller's drums on its horizontal axis and slowly rotated it. This resulted in sweet smelling compost being produced in three weeks. The compost was used on Jacobsen's plots with great success.

This was the birth of the Dano process and it is interesting to note that Jacobsen's 2 hectares were increased to 10 hectares and the original 20 plots rose to 200. It became so successful that it was too large to be administered by the Dano company and was taken over and is still flourishing under the guidance of the Corporation of Copenhagen and the State (Wylie, 1959).

The Dano Company continued with its experiments and now their largest drum is approximately 30 meters long by 4.5 meters in diameter and can handle up to 150 tonnes of waste per day (Fig. 4.1).

From 1922 to the present, a number of methods of turning household wastes and sewage into compost have been patented. Basically, they all shred the waste and provide optimum conditions for the bacteriological breakdown. This can be done in pits, on level ground, in towers, in rotating drums, or in static tanks, but

Fig. 4.1. Dano Compost Plant.

always, irrespective of the mechanical means employed, the conditions must be right for the degradation to occur in an economical period of time.

THE MECHANICS OF COMPOSTING

Before organic wastes and/or sewage sludge can be composted, certain criteria have to be met. A heterogeneous mass of multiple particle size has to be converted into a homogeneous mass of suitable particle size and suitable moisture content.

At this stage, we will take a "conducted tour" around an existing compost plant and see the operations involved. Later, we will study these operations in detail and consider the reasons or necessities for them.

As an example, we will use the Wanlip Compost Plant in Leicester, England. This is the largest Dano plant in the world that is used exclusively for composting. The windrow plant at Tel Aviv makes more compost and the Rome plant has more drums but the Leicester plant is ideal as an example as it embodies all the methods and problems that need to be understood.

The Wanlip plant was built by the Leicester City Council on a site adjacent to their Sewage Works. It was designed to treat domestic waste from the city of Leicester in the United Kingdom (pop. 290,000) and to utilize treated sludge from the Sewage Works to enhance the final product. Commissioned in sections during 1966 and early 1967, the plant was extensively damaged by fire in March 1968, and recommenced operations after rebuilding in 1969. In 1974, due to local government reorganization, the ownership of the Sewage Works was transferred to the Severn Trent Water Authority and the ownership of the Compost Plant was transferred to the Leicestershire County Council. During the next three years, extensive alterations were made and new methods were introduced which have resulted in an increase in the input of wastes and the output of compost. The maximum capacity of the plant is now approximately 1400 tonnes of solid domestic waste per week with an operational throughput of between 1000 and 1100 tonnes per week. The sequence of operations is given below.

Weighbridge

Two 30 tonne capacity weighbridges are installed and all laden vehicles weighed as they enter and leave the site. Domestic waste is collected in vehicles which make two journeys per day. The vehicles proceed round the one way system of roads to the reception hall to discharge their loads.

Reception Hall

This building has a steel portal frame and is 69 meters long by 27 meters wide by 7.6 meters high. It contains three steel hoppers 13.2 meters long by 2.8 meters wide at the bottom and 3.7 meters wide at the top, each having a capacity of 139.3 cubic meters. At the bottom of each hopper is a hydraulically operated moving floor that discharges the waste onto one of the three elevating conveyors, which, in turn, transfer the waste to the salvage hall. A static grate incinerator with a grate area of 5.6 square meters and a cyclonic dust extractor is situated in the reception hall for burning large items of waste.

Salvage House

This three story reinforced concrete building measures 68 by 18.6 by 13.7 meters. It contains the picking rooms for sorting the wastes, the main control room, the sludge equipment, the salvage balers and the welfare facilities of toilets and canteen. The central control room on the ground floor is the nerve center of the whole plant and a large mimic diagram clearly indicates which sections of the plant are operating. All the equipment is linked electrically so that the various

Fig. 4.2. Author in control room.

sections can only be started in the correct sequence. The intertripping circuits also ensure that in the event of a stoppage, no waste can be fed onto a dead section. Meters and recorders indicate flow quantities, temperature, and power consumption (Fig. 4.2).

Also on the second floor are the three picking rooms which house picking belts of 1.2 meters width, which move at 18 meters/min[1]. These rooms are provided with dust extractors. Textiles and non-ferrous metals are dropped, by the operators, down chutes on the conveyors slung beneath the first floor for distribution to the appropriate part of the ground floor for baling.

The wastes on the picking belt are split into two halves before passing under the magnetic overband extractors which remove the ferrous metal and drop them onto a conveyor belt for transport to the baling press. The residue is then passed into the stabilizers, where it is mixed with sludge to control the moisture content.

Four Komline-Sanderson coil spring vacuum filters are installed for de-watering digested sludge from approximately 95 to 80% water content. Each filter is 3.5 meters in diameter by 4.3 meters wide, giving 47 square meters of filter area. The auxiliary equipment for each filter is located on the first floor of the salvage house and includes storage tanks, dosing pumps, mixing drums, filtrate pumps, and vacuum pumps. On the ground floor of the salvage house are the baling presses for paper, ferrous metals, and textiles.

Stabilizers

Each of the six stabilizers is 25.6 meters long and 3.5 meters in diameter. Solid wastes and sewage sludge are fed into the stabilizers, which rotate at 1 rpm during the day and at 0.5 rpm outside of working hours. The tumbling action breaks down the material physically and ensures a homogeneous mix. The aerobic bacteriological decomposition is the start of the composting process. The biological degradation raises the temperature inside the drums to 55°C, which is sufficient to kill any pathogens present. The material slowly gravitates to the discharge end where large rejects are retained on a special screen. The retention period in the drums varies between two and four days. At the discharge end of the stabilizers, the material falls on to belts which transport it to the screen houses.

Screen Houses

The material from the stabilizers comes off the belt in the screen houses and falls onto specially designed, inclined, stepped dock vibrating screens. The compostable material, 4 centimeters in size and below, falls into the compost traps while the material over this size is rejected onto a belt which transports it to each of the screen houses for loading into lorries prior to tipping.

Windrowing

The compostable material is transported from the screen house bays by means of bucket loaders which place the material in windrows. These windrows are normally 90 to 140 meters in length and are roughly triangular in section—3 meters high and 9 meters across the base. Thermophilic biological degradation is brought about by protozoa, fungi, and actinomycetes, and the temperature rises to approximately 75°C. The windrows are turned by mechanical shovels every seven to ten days, depending upon the temperature. After three months, the process is complete and the windrows receive a final period in a drying shed before being finally screened to 1 centimeter, ready for bagging and sale.

General Refuse Handling Techniques

The receiving area for the plant should be enclosed to obviate wind blown debris and be large enough for vehicles to maneuver without danger.

If refuse collection is a one shift operation, then the normal pattern will be for all the collection vehicles to go out empty and to return to the plant full. Again, normally, dependent on routes, pick up points, and distances involved, this will mean that vehicles will probably return to the plant twice in the working day. Local conditions can vary this, but it must not be assumed, for example, that since the plant and 32 vehicles are operational for 8 hours and the vehicles make two trips each, the receiving area will have to deal with 8 vehicles per hour. It is more likely that the 64 trips involved will need to be handled in two 2 hour periods; i.e., 2 hours in the morning and 2 hours in the afternoon. This would require a reception area capacity for 16 vehicles per hour.

If receiving hoppers are used, they should be of sufficient size to hold one day's intake to prevent the need of storage outside the hoppers, which can occur in peak periods. If the receiving hoppers are unloaded from the top with cranes, then arching or bridging should not be a problem. If they are unloaded from the bottom by means of moving floors or conveyor belts, then arching and bridging can become a real problem. Careful design with at least one of the long sides being vertical will help (Fig. 4.3). The ideal would be a divergent hopper with the narrowest part being the entry point. This would ensure that the wastes are constantly dropping into an increasing cross sectional area and the likelihood of blockages will be minimal. The surest way of getting blockages is to design the bottom emptying pit with its long sides forming a convergent shape toward the bottom and to place leveling bars across the width to level the refuse. Leveling bars can practically be guaranteed to cause bridging.

As, at this stage, a load of hot ashes could inadvertently be emptied into the raw refuse, it is important that all the construction materials be fireproof (e.g., steel or concrete) and that any conveyor belting be fire retardant to the standards of the local mining industry.

Fig. 4.3.   Raw refuse in hopper.

A reverse drive on the moving floor would be an excellent aid to clearing blockages but it is unlikely that the cost would be considered justifiable. Strength of construction is obviously important, as a scrap engine block dropping 10 feet onto a piece of equipment, be it a conveyor belt or a feed chute, will obviously find any weaknesses that have been built into a system.

From the receiving area the wastes will be transported or conveyed to the size reduction plant. Hand picking, if required, can be catered for during this operation (Fig. 4.4). This is normally considered to be a degrading job but that is entirely dependent on one's viewpoint. The author is familiar with two cases where people from a local institution have been employed on this seemingly onerous task. Both of them had been in a hostel for a number of years because they were mentally slow and had no one to look after them. Their wages from the picking were invested for them by the hostel and small houses purchased for them and furnished. Both are now married and consider that refuse picking is the best job on earth.

Fig. 4.4. Hand picking.

In terms of plant economics, materials can be picked to obtain additional revenue—textiles, non-ferrous metals, etc.—or to prevent blockages and remove obviously non-compostable materials from the waste (e.g., large p.v.c. sheets or lumps of wood) (Fig. 4.4). Most modern plants in fact do not have hand picking.

After the hand picking has been carried out or bypassed, the ferrous metal should be removed. This can be done magnetically using overband or underband magnetic separators or magnetic end drums on the conveyors. At this stage, it is unlikely that more than 50% of the ferrous metal will be extracted due to the volume of the waste involved. After the waste has passed through the size-reducing process, further magnetic separators could be employed to remove the ferrous element from the now level bed of homogeneous wastes. If both systems are used, it is likely that 90 to 95% of the available ferrous material will be removed, thus ensuring maximum revenue from this source.

The next step for the wastes is for them to be reduced in size to produce particles suitable for composting. This is normally accepted to be from 4 to 6.5

centimeters. The only exception to this is the Van Maanen process, where un-ground refuse is used.

The purpose of grinding is to produce a homogeneous mass of suitable particle size to ensure maximum surface area compatible with the ability to pass air through it. The most common (though not necessarily the best) method of grinding refuse is to use a hammermill. These are machines which usually have swinging hammers revolving around a shaft. A system of grids ensures the particle size as the refuse is hammered through them. The normal particle size of this type of mill is 5 to 6 centimeters. Any material which is not grindable is rejected. The material is normally ground dry, with moisture having to be added at a later stage.

Such a mill would be powered by an electric motor of up to 500 hp capacity, and as they can operate up to 3500 rpm, the noise and vibration can be considerable. Very heavy hammer wear takes place and maintenance costs are heavy. Against this, the capital cost is usually fairly low when compared to other methods.

A wet pulper is another means of size reduction. These consist of a large bowl that has a rotatable steel plate covered in hardened steel teeth. The bowl is filled with water and when the plate is rotating at full speed, raw refuse is dumped in. The resultant shredded waste forms a slurry containing about 5% solids. The slurry must then be dried out to about 50% moisture before it is suitable for composting, and de-watering slurries or sludges of any type is expensive.

Refuse raspers can also be used for size reduction. This is basically a system whereby a surface covered in pins is moved against a perforated plate and the wastes fed between them. The pins rasp or shred the refuse which then falls through the perforations.

In the Dano drum, no pretreatment is required as the material is tumbling on itself for two to four days and the moisture can be added into the drum. The moisture weakens the fibers and the tumbling action breaks them down so that when the compostable material is exhausted from the stabilizers it has an appearance rather like peat.

This is not meant to be an exhaustive review of how to reduce refuse in size, but is indicative of the type of equipment in use.

Once the refuse is in a suitable state, moisture must be controlled to give a compostable mix of about 50% moisture. In most industrialized countries, this means adding moisture in the form of water, wet sludge, or dried sludge. In other countries (e.g., South America, Italy, and Malaysia), the refuse can be vegetable based with a moisture content of up to 80%. In these cases, moisture must be removed or the slurry diluted with dry material to achieve the 50% figure. In Beltsville, Maryland sewage sludge is successfully composted by mixing wood chips with it as a carrier to reduce the moisture content (Epstein, 1977).

In the Northern industrialized regions, the refuse is normally between 25 and

50% moisture and water must be added. If the prime intention is to compost refuse, then water or wet sewage sludge can be added. If the need is also to dispose of sewage, then the sludge can be dried and in this way more sludge can be disposed of. The effect on the compost, whether using water, wet sewage sludge, or dried sewage sludge, is minimal (see the chemical and biological sections of this chapter).

When refuse of the right particle size has been adjusted to the right moisture content, the composting process can start. This can be achieved in windrows with regular turning to ensure good mixing and aeration. Special machines can be obtained for windrow turning and for adding moisture during the turning process. Normally, this turning is done by front end bucket loaders as they remove problems that can be experienced with high cost specialist machines (Fig. 4.5).

Composting can also be carried out in windrows where air lines force oxygen through the mass and ensure that it remains aerobic. This method lessens the

Fig. 4.5. Windrow turning.

need for turning and throughout the windrow life it would probably only require one turn.

Other methods include:

1. Storing the mix in static towers where constant agitation and aeration occur.
2. Storing the mix in pits where agitation can be arranged by cranes and other mechanical means.
3. Storing the mix in tanks where mechanical agitation takes place.
4. Storing the mix in digestion tanks with movable conveyors for removing the compost.

All methods require aeration and usually agitation. The more sophisticated the mechanical means employed, the quicker (theoretically) the finished compost will be produced.

Following the composting process, the compost is usually moved to a shed where the self generated heat will dry it. In high humidity conditions, mechanical drying may be required. The compost is now ready for finishing even though further stabilization and curing may go on for a much longer period.

Finishing Techniques

At this stage, the compost will have the appearance of a good quality loam although it will still contain fairly large amounts of glass, bone, plastics, and other non-compostable items. These items must be removed before the compost can be considered suitable for sale.

The simplest way of finishing compost is to screen it using a rotary screen or flat bed screen down to 1 centimeter. This will leave a product that still contains glass and other non-compostable items but of insufficient size to make the product unacceptable.

Numerous methods have been tried to clean glass from the compost but with varying degrees of success. Ballistic separators, inclined conveyor separators, secators, crushing rollers, air separation, and fluidised beds are all in use and all methods have their protaganists. The choice of a final finishing system must take into account the additional expense associated with the product improvement and customer acceptability of the product.

Handling Compost

Compost requires special material handling techniques due to its sticky nature. It does not flow easily and if it contains excess moisture it can form into balls. At Leicester, the problems have been overcome by using steep angles on hopper sides and lining hoppers and chutes with a paraffin based plastic material.

THE CHEMISTRY OF COMPOSTING

The rate at which organic matter decomposition occurs is principally dependent upon the carbon-nitrogen ratio (C:N) of the material.

As the material decomposes, the carbon content declines due to its release as CO_2 while the nitrogen remains within the system, and as the composting process continues, the C:N ratio becomes lower.

The C:N ratio of organic wastes can vary between 20:1 and 115:1. Most household wastes are between 30:1 and 40:1, with items such as straw being 80:1 and sawdust 115:1.

Most soil micro-organisms have a C:N ratio of between 15:1 and 30:1, so it is generally accepted that if compost is added to the soil, the C:N ratio should be below 20:1. If it is above 20:1, at high rates of application, microbiological activity will immobilize the available nitrogen in the compost, causing a nitrogen deficiency.

This possibility of compost being applied with a C:N ratio above 20:1 is not a matter for concern, as adequate monitoring of the process will ensure that all the finished compost, prior to sale, will be below this figure.

Chemical Analysis of Compost

Nitrogen	1.33%
P_2O_5	0.83%
K_2O	0.36%
Humus	53.70%
Calcium	5.61%
Iron	2.1%
Zinc	285 ppm
Lead	575 ppm
Copper	65 ppm
Cadmium	5 ppm
Iron	21,250 ppm
pH	7.2

This analysis must be viewed with extreme caution as compost made from household wastes and sewage sludges can vary greatly in their analysis due to the differing constituents of wastes and sludges from different areas.

For instance, a sludge from a housing estate with a young population could be high in zinc due to the amount of baby powders used, while the domestic wastes from a ribbon development area bordering a major traffic route could be high in lead content due to the vacuum cleaner contents. The sewage from an area that has a big electroplating industry could be abnormally high in chromium and other industries can also affect the chemical contents of sewage and wastes.

Before any commercial or municipal composting arrangement is evolved, all these factors must be known and considered. For the majority of locations, there will be no problem at all.

Moisture

Moisture needs to be present during the composting process to provide the correct humidity for the decomposition of the organic wastes by the microbiological decomposers. The optimum conditions occur when the water content is between 50 and 60% by wet weight. An excess of water causes the compost to become more compact, the oxygen retaining voids to become filled, the mass to become anaerobic, and objectionable odors to be created. This will also increase the time cycle involved. Conversely, if too little moisture is present, the center of the mass will reach very high temperatures and once again, the rate of decomposition will be retarded.

Moisture can be added to the wastes in many forms. Water, wet sewage sludge, or dried sewage sludge are commonly used. Wet sludge or dried sludge being used will also solve the disposal problem of the sewage authorities. Raw sewage can be used but tends to give rise to offensive odors, and it is customary to use sludge that has been anaerobically digested. This is a process that is used in methane generation from the sewage, the methane being used as a fuel in dual fuel engines to power electrical generators.

Wet sludge has usually about a 96% moisture content (i.e., about 4% solids) and dried sludge has had some of the water removed by vacuum or centrifugal filters; its moisture content is usually about 80% (i.e., 20% solids). The amount of water removed to reduce sludge from 96 to 80% moisture is not 16%, as would appear: 1000 kg of sludge at 96% moisture will contain 960 kg of water and 40 kg of solids, and to reduce this to 80% water and 20% solids, the 40 kg of solids in the original mix will have to be 20% of the final dried mix; i.e., the weight of the final mix will have to be 200 kg—40 kg solids, 160 kg water. Therefore, as the original 1000 kg mix had 960 kg of water, 800 kg of this must be removed to get the 80% moisture final mix.

For a great many years, it was assumed that as dried sludge is rich in nitrogen, compost made from dried sludge would be of a higher quality than that made from wet sludge with its correspondingly lower nitrogen content. However, results of research carried out at the Wanlip Composting Plant have indicated that this is not so. Two windrows, each containing 150 tonnes of material, were made; one was composed of household wastes and anaerobically digested sludge at 96% moisture, and the other contained 150 tonnes of material made with household wastes and anaerobically digested sludge that had been dried to 80% moisture. The results in Figs. 4.6, 4.7, 4.8 and 4.9 showed that there was no difference between the two windrows throughout their life and chemical analysis

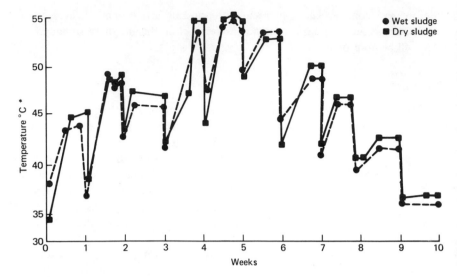

*Temperatures represent the mean over a depth of 60 cm, peak
temperatures are approximately 15°C higher than shown. Sharp
temperature drops are caused by turning.

Fig. 4.6.   Windrow temperatures during composting process.

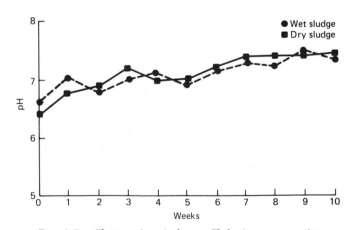

Fig. 4.7.   Changes in windrow pH during composting.

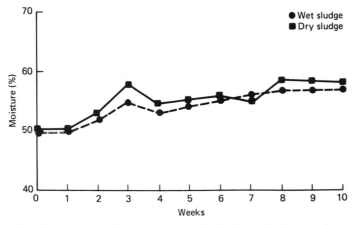

Fig. 4.8. Changes in moisture content of windrows during composting.

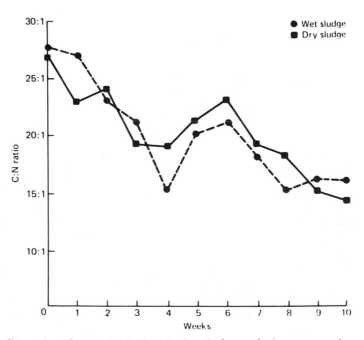

Fig. 4.9. Changes in C : N ratios in windrows during composting.

revealed that within experimental error, there was no difference in the finished product. Therefore, if the composting operation is being set up to dispose of solid organic wastes, the expense of drying sludge is not justified. If, however, the disposal of sewage is of equal or superior importance, then the cost may be justified. The drying of sludge for composting should be decided only on need and economics. Whether it is dried or not has no significance to the composting process.

pH

Household organic wastes will normally be between pH 5.5 and 7. The first few days of the composting process will usually show a drop in the pH of the mass, sometimes to below pH 5 and then, as the process gets under way, the acid content will be neutralized and the finished compost will have a pH above 7, probably between 7 and 8.

Heavy Metals—Toxic or Essential Trace Elements

It has long been recognized that all plants require not only nitrogen, phosphates, and potash to grow, but also small amounts of various metals: iron, magnesium, copper, zinc, sodium, manganese, chlorine, cobalt, iodine, molybdenum, and selenium, among others.

The use of municipal compost can increase the concentration of metals in the soil, but because the total metal is increased, it does not follow that plants will pick up this metal. What is important is the availability of the metal and the availability is dependent on the organic content of the soil and its pH. The higher the humus content and pH, the lower is the availability of some metals to the plants.

The fact that an analysis of municipal compost shows that lead and cadmium are present (for example) must be viewed in context. Let us take an actual case. One analysis of the compost with which this author is associated gave a figure of 160 parts per million of lead and 4 parts per million of cadmium.

This was a dry weight analysis; when applied to the soil, the compost would contain moisture and these figures would then drop to 84.8 parts per million of lead and 2.12 parts per million of cadmium. If 4 tonnes/ha of the compost were spread on the land and then mixed into the top 12 centimeters of soil, then the increase in the metal content of the soil would become 1.26 parts per million of lead and 0.03 parts per million of cadmium. Plants would remove some of this metal so the actual build-up would be slower. Putting this into the context that in the United Kingdom, soil will naturally contain between 2 and 200 parts per million of lead and 1 to 5 parts per million of cadmium, the powerful diluting effect can start to be understood.

TABLE 4.1.   Lead and Cadmium Content of Tomatoes Grown in Municipal
Compost, Compared with Shop Bought Tomatoes.

	Ring Culture 100% Municipal Compost	Tinned With Skins (Random)	Tinned Without Skins (Random)	Fresh Sample (Compost-free)
Lead ppm	0.4	0.3	1.5	0.14
Cadmium ppm	Not detected	Not detected	0.1	0.06

A further factor to be considered is that plants can be selective in their uptake of metals. To illustrate this, the author grew some tomatoes, by the ring culture method, using 100% municipal compost in the ring. (This is not recommended, as compost should always be mixed with soil, but was done in an attempt to obtain the highest possible uptake of metals.) The tomatoes were then analyzed and compared to other tomatoes. The results of the experiments can be seen in Table 4.1.

The only difference between an "essential trace element" and a "heavy toxic metal" is one of quantity and position. If zinc, lead, and boron are naturally present in soils, then they are referred to as essential trace elements. If they are present in sewage sludges or municipal composts, then the anti-sludge or anti-compost lobbies tend to refer to them as toxic metals.

In practice, a vast amount of research has been carried out on investigating the complex relationships between municipal compost and plants by highly respected universities and organizations and thus far none have found any cause for concern in the use of municipal compost.

THE MICROBIOLOGY OF COMPOSTING

The biological degradation process of composting can be broadly described in terms of four stages of micro-organism activity, characterized by different temperature ranges.

In the initial mesophilic stage (25 to 40°C), the decomposition starts with the digestion of the organic fraction of the wastes by fungi, bacteria, and actinomycetes. Fungal mycelia penetrate all parts of the mass, and in a windrowing process, early fruiting bodies of mesophilic fungi will appear. At the same time, mites (acarines), millipedes (diplopods), and sow bugs (isopods) will ingest the organic wastes. The soft tissue of decaying plants supports growth of nematodes and enchytraeids.

These consumers then attract and become food for the next level of consumers; i.e., collembolams eat fungi, ptiliids feed on fungal spores, nematodes ingest bacteria, and the protozoa and rotifers feed on bacteria.

After each consumer dies, all the nutrient constituents in its tissues, such as nitrogen, are again recycled into other organisms. The energy liberated during this conversion causes a rise in temperature to between 45 and 70°C.

At this higher temperature (the thermophilic stage), a specialized thermophilic flora of bacteria, fungi, and actinomycetes take over. Organic degradation is rapid at this point and it is during this phase that pathogenic organisms, fly larvae, and weed seeds are destroyed. In a windrow plant, at this stage, a 25 to 30 centimeter deep grey layer will form approximately 8 centimeters below the outer surfaces of the windrow. This is where the conditions are at their optimum and turning is required to ensure that all the material in the windrow passes through this layer (Fig. 4.10).

As thermophilic activity declines and the temperature falls to 30 to 40°C (the cooling stage), there is another increase in the number of organisms capable of growth at normal temperatures.

During the final maturing stage, the number of these micro-organisms declines slowly as the temperature drops to ambient level.

Fig. 4.10. Composting layer in windrow.

While the foregoing may appear complex, it is a naturally achieved process, as the wastes naturally contain the spores, eggs, and propagules of the decomposers, and it has been shown that the addition of various innocula is of no practical significance.

Composting Temperatures and Pathogen Survival

In a windrow of any enclosed system, biological activity will result in an increase in the mass temperature above ambient. Temperatures in windrows can easily rise to 77°C and can be held for comparatively long periods. Up to ten days is normal, and if, in a windrowing plant, turning is done when the temperature has reached a plateau, the temperature will drop sharply during the turning and then restore itself to a slightly lower peak. This peak will get progressively lower with each turning until the compost is finally matured and the peak temperature is only slightly above the ambient temperature (Fig. 4.6).

In some plants, where air is removed from the windrow by a suction fan, additional moisture may have to be applied to the windrow to keep the temperature of the composting mass down.

In an enclosed digestion system, the time/temperature graph will probably be a decreasing straight line.

The high temperatures present in the compost will have an adverse effect on the population of pathogens. If this high temperature can be sustained, then the possibility of pathogen survival is greatly diminished. For example, *Mycobacterium tuberculosis* is normally destroyed by the fourteenth day of composting if the temperature is at or above 65°C, while a temperature of 54°C for as short a time as 30 minutes will be fatal to the polio virus. Salmonella is deactivated by a temperature of 60°C for 30 minutes while *Ascaris lumbricoides*, a pathogenic worm found in the intestinal tracts of pigs, is killed by a temperature of 60°C for 60 minutes.

It is highly unlikely that compost is anything but safe and there are no recorded instances of compost plant workers or users of the compost being adversely affected.

USES OF COMPOST

Compost, although containing nitrogen, potash, phosphate, and trace elements, should not be looked upon as a fertilizer. It is a soil conditioner which will turn poor soil into a high quality loam.

The main difference between a heavy clay soil and a light sandy soil is one of particle size. The heavy clay soil is composed of very small particles which stick together, while the sandy soil has large particles which hold the structure apart and allow moisture to drain away.

Municipal compost has a very high humus content, up to 60%, and its action will be to break up the heavy clay soil or to tie the large particles of the sandy soil together. The compost should be applied to the land and mixed into the top 15 centimeters of soil. The humus in the compost will not only improve the structure of the soil but will also hold the water and nutrients where they are needed; i.e., around the roots of the plant. Leaching will be reduced, and as nutrients are being held where they are required, less fertilizers need to be applied. Most users report an increase in crop yield of up to 30% with a 50% drop in the fertilizer used.

The use of compost can have spectacular results on certain crops, with the 30% increase in crop yield often being exceeded. These crops include tomatoes, leeks, onions, broccoli, melons, courgettes, hops, and grass.

Some tests have been carried out using municipal compost on coal shale waste heaps which were being reinstated after open cast coal mining operations. The control plot with no compost added yielded 0.7 tonnes/ha of grass. When compost was applied at 4 tonnes/ha, the yield rose to 1.3 tonnes/ha; at an application rate of 8 tonnes/ha, the yield rose to 2.5 tonnes/ha. The normal application rate for most crops is between 1 and 3 kg/square meter, applied annually, preferably in autumn (although a spring application is quite satisfactory). When the soil has become a good, friable loam, no further applications need to be made for up to three years in moderate climates. In areas of high humic degradation, the applications will need to be made annually.

Soft fruits, vegetables, flowers, roses, and trees all benefit from the use of compost, although obviously, in the case of roses, shrubs, and trees, it will be used as a top dressing or mulch.

Apart from its use as a soil conditioner, compost has a multitude of other uses. In Europe it is used on vineyard terraces, not only to benefit the grapes, but also to minimize soil erosion. Erosion of vineyards with slopes of up to 30° has successfully been stopped by the use of compost. It is also used in Europe as litter in poultry houses, where the birds' droppings turn it into a high quality fertilizer. Municipal compost has also been marketed as a pig food addition. If pigs are kept in fields, they will ingest some soil when they feed on roots. In this way their bodies absorb the trace elements that they require. When they are farmed using intensive methods, these trace elements are denied them and they can suffer from nutritional diseases such as anemia and scouring. A handful of finely ground and screened municipal compost mixed in with the food restores the trace elements and anemia and scouring cease to be a problem.

The reclamation of disused areas and areas blighted by open cast mining are also uses to which compost can be put. Parks and terraced relaxation areas can be a powerful incentive to the establishment of a composting operation, and even where open site tipping is the order of the day, the use of composting techniques can rapidly turn an eyesore into an environmental advantage, even if the compost is only used as tip cover.

MARKETING

There is very little point in making compost if the end product is not used. The establishment of a marketing philosophy, if not a high powered commercial organization, is essential.

The decision must first be taken as to whether the marketing will be at plant level or whether an external organization will be brought into being.

The author's preference is for the marketing side to be done at plant level, as in this way the vagaries of production can be kept in line with orders. Overselling can be fatally easy, and a reputation for producing good compost but never seeming to have any for sale is something that can take a long time to overcome.

As compost selling is a seasonal occupation with very high peaks in the autumn and spring, it must be ensured that there is sufficient storage space for the out of season stockpiling. Markets such as tip covering and land reclamation are obviously round the year pursuits, but farmers, horticulturists, and domestic gardeners require their compost in the two main selling seasons. The domestic gardener will normally buy in bags in the spring just prior to planting. The professional grower, on the other hand, will buy in bulk in the autumn, treat his land, and allow it to stand over winter to gain maximum benefit. So immediately we have two separate sales peaks: loose bulk in the autumn, bags in the spring. Both of these peaks must be catered to in terms of stock. If bags are used, then they should be printed with instructions and trade name, etc., in the same way that any product requires packaging. Sales of compost in a plain bag will not equal the sales of the same compost that is attractively presented. Advertising must be carried out throughout the year, so that when the selling season starts all your potential customers are aware of the product.

It is unlikely that direct sales from the plant will dispose of all the compost produced, so a transporting system will have to be devised. This can be run by the plant, although it must be remembered that the main two selling seasons are unlikely to total more than six months of the year. For the other six months, the transporting lorries will not have sufficient work. It is therefore preferable that an outside contractor be employed on a tender basis, as he will have work for his transport in the compost non-selling season.

The distribution of the compost to customers in other areas can be satisfactorily done by existing wholesalers and commercial organizations. At the compost plant with which the author is associated, the compost is sold loose, in bulk, or in attractively printed plastic sacks. Orders received at the plant are delivered by a contractor up to 300 kilometers away. A distribution network has been set up by appointing one agent for each county. This agent works on a commission only basis.

Regular advertising ensures that the name of the product is constantly brought to the attention of the customer. A very healthy local trade has been built up with domestic gardeners, farmers, horticulturists, golf courses, allotment societies, and government departments. The result of this has been that all the compost gets sold, and very often in the selling season, deliveries have a two week waiting list. On occasions, there has been a queue of 60 vehicles at the plant waiting to buy compost.

From this it will be seen that if a good compost is being made, there is no reason why it should not be sold.

THE ECONOMICS OF COMPOSTING

It is not intended in this section to put monetary values on the points discussed, as these so quickly become out of date. Instead, general economic principles will be put forth.

The cheapest form of waste disposal is normally landfill, but the distribution of holes in the ground and tip covering material can make landfill unattractive. Certainly, aesthetic and environmental considerations and neighborhood objections are bringing landfill sites into disrepute. Most waste disposal authorities are looking toward mechanical treatment or recycling of wastes.

The present methods available are:

1. Incineration—already falling into disrepute because of high costs and airborne pollution.
2. Incineration combined with heat recovery or desalination projects.
3. Pulverization—pretreatment prior to landfill or composting.
4. Composting—pulverization followed by the composting process.
5. Pyrolysis—the reclamation of certain materials by burning refuse in an oxygen-free atmosphere; potentially useful but at present confined to small pilot plants.

The highest cost tends to be incinerators, with a lower cost involved if successful heat recovery provides an income. Desalination tends to increase the cost, as normally it is a state function rather than a commercial function. The cost of pulverization is probably half that of incineration but the material still has to be tipped. The gross cost of composting will be higher than that of pulverization, as pulverization has to be followed by composting. The income from the compost should significantly lower the net cost of a composting plant to below that of a pulverizing plant. Accurate costing for pyrolysis is not possible at the moment.

Composting has economic advantages that other forms of refuse disposal do not enjoy. For instance, the preparation of refuse into a state suitable for composting also renders materials with a possible market value into an easily recoverable state. These materials include ferrous metals, non-ferrous metals, textiles, paper, glass, and plastics. Before possible income from these sources is calculated, it is well to check that markets exist as they tend to fluctuate heavily. It is easy to employ staff and install machinery for, say, the recovery of paper, and then to find that the market has collapsed, leaving you with underemployed staff and bright new machinery that has to be paid for whether income is accruing or not.

These are some economic aspects to composting that are not so obvious. The rejects, for example, will weigh considerably heavier for a given volume than the refuse. This will not only extend the life of a landfill site but will also reduce the unit cost of the transport involved.

The income of a composting plant can rise to 50% of the gross cost, including capital cost repayment, and inflation works in favor of composting. As wages, electricity charged, and other variables increase, the value of the compost will also increase, unlike incineration, for example, where inflation will always send up the unit cost for disposal. However, it is wrong to look upon composting as a profit making process. In the United States, in particular, composting has had to shoulder this burden while other waste disposal systems have not. No waste disposal authority expects its landfill sites and incinerators to make a profit. Then why should composting?

If a waste disposal authority is prepared to credit a compost plant with an economic amount for the disposal of solid wastes, and the sewage authorities are prepared to do the same for the disposal of sludge, then this total income plus the income from the sales of compost and salvageable materials should exceed the gross cost.

THE POTENTIAL OF COMPOSTING

In general terms, at the moment, in industrialized societies, 1 tonne of solid wastes is generated per 1000 inhabitants per day. In a city of 100,000 people, the waste will amount to approximately 100 tonnes/day, and in a city of 1,000,000 people, it will amount to roughly 1000 tonnes/day.

Table 4.2 shows the analysis of raw refuse for a city of 280,000 inhabitants in England and for a city of 315,000 in North America. It will be seen that there are variations. For instance, the organic content of English wastes are 21.5%,

TABLE 4.2.   Analysis of Domestic Wastes.

	U.S.A.	England
Textiles (including rubber and leather)	9%	5%
Fines	4.6%	8%
Organic	18.5%	21.5%
Metal	8.2%	7.0%
Glass, plastic, and ceramics	24.2%	15.5%
Paper	35.2%	27%
Unclassified	0.3%	16%
	100%	100%

while those of American wastes are 18.5%. Theoretically, this will reduce the amount of compost that can be made, but this is offset by the higher paper content of American wastes.

On the average, 40 to 50% of the input of domestic wastes to a compost plant will finish up as saleable compost; i.e., for every 1000 tonnes fed into the plant, 400 to 500 tonnes of compost will be produced.

So, in a city of 100,000 inhabitants, if all the refuse is composted, the 100 tonnes/day of wastes generated will result in a compost yield (at 50%) of 50 tonnes/day, or (assuming a 5 day week) a total of 13,000 tonnes/year, which will treat 3200 ha at 4 tonnes/ha/year. This is not a great deal, and some communities may decide to compost their wastes for the benefit to the land and the ecological advantages.

It is hoped that this chapter has shown that composting is possible. Technically, all the knowledge is available. The author does not advocate composting as the be-all and end-all of waste disposal, but it is a powerful weapon in the waste disposal authority armory whose advantages can also benefit the community as a whole.

ACKNOWLEDGMENT: The author wishes to thank Leicestershire County Council for permission to publish and for the photographs of the Wanlip Compost Plant, Leicester, England.

REFERENCES

Epstein, E. 1977. "Sludge composting projects in U.S. cities." *Compost Sci.* **18**:5–7.

Howard, A. 1935. "The manufacture of humus by the Indore Process." *J. Roy. Soc. Arts.* **84**:26–59.

Rodale, R. 1962. "The intriguing history of composting." *Compost Sci.* **3**:30–33.

Wylie, J. C. 1959. *The Wastes of Civilisation*. Faber and Faber, London.

FURTHER SUGGESTED READING

Compost Science. Rodale Books, Emmaus, PA, U.S.A.

Golueke, C. G. 1977. *Biological reclamation of solid wastes*. Rodale Books. Emmaus, PA, U.S.A.

Davies, A. G. 1961. *Municipal Composting*. Faber and Faber, London.

USEPA, Solid Wastes Management Programs Office. *Composting of Municipal Solid Wastes in the United States*. Ref. EP 3.2: C 73/3 stock no. S/N5502-0033.

5

By-products from malting, brewing, and distilling

B. A. Garscadden, N.D.A., C.D.A.

Divisional Director, Brewers Grains Marketing,
Burton-on-Trent, U.K.

INTRODUCTION

The fermentation industry using cereal substrates in the United Kingdom is oriented to producing alcoholic beverages rather than alcohol for industrial purposes. During the process, marketable by-products occur, and as the cost of energy continues to escalate, the disposal of the by-products must have minimum energy input, as the markets for these are extremely competitive.

The limitations on the available supplies of energy and the consequent continual increase in its costs have been well publicized. It is, therefore, sufficient to say that all users of energy have to critically examine whether it is being used to the best advantage. Energy is also required for the transportation of raw materials in, and finished products and waste material out, and thus the location of the factory is invariably becoming an important factor in manufacturing and distribution costs.

The production of fermented and distilled drinks throughout the world is based on material that can be grown locally and is

TABLE 5.1.   Estimated Annual Flows of Malting Barley in the United Kingdom, 1971/72 to 1975/76 (million tonnes except where stated).

Production area (million ha)	2280
Yield (tonnes/ha)	3.9
Production	8.89
Barley sold off farm	5.52
Home grown barley used for malting	1.65
Malting barley imports	0.13
Total barley for malting	1.78

Source: Home-Grown Cereals Authority: "British Malting Barley: Supply and Demand" by I. M. Strugess and C.J. Knell.

best suited to the prevailing climatic conditions. In the United Kingdom, barley is the most preferred cereal; it likes a relatively high rainfall without extremes of temperature, and is widely grown. Therefore, the cycle of barley is from the farm, through a malting process, to the brewery or distillery. After extraction, it returns to the farm as a valuable ingredient for ruminant livestock production. The production area and quantities involved are shown in Table 5.1.

The methods of disposal of spent grains and other organic by-products produced in these processes, whether they result in revenue or cost, will also be reviewed.

MALTING

The object of malting is to modify the endosperm of the cereal by stimulating the production of naturally occurring enzymes, particularly alpha and beta amylases under controlled germination and kilning processes. This converts the endosperm of the grain into simple sugars which will later be partially converted by fermentation into alcohol. Malted barley is also used for the manufacture of malt extract as a sweetener or flavor enhancer for food and non-alcoholic drink manufacture. While raw materials selection characteristics may differ, the actual process of malting for both these purposes is similar to that for the material intended for use in a fermentation process. Details of the malting process and the by-products produced are summarized in Table 5.2.

In the main, malting in the United Kingdom is undertaken by the Brewer, Distiller, and Independent Sales Maltsters, and Table 5.3 shows the quantity undertaken by each category.

Barley Intake

The selection of barley for malting is dependent on the final use of malt. Brewers require malt to have a low nitrogen content to maximize the yield of extract and

TABLE 5.2.   The Malting Process and By-Products Produced.

Basic Stages	Details of Process	By-Products
1. Intake	Intake of barley Screening	Barley screenings
2. Premalting treatment	Drying to 12% moisture Cleaning and dust removal Grading Storage	Small corns and dusts
3. Malting	Intake of graded, dry barley Steeping (42 to 46% moisture) Germination Kilning (to moisture 2 to 4%) Screening	Steep liquor Malt culms Malt screenings

4. Storage and
uses
	Maturation
	Transportation
	Utilization of malt in:
	  Brewing
	  Distilling
	  Vinegar manufacture
	  Beverage and food manufacture
	  Malt extract manufacturer
		Spent grains

Source: Home-Grown Cereals Authority: "British Malting Barley: Supply and Demand" by I. M. Sturgess and C. J. Knell.

to reduce undesirable elements such as haze precursors in the final product. Until the developments in modern plant breeding, only certain varieties, such as Spratt and Plumage Archer, grown in selected areas of the country, were regarded as suitable for malting. With the introduction of Proctor, which although high yielding, still has the required malting characteristics, the "malting quality" price premium has tended to diminish. By virtue of their lower yield cost, non-malted

TABLE 5.3.   Estimated Annual Malt Production and Utilization in the United Kingdom, 1971/72 to 1975/76 (million tonnes).

	Brewer Maltster	Distiller Maltster	Sales Maltster	Other Uses	Exports	Imports
Barley used	0.33	0.38	1.07	—	—	—
Malt output	0.26	0.29	0.84	—	—	—
Malt usage	0.66	0.59	—	0.06	0.12	0.04

Source: Home-Grown Cereals Authority: "British Malting Barley: Supply and Demand" by I. M. Sturgess and C. J. Knell.

cereal adjuncts are used. They also have the added benefit of diluting the higher nitrogen levels in the malt, resulting from the greater usage of nitrogenous fertilizers by farmers to increase yield of barley per hectare. Suitable barley, therefore, is purchased at harvest time by agricultural merchants or maltsters and then transported to grain stores with drying facilities, where it is first predressed to remove any trash, weed seeds, and small (unripe, etc.) corns to avoid the high cost of drying.

The amount of material removed at this stage will very largely be dependent on the season. For instance, the report by the Home Grown Cereal Authority on the Quality of Wheat and Barley from the 1976 Harvest, stated that the amount of small corns passed through a 2.2 millimeter screen in 1975 was 8%, whereas in 1976 the amount was 27.7%. The rejected corns are either returned to the farmer-grower or sold on the open market at below market values for feed grade quality cereals. Thus it is in the interest of the farmer-grower to present the merchant or maltster with as clean a sample as possible to avoid the cost of delivery and grading, especially where material has to be moved over long distances.

Barley for malting is purchased on the basis of a minimum 16% moisture, and price adjustments are made for moisture contents above this figure. The drying of the grain from 16 to 12% moisture is designed to ensure it can be stored for long periods without any adverse effects on its viability, so that malting can be undertaken on an all year round basis. This operation must, however, be skillfully carried out by specialist driers to ensure that it is not dried too quickly or by too high temperatures that could kill the embryo of the grain and render the barley useless for malting.

Steeping

The raising of the water content of the grain to between 42 and 46% on the wet weight basis is to initiate the germination process. Traditionally, this is done by steeping the raw grain in water at 15 to 20°C which is changed two to three times. This process takes about 48 hours but technological changes such as barley abrasion and aerating the grain can shorten the duration of steeping by as much as 20 hours. The submersion of the grain allows any dust or awns to float to the surface and these may be skimmed off. The liquor from the first steep will have a B.O.D. in the order of 2,000 mg/l. With the present knowledge available, one cannot predict a useful conversion of this particular effluent to any form of cattle food or any other potentially valuable material. On the other hand, it is of sufficient strength to necessitate treatment prior to discharge.

In addition to being labor intensive, another very great problem of floor malting is that the environmental conditions (especially with regard to temperature and humidity) near the barley will be variable and not easily controlled since they will be affected by the prevailing climatic conditions. Consequently, the trend is towards highly mechanized and automated pneumatic maltings such as drum maltings and box maltings. When the desired degree of germination has taken place, the process is terminated by kilning.

Kilning

Kilning kills the germinating barley and produces material which has a longer shelf life, for use at a later date without reduction in quality.

The moist grain with roots is passed, in the traditional system, through the large kilns where the material is spread out on perforated floors through which hot air is passed. Kilning temperature and time will influence the flavor and color characteristics and will be varied according to the type of process for which the malt will be ultimately used. Some pneumatic malting systems allow the kilning to be done in the same container as the germination, which avoids another handling process. The dried roots which are screened off are termed "malt culms." These are sold, in the main, as a raw material ingredient for compounding into animal feeding stuffs. The quantities of malt culms produced, which are shown in Table 5.4, are in the order of 25 kg/tonne of malt produced. The analysis of malt culms is shown in Table 5.5. The value of malt culms to the animal feed compounder is dependent on the state of the market. In relation to feed grade barley, they carry a premium of between 5 and 10%, due to the relatively high protein content. Malt culms are very largely sold subject on approval of the sample, due to the variations of quality, especially from different maltings. A good malt culms sample should be light brown in color, contain no barley awns, husks, dust, or other extraneous matter and should be free from mold and have no unpleasant odors.

Dust extraction facilities at this production point are becoming more common, and in such instances better quality malt culms are produced. The dust which is collected separately is sold at a nominal price and is used as a filler ingredient in manufactured animal feeding stuffs.

TABLE 5.4. Estimated Quantity of Malt Culms from Malting in United Kingdom, 1976 (thousand tonnes).

Brewer maltster	6.500
Distiller maltster	7.250
Sales maltster	21.000
Total	34.750

TABLE 5.5. Barley Malt Culms Analysis.

Dry Matter Content	90%
Nutritive value on dry matter basis	
Metabolizable energy (MJ/kg)	11.2%
Dry crude protein	27.3%
D value	72%
DOM	20%
Chemical analysis on dry matter basis	
Gross energy (MJ/kg)	18.4%
Crude protein	27.1%
Ether extractives	2.2%
Crude fiber	15.6%
Nitrogen-free extractives	47.1%
Total ash	8.0%
Digestibility coefficients	
Crude protein	82%
Ether extractives	75%
Crude fiber	91%
Nitrogen-free extractives	73%

Source: M.A.F.F. 1975.

The low bulk density of malt culms presents problems in handling, storage, and transport. To overcome these problems, the bulk density may be increased by pelletizing. The relatively high fiber and crude protein content of malt culms, together with a low energy value, renders this material suitable for consideration for incorporation in specialized human foods. The requirement for natural fiber and additional bulk in the human diet is creating additional markets for this by-product and will command higher market values than for the animal feed usage.

The dry (2 to 4% moisture) malted barley can be stored ready for use by the brewer, distiller, or food manufacturer. The imperial weight of the old quarter of barley at 16% moisture is 448 lb, and for malt at 2 to 4% moisture, 336 lb. The difference in weight is not fully explained by the straight loss in moisture and the aforementioned by-products, but is due also to the chemical changes that have taken place during the malting process.

Malting the barley near to the growing area reduces the overall transportation costs to the processor.

BREWING PROCESS

The production of beer involves the solubilization of fermentable sugars, and, after the solution is bittered by boiling with hops, it is fermented to produce

THE BREWING PROCESS

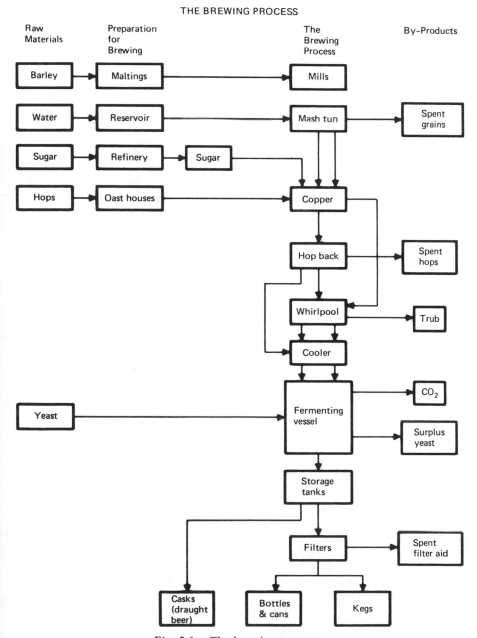

Fig. 5.1. The brewing process.

TABLE 5.6.   Brewery Raw Materials in the United Kingdom, Year Ending
             March 31, 1976.

United Kingdom beer production	39.2 million bulk barrels
Malt	707,210 tonnes
Unmalted corn	45,300 tonnes
Rice (grits and flaked)	34,641 tonnes
Maize (grits and flaked)	—
Sugar, etc.	133,435 tonnes
Hops	6,900 tonnes
Hop powder	974 tonnes

alcohol. The rough beer is then conditioned in various ways to prepare for dispensing. This operation is shown diagrammatically in Fig. 5.1, while Table 5.6
shows the raw materials used relative to the output of beer produced.

Mashing

Kilned malt, with or without the addition of relatively small quantities of unmalted cereal adjuncts, is ground and intermittently mixed with hot water so
that the final mash temperature is in the order of 65°C. This temperature is held
constant for a period to allow the starch in the endosperm to be converted to
soluble sugars by enzymes. The mashing vessels or mash tuns are cylindrical
vessels with drainage floors incorporated to allow the extract from the grains
known as "wort" to pass through leaving the undegradable products behind.
Washing or sparging with hot water at 55 to 80°C removes any residual soluble
material. The undegraded solid material left in the mashing vessel is termed
spent grains.

Hop Bittering

The sweet wort is boiled with hops which impart bitterness and aroma, the by-
product from which is spent hops. Recent technological developments have
enabled hops to be ground and pelletized, the residue from which can be recovered along with another by-product—trub (largely coagulated protein), where
whirlpool separation is available. A less common practice is the use of hop extract, from which there are no by-products at the brewery.

Fermentation

The bittered wort is then cooled, to around 15°C. Yeast is added to convert
simple sugars into alcohol and carbon dioxide. The temperature rises during

fermentation. During this process, the yeast utilizes the nutrients in the wort and multiplies, and by the end of the process, the yeast crop has multiplied by between three and six times. The yeast can be separated from the beer by centrifuge or filter press and the yeast surplus to the reuse requirement is sold as a by-product. The other by-product of this process is carbon dioxide.

In the production of cask conditioned beer, the green beer is racked directly into casks, perhaps with a few hours sedimentation period in an intermediate vessel. Finings are added to the cask immediately before being dispatched to the customer and are occasionally used to assist sedimentation before racking. In the production of bright or keg beers, the beers can go through a "conditioning" stage when more of the residual fermentable matter is changed into alcohol; it is filtered usually using kielselguhr, perlite, or mixtures of the two filter aids. During filtration, the filter aid becomes contaminated with separated suspended matter and the spent material presents a major disposal problem.

BREWERY BY-PRODUCTS

Table 5.7 shows the estimated quantities of the various by-products produced during the brewing process.

Brewer's Grains

It will be seen from Table 5.7 that spent brewers' grains represent the largest quantity of all the by-products from the brewing industry, the majority of which are discharged undried. Virtually the whole output of spent grains, whether wet or dried, is used for ruminant livestock feeding, and Table 5.8 shows the average analysis of fresh material as it leaves the brewery, after ensiling on the farm and after drying. Regardless of whether the material is dry or wet, the energy and protein values are expressed on a dry matter basis. As spent grains can be re-

TABLE 5.7. Estimated Quantity of By-Products Produced from Brewing in the United Kingdom, Year Ending March 31, 1976.

Spent grains	
Fresh, 22% dry matter	800,000 tonnes
Dried, 90% dry matter	9,000 tonnes
Yeast (wet/pressed)	50,000 tonnes
Carbon dioxide (50% recoverable)	124,727 tonnes
Trub (potential dry matter)	3,536 tonnes
Beer and wort losses	Unknown
Filter aid material (unused, dry)	6,000 tonnes

TABLE 5.8.   Average Analysis of Fresh Material After Ensiling on Farm and After Drying.

	Dry Matter (g/kg)	Metabolizable Energy (MJ/kg DM*)	Dry Crude Protein (g/kg DM)	Total Ash (g/kg DM)	Gross Energy (MJ/kg DM)	DOMD (%)
Brewers' grains (fresh)	220	10.0	149	45	19.6	59
Brewers' grains (ensiled)	280	10.0	149	43	19.7	59
Brewers' grains (dried)	900	10.3	145	43	19.8	60

Source: M.A.F.F. 1975.
*Dry matter.

garded as a succulent feed in an undried state, comparisons of analysis have been made with other succulent and farm produced feed materials. These have been extended to show their feed value on an "as fed" basis, after taking the moisture content into consideration (see Table 5.9). Brewers' grains, in their undried form, contain a considerable amount of free water, which is a very dilute form of sweet wort. This is of negative value to the brewery as it contains unwanted matter such as tannin, which would ultimately affect beer quality. To increase the dry matter content of spent grains, treatment is necessary by a de-watering process, but to do so creates added effluent problems. The expressate after mechanical de-watering of spent grain contains high levels of B.O.D., C.O.D., and suspended solids, as shown by typical figures in Table 5.10.

Brewers require a guaranteed reliable and efficient collection service for the removal of their by-products, and as a result the grains are sold to specialized grains contractors. Specialized vehicles incorporating moisture separation floors have been developed to ensure that high moisture grains are not unstable during transit and cause seepage of effluent on the highway, as well as to improve the consistency of the product on delivery to the farm.

The analysis of spent grains, as previously stated, does not fully demonstrate their full ruminant feed value. The investigation into the site of digestion of brewers' grains in the ruminant carried out at Nottingham University School of Agriculture (Swann, 1977) found that the protein is protected, not degraded, in the rumen. Degradation in the rumen results in losses in methane and ammonia, which are therefore not available for utilization by the animal. The absorption of nutrient into the bloodstream through the wall of the hind gut affords a greater efficiency of utilization for both liveweight gain and milk production. It is believed that this is due to the heat treatment in malting and mashing, which renders the protein insoluble. Mashing fractionates the soluble

TABLE 5.9.   Analysis of Brewers Grains and Farm Produced Feeds on a Dry Matter And "As-Fed" Basis.

	Dry Matter (g/kg)	Metabolizable Energy (MJ/kg DM)	Dry Crude Protein (g/kg DM)	"As Fed" Metabolizable Energy (MJ)	"As Fed" Dry Crude Protein (g)
Brewers' grains (fresh)	220	10.0	149	2.2	32.78
Brewers' grains (ensiled)	280	10.0	149	2.8	41.72
Brewers' grains (dried)	900	10.3	145	9.27	130.50
Kale	140	11.0	133	1.54	18.62
Swedes	120	12.8	91	1.54	10.92
Potatoes	210	12.5	47	2.63	9.87
Straw	860	5.3	8	4.99	6.88
Hay (mod.)	850	8.4	39	7.14	33.15
Silage grass (mod.)	250	8.8	102	2.20	25.50
Silage maize	210	10.8	70	2.27	14.70
Barley	860	12.9*	82	11.09	70.52
Wheat	860	14.0	105	12.04	90.30

Source: M.A.F.F. 1975.
*Rowett Research Institute. 1975.

and insoluble components at a pH similar to that of the rumen. In addition, the quality of the protein by amino acid evaluation was found to be much greater than hitherto believed.

By virtue of improved malting and mashing techniques, resulting in improvement in extracts, the proportion of residual and insoluble constituents in the by-products is increased. An increase in concentration of the oil fraction leads to an increase in metabolizable energy value. From the farmer's point of view, the incorporation of brewers' grains into his feeding system not only improves the nutrient feed value of the succulent part of the ration, but there is less

TABLE 5.10.   Analysis of Spent Grains Press Liquor.

	mg/l
B.O.D.	13,000
C.O.D.	20,000
Suspended solids	5,000

dependence, both in terms of quantity and quality, on the availability of farm produced succulents. Buying in succulent feed allows greater stocking rates per forage hectare to be achieved.

As most of the spent grains are discharged in an undried form, it is worth briefly examining the reason for this. Wet disposal relieves the brewer of the operation of drying and all its attendant costs, such as capital investment, depreciation, use of expensive energy, and the increased effluent concentration due to mechanical de-watering of the grains before drying. Recent studies by Garscadden (1977) have shown that the drying cost per dried tonne during 1977 ranged between £30 and £40 per tonne. Four and one-half tonnes of undried spent grains, which could otherwise be sold as such, are required to produce one tonne of dried material, and this raw material content, together with the drying charges, means that the break-even value of dry disposal is generally in excess of the value of feed-grade barley. In addition, by virtue of the seasonal production of beer, the majority of spent grains are produced during the summer months, at a time when demand for bought-in raw materials for compounding and ruminant feeding is at a minimum. A significant price allowance must therefore be made for storage in an assessment of true costs. This problem is overcome with wet disposal, since the spend grains which are produced during the summer can be efficiently stored under anaerobic conditions on the farm, where they will be ultimately used during the following winter. The conserved supplies balance out the reduced supplies during the critical feeding period.

Surplus Yeast

The analysis of brewers' yeast after drying is shown in Table 5.11, and for yeast extract, in Table 5.12; Table 5.13 shows the amino acid composition and typical vitamin B contents for both yeast extract and brewers' yeast; and Table 5.14 shows the comparative vitamin B potency of brewers' yeast with common foodstuffs.

TABLE 5.11.   Analysis of Brewers' Yeast After Drying.

Dry matter	900 g/kg
Crude Protein	44.3%
Dry Crude Protein	38.1%
Gross Energy	18.3 MJ/kg DM
Metabolizable energy	11.7 MJ/kg DM
Total ash	10.2%
Crude Fiber	0.2%
DOMD	75%

Source: M.A.F.F. 1975.

TABLE 5.12.   Typical Analysis of Yeast Extract.

Moisture	27%
Protein (N × 6.25)	44%
Salt (as NaCl)	10%
Ash (excluding NaCl)	13%
Carbohydrate (by difference)	6%
Fat	Trace

Note: Approximately 50% of the total nitrogen is present as NH_2 nitrogen.

Source: Ellison. 1973. Reproduced by permission of the author.

TABLE 5.13.   Amino Acid Composition and Typical Vitamin B Contents of Brewers' Yeast Extract and Brewers' Yeast.

	Yeast Extract (μg/g)	Yeast (μg/g)
Amino Acid		
Alanine	7.2	6.0
Arginine	1.7	7.0
Aspartic acid	10.2	15.3
Cystine	trace	trace
Glutamic acid	11.5	14.4
Histidine	2.3	2.3
Isoleucine	4.7	4.9
Leucine	7.0	8.3
Lysine	7.0	10.0
Methionine	1.5	1.3
Phenylalanine	3.8	4.9
Serine	4.5	2.0
Threonine	0.5	4.9
Tryptophan	1.7	1.3
Tyrosine	3.0	4.0
Valine	6.0	6.5
Vitamin		
B_1 Thiamine	20–70	50–360
B_2 Riboflavine	55–100	36–42
B_6 Pyridoxine	15	9–102
Niacin	250–700	310–1000
Pantothenic acid	90	100
Inositol	3800	2700–5000

Source: Ellison. 1973. Reproduced by permission of the author.

TABLE 5.14.   The B Vitamin Potency* of Dried Brewers' Yeast Compared with Some Common Feedstuffs.

	Brewers' Yeast	Bakers' Yeast	Wheat Germ	Flour	Beef, Average, Lean	Fish, Average	Milk	Eggs
Aneurin	13.3	1.0	2.62	0.33	0.20	0.06	0.04	0.15
Riboflavin	5.3	2.5	1.00	0.13	0.22	0.20	0.15	0.25
Nicotinic acid	52.5	30.0	5.20	1.06	6.00	2.30	0.10	0.10
Pyridoxin	3.5	1.6	1.15	0.31	0.40	0.10	0.01	0.02
Pantothenic acid	15.0	–	0.85	0.51	0.49	0.66	0.29	0.83
Folic acid	0.8	–	–	–	–	–	–	–
Inositol	28.0	–	–	–	–	–	–	–
Biotin	0.07	–	–	–	–	–	–	–

Source: Ellison. 1973. Reproduced by permission of the author.
*Potency in mg/100g.

The demand and consequent market value of brewers' yeast for animal feeding stuffs has declined considerably over the past ten years. Limited quantities are still dried for pharmaceutical uses or health foods, but the majority is now used for the manufacture of yeast extract, with a small quantity for use in the distillery industry. A process flow diagram for the manufacture of yeast extract is shown in Fig. 5.2. Yeast extract has a variety of uses in the food industry as a food flavor enhancer and meat extract substitute, used mostly in processed foods to adjust flavor and color to maintain consistency in finished products.

Yeast for manufacturing is collected either in bulk, after being slurried by the addition of controlled amounts of water to produce a dried solids content of 12 to 15%, or as pressed in galvanized steel containers at 23 to 27% dried solids content. Partial treatment of yeast at the brewery by autolysis to assist in the mechanical handling is not favored by the yeast extract manufacturers, as this part of the manufacturing process is outside the control of the specialist processor. Yeast destined for reuse in pitching in the distillery industry is now normally pressed and then packed and well-consolidated in woven polypropylene bags to eliminate pockets of air which otherwise can cause a rise in temperature and accelerate the commencement of autolysis. Prior to transit and during transit, distillers' yeast should be kept in cold store.

Breweries sited nearer to the distilleries were formerly used to liquify the yeast by the addition of water in conjunction with mechanical stirring and put into barrels or bulk road tankers for transportation. This practice is losing favor due to the higher transportation costs and the lack of control of the

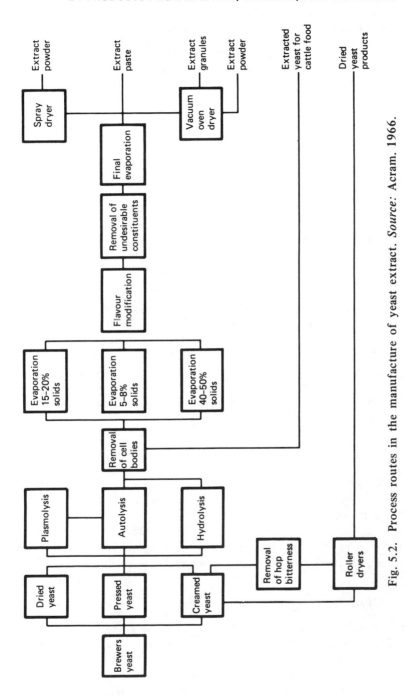

Fig. 5.2. Process routes in the manufacture of yeast extract. *Source:* Acram. 1966.

TABLE 5.15. Composition of Spent Hops.

	Range (%)	Mean (%)
Moisture	9–87	73
Nitrogen (N)	0.6–5.7	1.1
Phosphorus (as P_2O_5)	0.2–3.4	0.3
Potassium (as K_2O)	0.006–2.6	0.1
Magnesium (Mg)	0.12–0.11	—

Source: Berryman. 1970.

environment conditions during storage and transit, which adversely affects the viability of the yeast.

Spent Hops

The by-product from fresh spent hops is discharged in an undried state, and, where possible, supplied to nurserymen and market gardeners as a humus provider. Table 5.15 shows the analysis of spent hops. By virtue of their selection and the boiling treatment by the brewer, they are virus- and disease-free, and the resultant hop by-product is suitable, after composting, as an organic fertilizer in glass-house production. Usage is limited due to increasing labor costs, since commercial growers prefer products such as peat moss, which require no composting prior to use. The residue from ground and pelletized hops can be recovered as an admixture with trub.

Trub

Trub is the protein fraction resulting from the vigorous boiling of the sweet wort with hops. Table 5.16 shows the feeding value of this material. It also contains hop resins, polyphenols, and ash. With the increasing popularity of whirlpool separators, trub and the residue from pelletized hops are separated from the wort in a cone at the bottom of the whirlpool. The mixture is slurried with water and pumped onto the spent grains in the grains silos.

Carbon Dioxide

It can be seen from Table 5.7 that the anticipated quantities of CO_2 produced by the brewing industry in 1976 amounted to about 125,000 tonnes on the basis of 3.2 kg per barrel, though only about half of this is recoverable. However, in 1975, the beer and soft drinks industry used around 45% of the total conventionally produced CO_2, and therefore it is appropriate to question

TABLE 5.16.   Trub: Results of Analysis for Feed Value on Samples Taken
From One Brewery Using Whirlpool Separation, Dry Matter (%).

	Sample			
	1	2	3	Mean
Crude fiber	1.82	2.78	2.28	2.29
Oil	1.08	1.78	1.55	1.47
Ash	8.73	7.61	5.89	7.41
Crude protein	36.5	35.6	34.1	35.4
Dry crude protein	21.1	22.2	22.5	21.9
True protein	35.7	31.5	34.8	34.0

Source: Chapman. 1976.

methods of CO_2 disposal within the brewing industry. Prior to the rationalization of beer production in the United Kingdom, resulting in the closure of small, uneconomic production units within the main brewing groups, fermentation took place in relatively small quantities in open vessels, which was not conducive to the recovery of CO_2. But with the advent of new and larger production units, using larger, deeper, enclosed fermentation vessels, CO_2 collection becomes a practical possibility. The collection and use of CO_2 is not a common practice today because CO_2 is available at a relatively low cost from two major suppliers as a by-product from other processes. The large grain whisky distillers are one of the major sources, being well established and with operational plants set up before high inflationary building costs became prohibitive, and for them recovery costs for this by-product are economic. The other major source is as a by-product from the non-fermentation chemical and petrochemical industries.

The other large-scale users of this product, besides the beer and soft drinks production industries, are the nuclear and foundry industries. Specialist users, such as the medical profession, account for only a small percentage. CO_2 production, therefore, is not a rapid growth industry with potential supply exceeding demand, which results in relatively stable market values. The brewer thus has had little incentive to recover what financial revenue there is to be obtained. But with the increasing prevalence of industrial action and more stringent health and safety regulations being introduced, the large-scale brewing operations are now more likely to make provisions for CO_2 recovery to ensure a continuity of supply.

CO_2 is used in the movement of beer and the carbonation of certain final products within the brewery, as well as for beer dispensing in public houses. As CO_2 is evolved during the fermentation stage of the beer production process, associated treatments such as drying and purification are necessary. A typical

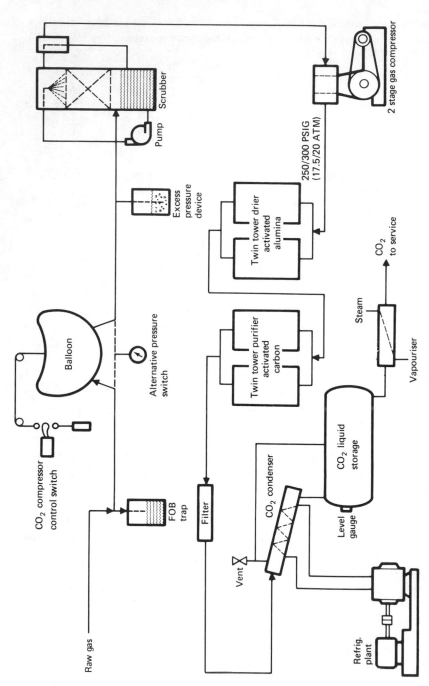

Fig. 5.3. Typical CO$_2$ recovery plant flow diagram. *Source:* Turvill. 1973. (Reprinted from an article by W. R. Turvill, Hall Thermotank International Engineering, Ltd.)

CO_2 recovery plant flow diagram is shown in Fig. 5.3. The siting of large conical fermenters, unhoused in the production precinct, avoids the necessity of using extraction fans in housed containers in order to conform with health and safety regulations. Surplus on-site CO_2 production may in the future become an attraction as a neutralizing agent for effluent.

The United Kingdom brewer is a more fortunate position than some overseas breweries which have no locally produced supplies of CO_2. This situation means that CO_2 recovery must be a regular practice and in many instances used in closely sited soft drink plants.

Spent Filter Aids

The two most common materials used as aids for filtration are kieselguhr and perlite. Both are imported, though the former is already processed and graded, while the latter is imported in its unrefined state and is appreciably cheaper. Particle retention and no contamination of the filtrate are the two most important properties of filter aids. By and large, the spent material is either washed to the drain or dumped. In each case, a disposal cost is incurred.

Attempts have been made (Garscadden, 1976) to recover kieselguhr by re-kilning at high temperatures to burn away all the organics. Unless the material has a high dry solids content, transport and rekilning costs make the process uneconomical. The yield was found to be low and the kilning losses very high. Indeed, with odd mixtures of kieselguhr and perlite, rekilning to an acceptable specification is virtually impossible due to the difference in fusion temperatures of the respective materials. Acid treatment of spent kieselguhr is feasible on a laboratory scale, but not so in practice, due to the problems of treatment and disposal costs of the used acids on a large scale.

Although the small amount of material retained in the spent filter aid has a high animal feed value, the safe tolerance levels for silicon materials is very low. Providing waterways are not polluted, disposal by spreading on agricultural land may possibly offer the least cost solution, and some intangible benefits as a soil improver may also ensue.

DISTILLING

The Process

Several parts of the distilling process are common, in general terms, to the brewing process. There are two distinct types of whisky production: malt and grain. Malt whisky is made from malted barley only, while grain whisky is made from a mixture of malted barley and unmalted maize or other cereals (see Figs. 5.4 and 5.5 for the traditional processes of both products). In malt distilleries, the

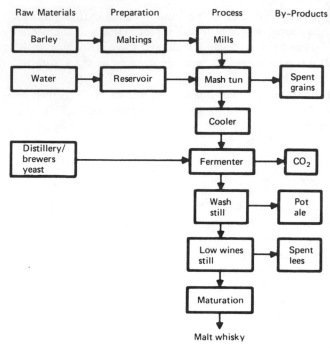

Fig. 5.4.   Traditional malt whisky process.

malted barley, with no cereal adjuncts, is mashed in conventional brewing mash tuns, with the resulting sweet wort fermented in vessels similar to those used in a conventional brewery. Spent grains, or draff, is the by-product of mashing, and CO_2 is the by-product from the fermentation. There is no yeast recovery as a by-product at this stage.

From this point, the process differs. Malt distilleries use no added flavors (e.g., hops) but rely on the concentration of the alcohol by double distillation in specially designed and constructed pot stills to provide the characteristic flavor and aroma in the finished product. The residues from this process are pot ale and spent lees. The spent lees are not recovered. If the pot ale, which contains yeast residue, is dried, it is known as distillers' dried solubles. If pot ale is evaporated to a syrup and then admixed with draff followed by drying, the product is malt distillers' dark grain (MDDG). The siting of malt distilleries in areas of plentiful supplies of naturally occurring clean water is said to also impart certain flavor characteristics in the final product, malt whisky.

After a lengthy maturation period in wooden casks of between 4 and 20 years, the whisky is either blended and bottled or sold as a bottled straight malt whisky. Consequently, malt distilleries are to be found in the remoter areas of the High-

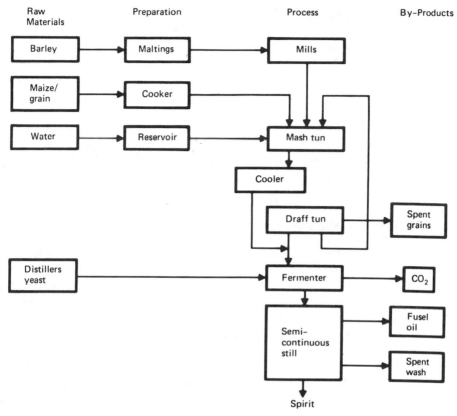

Fig. 5.5. Traditional grain whisky process.

lands around the River Spey, or in the Lowlands, or even on the Islands of Islay, Jura, Orkney, and Skye.

Grain distilleries are usually very much larger units and are normally sited nearer to the industrial areas. Unmalted maize, which is precooked to solubilize the starch, and small quantities of high diastatic-power malt, are conventionally mashed. Spent grains or draff are again the by-product of this process. Distillation concentrates the alcohol, and the by-products left are fusel oil and spent wash. Feints are also produced, but these are recycled through the process. Spent wash, if dried, is known as distillers' solubles. By virtue of the size of the operation, there are many opportunities to make the production savings necessary, by development of new technology. The development of a single vessel for the vacuum cooking of the maize and mashing with the malt has enabled higher concentrated worts to be fermented, thus reducing heat energy requirements. The fermentation and distilling of the whole mash, including the spent grains, is another example of the technological developments of the process. The by-products previously mentioned are produced as an admixture from that process technique. Grain spirit can also be used in the production of gin and vodka.

By-Products

CO_2. Although CO_2 is produced later in the process, its method of disposal is being dealt with first, due to the interaction of spent grains (draff) and pot ale/spent wash. In the small malt distilleries, the CO_2 is allowed to go to waste. In the large grain distilleries, CO_2 recovery has long been practised and sold to the brewing and soft drinks industries, where CO_2 from natural sources is welcomed. The flow diagram in Fig. 5.3 is identical to the CO_2 recovery in breweries.

Spent Grains, Pot Ale/Spent Wash. Traditionally, spent grains (or draff) were discharged and sold in their undried form to local farmers but without any organization for orderly marketing. Consequently, prices obtained were below their true market value. The feeding values of distillers' draff are shown in Table 5.17, and it will be seen that they are not dissimilar to brewers' spent grains. Pot ale and spent wash with a dry solids content of no more than 3 to 4%, but with the high B.O.D. content of 30,000 mg/l, have been traditionally disposed of via the drain and main sewer or directly out to sea.

TABLE 5.17. Nutritional Value of Distillers By-Products and Dried Skimmed Milk for Animal Feeds.

	Dry Matter (%)	Crude Protein (%DM)	Ether Extract (%DM)	Nitrogen-free Extract (%DM)	Crude Fiber (%DM)	Ash (%DM)
Draff (distillers' grains), fresh	23.8	19.8	8.1	51.5	17.3	3.34
Barley (brewers' grains), fresh	22.0	22.4	6.8	46.4	20.1	4.1
Barley (distillers grains), dried	90.9	19.8	9.3	49.5	18.1	3.34
Barley (brewers' grains), dried	85.6	19.2	6.9	50.4	16.3	3.3
Distillers' dried solubles	94.9	26.5	0.31	46.7	0.26	26.6
Barley dark grains	94.8	27.7	9.4	54.6	11.3	4.7
Dried skimmed milk*	97.0	36.1	4.1	51.5	0	8.2

Source: Leitch and Boyne. 1976.
*M.A.F.F. 1975.

However, with the introduction of the Rivers (Prevention of Pollution) (Scotland) Acts, 1951 and 1965, to control and define the quantity and quality of effluent into the main inland waterways, and more recently at certain coastal regions, the method of distillery by-product disposal has altered considerably. Basically, the changes have been centered around pot ale/spent wash. The enormous volumes of this material with a low solids concentration is the main problem. It can be seen from Table 5.17 that the nutritive feed value is reasonably high, being similar in its dried form to that of dried skimmed milk. The low fiber content but relatively high protein content of fairly high biological value, together with unidentified "growth factor" characteristics, enables it to be suitable for inclusion in small quantities in rations for the monogastric species, such as the pig, poultry, turkey, and mink. However, the high ash content and low carbohydrate value limit the amount that can be incorporated. The high cost of concentration and subsequent spray or roller drying is inhibitive, relative to the realizable price on the market, which could most likely only absorb a small percentage of the total potential output. The process flow diagram for the production of spray dried distillers' solubles is given in Fig. 5.6.

It is, therefore, a generally accepted practice to process a mixture of draff and pot ale/spent wash to produce dried distillers' dark grains, the process flow diagram for which is shown in Fig. 5.7. The average analysis for distillers' dark grains is shown in Table 5.18. It is estimated that the total potential output of distillers' dark grains in 1976 was 340,000 tonnes, but during that year, it was estimated that 370,000 tonnes of undried straight draff were sold. The United Kingdom production of compound cattle foods, excluding calf starters, balancers, and concentrates, was 3.9 million tonnes in 1975 and the relative potential output of distillers' dark grains represents 8.7% of this figure. An example of the place of dried distillers' dark grains in milk production ration is shown in Table 5.18.

The developments in ruminant nutrition indicate that the role of the animal food compounder will significantly change in the next few years, with the emphasis on the manufacture of fewer but higher energy concentrated feeds, for feeding during the limited period in the first stage of lactation. Production requirements for the remainder of the lactation, between lactations and during rearing (the dairy replacements), and during rearing and fattening of the beef animals, can be more economically obtained from on-farm produced feeds and non-processed by-products. It is estimated that a fairly high proportion of the distillers' dark grains output is exported, which has been lucrative when the overseas demand has been high and relative sterling values low. In the latter part of 1977, when overseas demand fell and sterling improved, the value of distillers' dark grains was reduced by as much as 50%. With these lower values, at a time of spiralling drying costs, the attractions of the wet draff market were very great. However, problems of pot ale/spent wash disposal, other than by processing with draff, have yet to be solved. Limited quantities can be spread onto land with the benefit of improving the quality of grazing of marginal land. Island distilleries

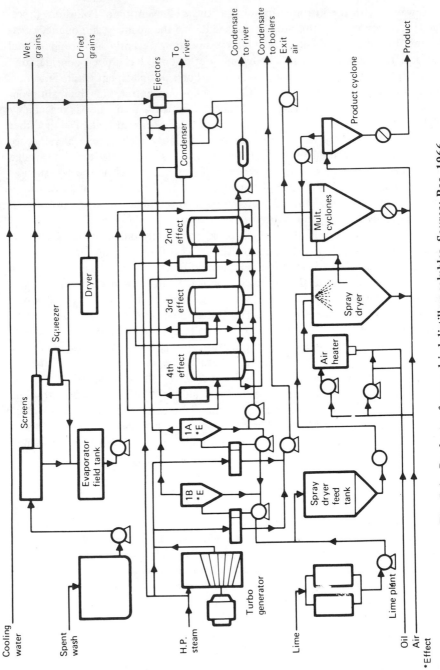

Fig. 5.6.   Products of spray dried distillers solubles. *Source:* Rae, 1966.

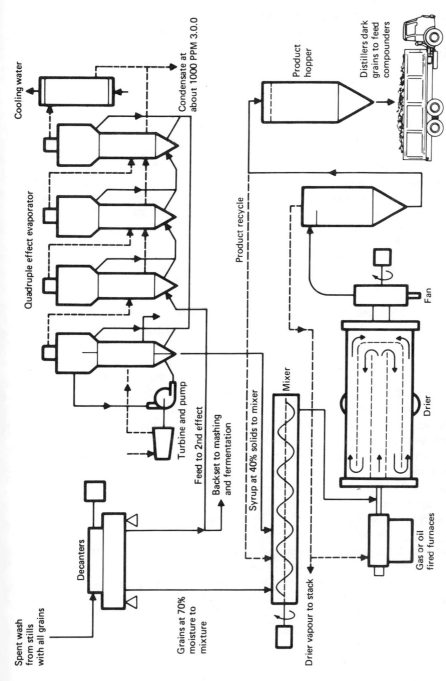

Fig. 5.7.   Process flow sheet for production of Distillers' Dark Grains. *Source:* Rankin. 1977. (From an article published originally in *The Brewer.*)

TABLE 5.18.   An Example of the Use of Distillers' Dark Grains in the
Formulation of a 13% Protein Daily Ration.

	As Fed Basis				
	Metabolizable Energy (MJ)	Crude Protein (g)	Dried Crude Protein (g)	Calcium (g)	Protein (g)
3½ parts barley	38.8	325	247	1.5	11.4
1½ parts wheat	18.1	160	135	0.4	5.2
2 parts sugar beet pulp	22.0	191	110	11.3	1.3
2 parts distillers' dark grains	20.4	500	330	3.0	16.0
2 parts dried grass	19.0	336	244	5.3	4.4
Calculated nutritional value of mixture	10.8	137	97	2.0	3.5

Note: The above ration, when fortified with additional minerals, would be suitable for average yielding dairy cows either grazing on average/good pasture or self-fed grass silage.

still discharge direct to sea, but proposed legislation will prohibit this in the future. If these problems could be solved without a negative cost to the distiller, the enhanced value of the by-products could provide a valuable contribution to reducing the cost of distilling raw materials inputs.

The solids in pot ale/spent wash contain yeast cells, amino acids, minerals, and unfermented carbohydrates. In addition to riboflavin, other substances classified as "unidentified growth factors" promote growth and lead to an improvement in efficiency of food conversion. Pot ale/spent wash, or distillers' solubles, in its dried form is similar in feed value for the monogastric species to dried skimmed milk, which is used extensively in the processing of human foods. The value of the protein as defined by the amino acid content is shown in Table 5.19. This nutritionally-attractive material would therefore appear to be suitable for upgrading to human food processing.

Hawthorn and Anderton (1977) found that conditions at distilleries and by-products plants give rise to unacceptable variations in some organoleptic characteristics of the product as produced at present, most notably acidity and flavor. Copper contents in dark grains and dried solubles, which are of the order of 50 parts per million and 150 parts per million, respectively, are also variable and above the limit of 20 ppm considered acceptable for most human foods in the United Kingdom. Such problems associated with the present by-products may in time be solved by product and process development research.

Laboratory and trial scale use of pot ale/spent wash as a substrate for microfungal growth has demonstrated the technical feasibility of such a process. The mycelium produced in a continuous culture is relatively easy to separate from the culture liquor and is compatible with the continuous availability of supply of

TABLE 5.19.   Average Amino Acid
                          Content of Malt
                          Distillers' Dried
                          Solubles.

	g/kg
Argenine	4.5
Cystine	0.8
Glycine	11.2
Histidine	4.3
Isoleucine	15.2
Leucine	20.8
Lysine	11.9
Methionine	3.7
Alanine	7.0
Serine	
Threonine	9.7
Tryptophan	0.7
Tyrosine	2.1
Valine	15.2
	107.1

Source: Fullbrook. 1977.

the raw material. The resultant end product, however, still requires drying and is therefore another demand for heat energy.

A limited consumer acceptability for novel proteins has been attained for products mainly derived from material of cereal origin. Yeast protein from hydrocarbons and wood sulfite liquors have generally yielded animal feed grade material, and it is in this category that the product from pot ale/spent wash substrate is to be found. In the world of animal nutrition, maximum protein requirements, especially for the ruminant, have recently been found to be somewhat less than has been traditionally fed. There is, however, an ever-increasing demand for higher energy values in feeds for livestock, to accelerate growth and yield. This is in direct contrast to the trend for human nutrition, where the demand is for less starch and more protein. The demand for animal feed grade single cell protein is thus limited, and the market at present cannot afford the costs required for economic production by the processors. Consequently, the progress in this direction has virtually ceased.

Liquid and block feeding to ruminant livestock have developed since it has been found that the ruminant has the ability to increase the population of digestive bacteria. This ability can be emulated by higher intakes of non-protein nitrogen, which block and liquid feeds can more easily incorporate, with the result that an improved conversion of high cellulose feeds can be obtained. As molasses has so far been used as the carrier, supplemented with urea for the liquid products, it may well be that pot ale/spent wash can be substituted to advantage. Although the ruminant has the ability to synthesize its own vitamin B, the use of concentrated pot ale/spent wash has been found to be a means of increasing the rumen micro-flora to enable it to consume and more efficiently utilize low value cellulose material such as straw and other similar forage materials of low digestibility.

The transportation of the material over long distances necessitates its concentration, and at the present time this can only be done by evaporation using heat energy. As this will be conveyed and stored, and dispensed at ambient temperatures, concentration is limited to no more than around 40% dry matter without seriously affecting its bulk handling properties. Table 5.20 shows the composition of evaporated spent wash. As the already well established dark grains process has end-product acceptability, although not always with full cost recovery price, this does not give distillers much encouragement to explore new processes and markets. This, however, could prove untrue if means were to be found to handle and market an undried distillers' dark grain without the need for heat energy to concentrate pot ale/spent wash by evaporation, since the greater part of evaporation costs are in the first stages of concentration.

Fusel Oil.   This may contain isopropyl, *n*-propyl, isobutyl, *n*-butyl, isomyl, and *d*-amyl alcohols. It is a source of amyl alcohol and also has a ready market in the cosmetics industry.

TABLE 5.20.   Composition of Evaporated Spent Wash.
(Range of Values Obtained with Four Batches.)

Dry matter	29.0–32.0%
on a dry weight basis	
Protein	18.0–28.0%
Lipid	0.6–11.6%
Fiber	0 –0.3%
Ash	6.1–8.1%
Soluble carbohydrates	28.0–50.0%
Ca	0.3–0.6%
Mg	0.4–0.6%
K	0.7–1.5%
Na	0.03–0.04%
P	1.0–1.2%
Cu	192.0–284.0 ppm
Mn	16.0–69.0 ppm
Zn	52.0–1012 ppm

Source: Miller. 1973.

CONCLUSION

This chapter has dealt with organic waste conversion from malting, beer, and whisky production in the United Kingdom. Conversion methods in the United States have until recently followed a different line, with the majority of the by-products being dried to facilitate transportation over greater distances. However, with the escalating cost of heat energy, and the increasing cost of complying with anti-pollution legislation, progress toward wet disposal, especially from breweries, is being made.

This discussion may pose some seemingly insurmountable problems concerning achieving acceptable and economic methods of disposal of by-products from the malting, brewing, and distilling industries. However, certain widely accepted disposal practices must surely be the envy of other manufacturers in the food and drinks sector. The maltster, brewer, and distiller are direct or indirect outlets for some of the farmer's products; therefore, they—the by-products producers—take a more understanding view when considering the disposal of by-products back to the farmer to repeat the cycle.

REFERENCES

Acram, A. R. 1966. "Processing brewery yeasts." *Proc. Biochem.* **1**:313–317.

Berryman, C. 1975. "Composition of organic manures and waste products used in agriculture." *A.D.A.S. Advisory Papers No.* **2**.

Chapman, J. 1976. "Disposal of a brewery waste with special reference to the whirlpool separator." Ph.D. Thesis, Newcastle University.

Ellison, J. 1973. "The commercial utilization of waste brewery yeast." *The Brewer* **59**:601–606.

Fullbrook, P. 1977. "By-products processing with the aid of enzymes." Portland Symposium, *Distillers By-Products*, April 1977.

Garscadden, B. A. 1976. Unpublished research.

Garscadden, B. A. 1977. Unpublished research.

Hawthorn, J. and Anderton, A. 1977. Unpublished research.

Leitch, I. and A. W. Boyne. 1976. "Composition of British feedingstuffs." *The Nutrient Requirements of Farm Livestock No.* **4**. A.R.C.

M.A.F.F. 1975. "Energy allowances and feeding systems for ruminents." *Tech. Bull.* **33**.

M.A.F.F. 1977. "Output and utilization of farm produce in the U.K." 1969/70–1975/76.

Miller, T. B. 1973. *Malt Distillers Grains*. Symposium arranged by the Agricultural Industries Training Services, June 1973.

Rae, I. J. 1966. "Distillery by-product recovery." *Proc. Biochem.* 1:407–411.

Rankin, W. D. 1977. "New production methods in distilling." *The Brewer* **63**: 90–95.

Rowett Research Institute. 1975. First report, Feedingstuffs Evaluation Unit, p. 36.

Sturgess, I. M. and C. J. Krell. 1978. *British Malting Barley: Supply and Demand*. Home-Grown Cereals Authority, London.

Swan, H. 1977. Unpublished research.

Turvill, W. R. 1973. "The process and economics of carbon dioxide collection." *The Brewer* **59**:607–611.

FURTHER SUGGESTED READING

Chapman, J. and J. R. O'Callaghan. 1977. "An evaluation of brewery wastes." *The Brewer* **63**: 209–215.

Simpson, A. C. 1977. "Gin and vodka." *Alcoholic Beverages* A. H. Rose (Ed.). Academic Press, New York, pp. 537–593.

Underkofler, L. A. and R. J. Hickey. 1954. *Industrial Fermentations*. Chemical Publishing Co., New York.

6

Wastes from the fermentation industries: their use as feedstuffs and fertilizers

M. W. M. Bewick, Ph.D., M.I.Biol.

Department of Applied Biology, University of Cambridge, Cambridge, England

INTRODUCTION

There are a large number of compounds which are capable of being synthesized by microbial fermentation, but relatively few are manufactured on a large scale by fermentation, as in many cases the synthetic pathway to a particular compound is more economical than a fermentation process. Several books have been written on the subject of the microbiology of industrial fermentations and include those by Rose (1961), Casida (1968), Smith (1969), Miller and Litsky (1976), and Yamada (1977). This chapter deals with the solid wastes which remain after the fermentation process is complete and examines their potential as fertilizers and feedstuffs. In the brewing and distilling industry, virtually all the wastes produced from the actual brewing operation have been utilized for many years, and these are dealt with in Chapter 5 to illustrate what can be done with the waste of an industrial fermentation if the will, and a suitable market, is there.

Of the fermentation wastes considered here, the ones which will be examined in the most detail are those from the antibiotic industry and wine and cider production. Wastes from other industrial fermentations tend to be produced on a smaller scale and companies tend to be very security conscious with respect to the quantities of wastes produced. Very few studies have been carried out into the potential uses of these materials.

ANTIBIOTIC FERMENTATION WASTE

Antibiotics were defined by Waksman (1947) as "chemical substances that are produced by micro-organisms and have the capacity, in dilute solution, to inhibit the growth or even destroy other micro-organisms." They are produced today on a large scale using the fermentation process. The exception to this is chloramphenicol, which has been produced synthetically since 1948 (Olive, 1949). The wastes with which this section is concerned are the solids remaining after the liquid containing the antibiotic has been extracted from the fermentation broth. This is usually done by filtration. These solids are a heterogeneous mixture of hyphae of the producing organisms, unused media, defoaming oils, and metabolic products of the micro-organisms, the most important of which is the residual antibiotic. Levels of residual antibiotic can vary greatly, and recent studies have indicated that tylosin fermentation waste contains 2000 parts per million residual antibiotic, while oxytetracycline waste contains 1500 parts per million residual antibiotic.

Quantities Produced

Total quantities of waste produced are difficult to obtain, as pharmaceutical companies are naturally reluctant to divulge the quantities of organic waste which they are discharging into the natural environment. Personal communications with producers of antibiotics in the United Kingdom suggest that total United Kingdom production of antibiotic fermentation waste amounts to some 150,000 to 200,000 tonnes/annum. Figures for the production of antibiotic fermentation waste in the United States can be estimated from figures obtained by Howe (1960), who found that 2.2 kg of antibiotic produced 13,500 to 16,000 liters of waste water. Based on estimated production figures of antibiotics in the United States, this indicates a total waste volume from antibiotic processes of 140 to 160×10^6 liters/day. Howe (1960) quoted the average solids content of these wastes at about 40,000 parts per million, and this suggests a production of 6,350 tonnes of solid wastes per day or 2.3×10^6 tonnes/year. This figure is similar to that derived from the figures of Lederman et al (1975), who reported that an average pharmaceutical plant in the United States consumes 2.36×10^6 kg/month of raw material for fermentation processes. Approximately 50% of this will be lost as CO_2 and compounds dissolved in the medium, and therefore the solid waste produced will be approximately 1.18×10^6 kg/month, equivalent to 14.2×10^3 tonnes/annum from one average plant. Lederman $et\ al$ (1975) also state that 115 pharmaceutical companies in the United States account for 95% of the industry's sales volume and thus the overall level of antibiotic waste production in the United States, based on these figures, is approximately 1.6×10^6 tonnes/year. In terms of total world production of this waste, one must also include the waste produced by the European, Japanese, and East European antibiotic manufacturers, of which nothing has been published. Berdy (1974) states that total world production of antibiotics in the early 1970's was 49,000 tonnes/annum. If one assumes that the figures produced by Howe (1960) for the United States are applicable worldwide, then total world production of antibiotic fermentation waste would have amounted to approximately 13.4×10^6 tonnes/annum in the early 1970's.

Disposal

Most present day methods of disposal involve biological treatment or the dumping of the waste untreated. Some pharmaceutical companies in the United Kingdom discharge their waste—which can have a B.O.D. of up to 32,000 parts per million (Hilgart, 1950; Hurwitz $et\ al$ 1952)—directly into rivers. In the United States, in 1970, Pfizer was prevented from dumping its waste into Long Island Sound, Connecticut, by the Environmental Protection Agency (Anon, 1973a), and now, like many other pharmaceutical companies in the United States, Pfizer dumps or buries waste on the land (Melcher, 1962; Colovos and Tinklenberg, 1962). Several pharmaceutical companies have investigated the possibility of biological treatment as a method of reducing the B.O.D. of the fermentation wastes. Brown (1951) looked at the possibilities of using sewage filters to treat antibiotic wastes, and Brown and Niedercorn (1952) found that the disposal of antibiotic containing fermentation wastes via sewage treatment plants could adversely affect the micro-flora of the filter beds. Hilgart (1950) appeared to overcome these problems in a treatment plant for penicillin and streptomycin wastes at the Upjohn Company, Michigan, U.S.A., when B.O.D. reductions of 97% were achieved. Barker $et\ al$ (1958) examined the possibilities of speeding up the breakdown of antibiotic wastes using compressed air, while Lederman $et\ al$ (1975) produced a reduction in B.O.D. using pure oxygen in the lagoons of the treatment plant. Several other pharmaceutical companies in the United States have also set up biological treatment plants, including Merck, Sharpe, and Dohme (Liontas, 1954) Upjohn (Tompkins, 1957), and Lederle (Molof, 1962). Attempts have been made to recycle the waste and use it as a base for a fresh batch of medium (Lawrence, 1974), but the inclusion of this material into a fresh batch caused a reduction in the yield of antibiotic.

Use as Animal Feedstuffs

In the 1950's and 1960's, much of the waste from antibiotic fermentations was used as animal feedstuffs. Its use as a feedstuff was first suggested by Stokstad *et al* (1949) when they observed a growth promoting effect in chickens fed with crude extracts of *Streptomyces aureofaciens* waste. Similar growth responses were reported when aureomycin was fed to chicks (Whitehill *et al*, 1950) and pigs (Jukes *et al*, 1950; Stokstad, 1954).

Reviews by Stokstad (1954), Cunha (1955), Robinson (1962), Bird (1969), and Jukes (1973) indicate that growth responses have been observed in many studies of swine, ruminants, turkeys, and chicks. Optimum levels of antibiotic were found to range from 2 to 50 parts per million, depending on the antibiotic. A report from the Office of Health and Economics (O.H.E., 1969) indicated that antibiotics stimulated the appetite, increased the efficiency of food conversion, reduced vitamin requirements, and increased the growth rate. These effects appear to be due to the action of the antibiotic on the intestinal micro-flora of the animals (Kellogg *et al*, 1966; Linton *et al*, 1975). Animals raised in a sterile environment showed no growth response (Lev and Forbes, 1959; Forbes and Park, 1959). Antibiotic fermentation waste would therefore appear to be an ideal feedstuff, as not only has it an intrinsic nutritive value, but the residual antibiotic stimulates growth. Reports of the use of antibiotic fermentation wastes as feedstuffs include those of Edwards *et al* (1953), Summers *et al* (1959), Young (1969), Johnson and Arscott (1974) and Struzeski (1977).

There are potential problems with the feeding of feedstuffs containing antibiotics routinely to animals.

1. In the presence of continuously low levels of antibiotic, resistance levels in pathogenic organisms could increase.
2. Antibiotics in animal feeds could be transferred to man via animal products.

Cases of dermatitis have been attributed to the presence of penicillin in milk (Borrie and Barrett, 1961). But in these cases the presence of penicillin in the milk resulted from the administration of relatively high levels of penicillin to treat mastitis and not from the feeding of sub-therapeutic levels of penicillin as growth promoters. Stokstad (1954) reported that levels of antibiotics used as growth promoters ranged from 2 to 50 parts per million, 0.2 to 5% of the therapeutic dose needed to maintain effective blood levels. Even at relatively high levels (200 parts per million) of aureomycin administered to chicks, Stokstad (1954) found that the antibiotic could not be detected in the serum, liver, and muscle. Jukes (1973) reported that tetracyclines could not be detected in soft tissues and were rapidly excreted in the urine, while erythromycin, tylosin, and bacitracin had not been found in animal tissues after feeding with low levels of antibiotics.

In 1962, in the United Kingdom, the Netherthorpe Committee (Netherthorpe, 1962) examined the practice of supplementing feedstuffs with antibiotics and found that the economic benefits outweighed the theoretical hazards. They also recommended that the practice, previously restricted to pigs and poultry, be extended to calves. But after 1962 there was an increased awareness of microbial resistance to antibiotics, and resistant bacteria were found to be becoming more common in the human intestinal tract (O.H.E. 1969; Moorehouse, 1969). There was concern over the transfer of resistance from non-pathogenic to pathogenic organisms, and in 1969, the Swann Report (Swann, 1969) was published, casting doubts on the wisdom of adding antibiotics to animal feedstuffs. As a result of this, from September 1, 1973, feedstuffs containing medicinal additives came under the control of the Medicines Act, 1968 (H.M.S.O., 1969) and control over the incorporation of antibiotics in feedstuffs and the sale of feedstuffs containing antibiotics is now exercised under the Medicines Act licensing provisions. As a result of these restrictions, the market for the use of antibiotic fermentation waste as an animal feed collapsed in the United Kingdom. In the United States, the Antibiotic Task Force Report (FDA Task Force, 1972) arrived at the same conclusions as the Swann Report. The implementation of this report was delayed, however, and only in 1977 was a ban on the inclusion of penicillin in animal feedstuffs announced with an imminent ban on tetracyclines (Anon, 1977).

There is still much controversy over the effects of sub-therapeutic levels of antibiotics on the resistance patterns of pathogenic micro-organisms. A review by Jukes (1973) strongly stated the case for the inclusion of low levels of antibiotics in animal feed and suggested that experiments carried out over a period of 20 years had not demonstrated a significant increase in the resistance of micro-organisms to the antibiotics used. Increasing levels of resistance were attributed to the widespread use of antibiotics in hospitals and general medicine. Therefore, in view of the problems which at present surround the use of antibiotic fermentation waste as an animal feedstuff, the rest of this section will be devoted to an examination of the potential of this material as a fertilizer, together with an examination of its potential effects on the soil ecosystem.

Fertilizer Value

Recent analyses of oxytetracycline, tylosin, and penicillin fermentation wastes carried out in the author's laboratory are presented in Table 6.1. Nitrogen contents ranged from 6.34% in tylosin waste to 2.8% in oxytetracycline waste, on a dry weight basis. Zvara et al (1960) reported that a penicillin residue contained 4.59% N; this is less than the figures obtained from United Kingdom penicillin waste and probably reflects the different media constituents in the two processes.

TABLE 6.1. Composition of Antibiotic Fermentation Wastes.

	Tylosin	Oxytetracycline	Penicillin
Water (%)	83.6	65.7	80.8
pH	6	2.2	NA*
Organic matter (as % of dry weight)	85.5	50	90.8
Organic C (as % of dry weight)	43.7	26.5	40.7
N	6.34	2.80	5.92
P	2.00	0.43	1.36
K	0.59	0.07	0.38
Ca	0.30	0.71	0.72
Na	0.21	0.35	0.25
Microelements (ppm):			
Mg	500	60	550
Fe	275	460	60
Mn	7	11	6
Cu	245	3.5	5
Pb	15	5	5
Cd	1.5	1.5	1
Antibiotic content (ppm)	2000	1500	NA*

*Figures not available.

To date, there have been very few reports in the literature on the use of anti-biotic fermentation wastes as fertilizers. Molof (1962) reported that Lederle in the United States mixed their antibiotic wastes with sawdust and animal manure to produce a compost which was sold as a fertilizer. But no further reports of this process have been seen in the literature. Uhliar and Bucko (1974) in Czecho-slovakia, reported an increase in fresh matter, dry matter, and numbers of root nodules in lucerne sprayed with penicillin and streptomycin wastes, although no indication was given as to the nutritional status of the wastes, the levels at which they were applied, or their antibiotic content. There have also been newspaper reports in Connecticut, U.S.A., of investigations by Pfizer of Groton, Connecticut into the use of fermentation waste as a fertilizer for potatoes (Anon, 1973a, b; 1974a, b; Warner, 1973). Although the type of antibiotic waste is not named, the experiments appear to have been successful. Should the use of antibiotic fermentation wastes as fertilizers become widespread, one of the potential problems, apparently ignored in the work of Pfizer, would be the presence of the residual antibiotic. The addition of antibiotics to the soil could affect the micro-organisms in the soil as well as plants growing in the soil. Therefore, it is important to review the potential effects of antibiotics on these aspects of the soil ecosystem, bearing in mind that the levels of residual antibiotics will be low.

Behavior of Antibiotics in Soil

A review by Pramer (1958) showed that antibiotics can be inactivated in soils, and this inactivation may be as a result of one or more of three processes:

1. The chemical instability of the antibiotic molecule.
2. Adsorption onto clay minerals and organic matter.
3. Microbial degradation.

Jefferys (1952) and Wright (1954) attributed the inactivation of penicillin, viridin, gliotoxin, frequentin, and albidin to their chemical instability. Using sterile soil, Gottlieb et al (1952) and Jefferys (1952) found that cyclohexamide, gladiolic acid, and penicillin were rapidly inactivated in soils, even under conditions favorable to their stability. The adsorption of antibiotics by clays and organic matter has been studied extensively by many workers, including Gottlieb et al (1952), Gottlieb and Siminoff (1952), Jefferys (1952), Martin and Gottlieb (1952, 1955), Pinck, Holton, and Allison (1961), Ghosal and Mukherjee (1970), and Bewick (1979). Basic antibiotics—e.g., streptomycin, neomycin, erythromycin, and tylosin—are adsorbed by the clay minerals montmorillonite, kaolinite, bentonite, illite, and vermicullite, although different clays adsorb different amounts of the same antibiotic (Pinck, Holton, and Allison, 1961; Bewick, 1979). Amphoteric antibiotics—e.g., bacitracin, aureomycin, and terramycin—can act either as acidic or basic antibiotics depending on their isoelectric points and the pH of the soil. As the pH of the soil is usually below the isoelectric point of the antibiotic, they behave as bases and are adsorbed onto clay particles. Acidic or neutral antibiotics—e.g., penicillin and chloramphenicol—are adsorbed by clays only in small amounts. Some antibiotics adsorbed by clays can be released by phosphate and citrate buffers. This release is more prevalent with amphoteric than with basic antibiotics (Soulides et al, 1961; Pinck, Soulides, and Allison, 1961; Bewick, 1979). Ghosal and Mukherjee (1970) found that different cations can produce different rates of release of streptomycin from clays. Microbial degradation can also play an important part in the disappearance of many antibiotics in soil. Jefferys (1952) found that 72 parts per million penicillin was rapidly degraded in soil, and Gottlieb and Siminoff (1952) showed that of 50 parts per million chloramphenicol added to non-sterile soil, 85% had disappeared after four days. Even with streptomycin, a basic antibiotic rapidly adsorbed by clay minerals, Pramer and Starkey (1951) found that disappearance of 1000 parts per million was more rapid in non-sterile than in sterile soil and that second and third additions disappeared more rapidly than the first addition. This observation was also reported when repeated additions of 20 parts per million griseofulvin were made to soils (Wright and Grove, 1957).

Studies on the release of antibiotics from their fermentation wastes when added to soils have been initiated by Bewick (1979) using tylosin fermentation waste. The fermentation waste was added to soil at levels of 3, 8, and 13 tonnes/ha. This resulted in the addition of 3.5, 11.2, and 18.7 parts per million tylosin, respectively, to the soil. These additions were carried out at 4, 10, 15, and 23°C. Increasing temperature was found to have little effect on the uptake of tylosin by the soil, while with the addition of increasing amounts of waste, increasing amounts of tylosin were adsorbed by the soil (Table 6.2). Using the means of tylosin uptake and percentage uptake at the four temperatures examined for each concentration of fermentation waste in the soil, it was found that with the addition of 2 tonnes of fermentation waste per hectare, the adsorption capacity of the soil for tylosin had not been reached and the soil had an adsorption constant for tylosin of 0.72 (Fig. 6.1a). Higher amounts of tylosin fermentation waste were added to a limestone waste to examine tylosin adsorption in a mineral soil containing no organic matter. With additions of 13, 25, and 38 tonnes of fermentation waste per hectare, 13.1, 26.2, and 39.3 parts per million tylosin were added, respectively. The adsorption of tylosin at these levels is shown in Fig. 6.1b. The adsorption capacity of the limestone waste was not reached with additions of 39.3 parts per million tylosin and the adsorption constant for tylosin and limestone waste was found to be 0.46. The total amount of tylosin released by the soils in these experiments was determined by leaching soil samples amended with different amounts of tylosin fermentation waste at fortnightly intervals over a 12 week period. Over this period of time, maximum release of tylosin occurred immediately after addition of the fermentation waste to the soils (Bewick, 1979). In both soils, the initial levels of tylosin detected ranged from 0.4 parts per million with the addition of 3 tonnes of fermentation waste per hectare, to 13.8 parts per million at 38 tonnes of fermentation waste per hectare. Although these levels are above the minimum inhibitory concentration for many

TABLE 6.2. Uptake and Release of Tylosin from a Modified John Innes Compost Amended with Tylosin Fermentation Waste. [Reprinted from *Plant and Soil* (Bewick. 1979.).]

| Concentration of Tylosin (ppm) | Temperature (°C) | | | | | | | | | | |
	4			10			15			23		
Added	3.5	11.2	18.7	3.5	11.2	18.7	3.5	11.2	18.7	3.5	11.2	18.7
Total release	1.1	3.4	7.2	0.6	2.7	3.9	1.1	3.2	6.1	0.8	3.6	3.9
Uptake	2.4	7.8	11.5	2.9	8.5	14.8	2.4	8.0	12.6	2.7	7.6	14.8
Uptake (%)	68.6	69.6	61.5	82.9	75.9	79.1	68.6	71.4	67.4	77.1	67.9	79.1

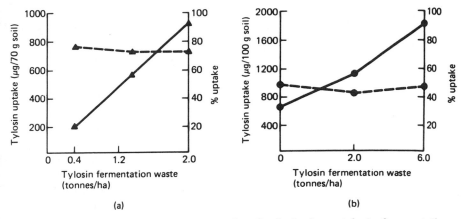

Fig. 6.1a.   Uptake and percentage uptake of tylosin from tylosin fermentation waste by modified John Innes compost. *Source:* Bewick. 1979. (Reprinted from *Plant and Soil*.)

▲———▲ uptake
▲-----▲ % uptake

Fig. 6.1b.   Uptake and percentage uptake of tylosin from tylosin fermentation waste by limestone waste. *Source:* Bewick. 1979. (Reprinted from *Plant and Soil*.)

●———● uptake
●-----● % uptake

sensitive organisms, as reported by McGuire *et al* (1960), it must be borne in mind that had these initial concentrations of tylosin been left in the soil, and not removed by leaching, they would have been subject to chemical and micro-biological breakdown in the soil, as well as adsorption onto organic matter and clay minerals. The results show the release of tylosin into the soil water over specific periods of time, i.e. the intervals between leachings, and not the subsequent fate of that tylosin in the soil system. Experiments are at present being carried out on the persistence of tylosin left in contact with soils for different periods of time.

The Effects of Antibiotics on Microbial Processes in Soil

There have been several studies on the effect of pure antibiotics on microbial populations and processes in the soil. Several of these studies were carried out on pure cultures of soil micro-organisms in laboratory media (Waksman and Woodruff, 1941; Trussel and Sarles, 1943; Casas-Campillo, 1951; Pramer and Starkey, 1962). However, the activity of an antibiotic in the soil may be very different from that in laboratory media. This is illustrated in a study by Pramer and Starkey (1952) on the influence of streptomycin on nitrification. Nitrification was found to be completely inhibited in solution medium by 1 part per million streptomycin. In sand, a comparable inhibition required 1,000 parts per million, while in soil, a concentration of 10,000 parts per million was required. The results of this and other studies indicates that relatively few antibiotics influence the development of micro-organisms in the soil. Clavicin at 50 parts per million and Actidione at 5 parts per million were found by Gottlieb *et al* (1952) to inhibit sensitive organisms in the soil, while antimycin was found to inhibit *B. subtilis* growth at 6 parts per million in soil (Martin and Gottlieb, 1955). More recent experiments using the antibiotic tylosin (Bewick, 1978) have shown that additions of the pure antibiotic at concentrations of 41, 184, and 368 parts per million caused a temporary inhibition in overall microbial activity in soil, as measured by O_2 uptake and CO_2 release, for six weeks after addition. Nitrogen mineralization in the soil amended with tylosin was found to be inhibited throughout the ten weeks of the experiment.

In contrast to this, experiments carried out using tylosin fermentation waste (Bewick, 1978) showed that at all the concentrations of waste used, there was always an increase in microbial activity, as measured by O_2 uptake, CO_2 release, and N mineralization, compared with the unamended soil. Similarly, the results from research yet to be completed suggest that the addition of penicillin and oxytetracycline fermentation wastes to soil does not cause a decrease in microbial activity. But the results of the effect of pure tylosin on microbial activity suggest that it is possible that the concentrations of tylosin and other antibiotics present in high rates of application of their fermentation wastes to soils could exert a restraining influence on the microbial activity in the soil, and therefore the full potential of the material would not be realized. In the case of tylosin, the amounts of pure tylosin added to soil which resulted in a temporary decrease in microbial activity were those which would be present in additions of 25, 125, and 249 tonnes of dried fermentation waste per hectare, equivalent to 152, 762, and 1524 wet tonnes/ha. It is, therefore, important that the addition of fermentation waste containing residual antibiotic does not permanently affect microbial populations and processes in the soil and does not increase levels of resistance in soil microbial populations. These aspects of antibiotics fermentation waste are at present under investigation in the author's laboratory.

Antibiotics in Relation to Plants

In this section, the possible uptake by plants of the low levels of antibiotics present in antibiotic fermentation waste are discussed. There have been a number of studies on the uptake of antibiotics by plants and the subject was reviewed by Brian (1957) and Goodman (1962). Studies have been mainly concerned with the uptake of streptomycin and its effect on plant growth and include recent investigations by Mukherjee and Bag (1974) and Mukherjee *et al* (1975). Other antibiotics have also been examined and include griseofulvin (Srovastra, 1966), chloramphenicol (Crowdy *et al*, 1955) and aureomycin, neomycin, and terramycin (Pramer, 1953). Streptomycin at concentrations greater than 100 parts per million were found to inhibit root and shoot development in many plants (Gray, 1955). Griseofulvin and chloramphenicol were also found to be taken up by plants, while aureomycin, neomycin, and terramycin did not appear to be taken up by cucumber seedlings (Pramer, 1953). There have been no reports in the literature concerning the uptake of antibiotics by plants fertilized with antibiotic fermentation residues. Recent preliminary investigations carried out in the author's laboratory suggest that tylosin and oxytetracyline are not detectable in tomatoes and potatoes fertilized with fermentation residue, but further investigation is necessary before firm conclusions can be drawn.

Antibiotics are also used in plant pathology, and the subject was reviewed by Goodman (1967) and Crosse (1971). Many of the studies by these authors concerned antibiotics used in human medicine. Berdy (1974) reported that a series of antibiotics active specifically against plant pathogens are manufactured on a large scale, mainly in Japan. These include blasticidin S, kasugamycin, polyoxin, and validamycin. He found that 25,000 tonnes of antibiotics active against plant pathogens were manufactured annually in the 1970's. Berdy also reported that 10,000 tonnes of blasticidin S, an antibiotic highly effective against *Piricularia orzae* (rice blast disease) were manufactured in 1967. It is interesting to compare the 10,000 tonnes of blasticidin S, an antibiotic used against one disease, with the world production of 10 tonnes of streptomycin produced in 1966 (Perlman, 1968). In terms of the total world antibiotic production of 49,000 tonnes/annum in the early 1970's, Berdy (1974) reported that 25,000 tonnes, or 58%, were used as plant protecting antibiotics. Therefore, the use of fermentation waste from antibiotics produced specifically for use in plant pathology should pose no problems in terms of the uptake of the residual antibiotics by plants, as these plants are already being exposed to much higher concentrations of antibiotic than would ever be present in applications of fermentation waste. It should, however, be borne in mind that although questions have been raised concerning the presence of residual antibiotics in animal products consumed by man, no questions appear to have been put forth concerning residual antibiotics in plant tissues consumed by man, and this could be a problem which will arise in the future. Therefore, it would be advisable for the subject of the uptake of antibiotics by plant tissues to be investigated before questions similar to those asked in the Swann Report (Swann 1969) arise.

WINERY BY-PRODUCTS

The major wastes produced by the wine-making process are stems, pomace, and lees. The quantities of stems and pomace produced are large, especially in Europe. Amerine and Joslyn (1951) reported that 1 tonne of grapes gave rise to approximately 746.7 liters of new wine, 87.6 kg pomace, and 33 kg stems. Figures given by Wicks (1976) indicate that total world wine production in 1973 was $36,410 \times 10^6$ liters 76.4% of which were produced in Europe and 3.8% in the United States. Using these figures, it can be seen that total world production of grapes in 1973 was 48.8×10^6 tonnes, which gave rise to 4.3×10^6 tonnes of pomace and 1.6×10^6 tonnes of stems. As United States wine production was 3.8% of total world production, waste production from wine-making in the United States in 1973 was 163,000 tonnes of pomace and 60,000 tonnes of stems.

Stems

These are the stems and leaves of the vines and represent between 2 and 6% of the total weight of the fruit. There have been few reports on the potential uses of this material, although Schreffler (1954) suggested that the stems could be ground up in water and the resulting liquor distilled for its alcohol content. Potential commercial utilization of the stems was examined by Raback (1921), while Amerine and Cruess (1960) reported that in the United States the stems are usually disposed of by scattering on the land, and in Europe, crushers are used to break the stems up.

Grape Pomace

This material is a mixture of the skin, seeds, and pulp of the grapes. When the juice is pressed out for wine production, the solids left represent between 10 and 20% of the weight of the fruit, with a moisture content of 35 to 70%. Reports by Fuchs (1911) and Amerine and Cruess (1960) suggest that grape pomace contains approximately 11% crude protein, 6% fat, 40% fiber, and 8% ash. The value of grape pomace as a feedstuff was first suggested by Semichon (1907) and the feeding value was considered to be equal to meadow hay, but experiments by Folger (1940) suggested that the material was no better than wheat straw in feeding value. He found that although sheep would eat pomace readily, dairy cows found it unpalatable. Amerine and Cruess (1960) also suggested that the material could be useful as a feed supplement. Before feeding it to animals, the pomace is dried in rotary, direct-fired drum driers heated by gas or oil to temperatures of approximately 1000°C. The dry pomace is crushed and molasses may be added to improve its palatability. The moisture content must be kept below 6% to prevent spoilage. Agostine (1964) reported that the feed value of grape pomace could be improved by the addition of lime to raise the pH. The use of grape pomace as an animal feed does not appear to be widespread and recent reports in the literature (Walter and Sherman, 1976) suggest that the material may be of use as a fuel.

The potential of grape pomace as a fertilizer has also been investigated. Jacob and Proebstring (1937) reported that an analysis of grape pomace showed levels of 1.5 to 2.5% N, 0.5% P, and 1.5 to 2.5% K on a dry weight basis. They gave its fertilizer value as equivalent to farmyard manure and suggested that it could be used to improve the texture of heavy soils. It was also noted that heavy applications gave rise to phytotoxic effects due to substances formed during the early stages of decomposition. After this initial period, it was found that the toxic effects disappeared and areas where waste was dumped supported luxuriant plant growth. Perrone (1958) advised that the material should be composted before spreading and this would overcome the toxic problems noted by Jacob and Proebstring (1937).

Lees

These are the spent yeast cells which remain after fermentation of the wine is complete. Lees contain potassium bitartrate in large amounts, wine-making representing the largest single source of tartrates in the world. Tartrate is also recovered from grape pomace and can often be scraped off the walls of the wine storage tanks as almost pure cream of tartar. In the form in which they are found in wineries, lees are unsuitable for animal feed, as the potassium bitartrate which they contain is a potent laxative. Lees do find a ready market in the distilling industry, where they are mixed with water and the residual alcohol is distilled off. The wastes remaining after this operation are then the concern of the distilling industry (see Chapter 5).

WASTES FROM CIDER PRODUCTION

The main waste products from cider production are apple pomace, yeast, and filter aid.

Pomace

As with grape pomace, cider pomace consists of the remains of the skins, pulp, core material, and seeds of apples from which most of the juice and soluble constituents have been extracted in the making of cider. Much of the pomace produced is further treated to extract the pectin, which is subsequently sold. The composition of these two types of pomace is given in Table 6.3. Levels of production of pomace are difficult to estimate since there are no recently published figures and not all cider producers are prepared to release figures of the wastes which they discharge into the environment. Barker and Gimingham (1915) estimated the annual production of pomace in the United Kingdom to be between 60,000 and 75,000 tonnes, while recent figures obtained suggest that an estimate of pomace production today would be in the order of approximately 15,000 to 20,000 tonnes/annum. In the United States a report by Walton and Bidwell (1923) estimated pomace production at about 150,000 tonnes/annum, and there do not appear to be any more recent estimates.

TABLE 6.3.   Analysis of Three Types of Apple Pomace.

Percentage Composition	Dried Pomace	Dried Pomace (After Removal of Pectin)	Wet Pomace (After Removal of Pectin)
Dry matter	87.5	96.3	21.3
Crude protein	6.2	10.4	10.5
Crude fiber	20.9	36.9	37.8
Ether extract	2.5	4.6	4.7
Total ash	2.0	0.9	1.0
pH	—	—	3.0

In the United Kingdom, much of the pomace, after being depecturized, is dried and sold as a feed additive. Many of the reports dealing with cider pomace as a feed date from the first half of this century, and in most of these it is suggested that the wet pomace be ensiled before feeding to prevent decomposition. (Barker and Gimingham, 1915; Atkeson and Anderson, 1927; Barker, 1933; Charley *et al*, 1942; Smith, 1950). Smith (1950) states that the drying of pomace requires a special plant, whereas ensiling can be easily carried out on the farm. In the first half of this century, cider-makers tended to be small organizations and could easily distribute their wet pomace locally, but with the formation of large cider manufacturers with large plants, it became possible to dry the pomace, on site if necessary, and then distribute it as a feedstuff. All the reports in the literature point out that cider pomace is readily palatable; reviews of feeding trials were carried out by Atkeson and Anderson (1927) and Smith (1950). Today, in the United Kingdom, the material is marketed as a source of a hemicellulose-type material, suitable for blending with oils and proteins in cattle cake.

Barker and Gimingham (1915) and Barker (1933) suggested that apple pomace could be used as a manure. Barker reported that the material had been used in Devon as a manure for grass, after being composted with phosphate. Barker and Gimingham (1915) have an analysis of pomace used as a manure as 0.2 to 0.6% K_2O, 0.4 to 0.7% H_3PO_4, and 1.6 to 1.7% N. They pointed out that in France, the material was mixed with bone or rock phosphate. The acids solubilized the phosphate, increasing its availability. Today, however, it appears that most of the pomace produced by the cider manufacturers is used as a feedstuff rather than a manure.

Yeast

The yeast produced in the fermentation of cider, approximately 160 tonnes/annum in the United Kingdom, is usually mixed with the pomace and sold for animal feed. Although there are no details concerning the situation in the United States, it is likely that the yeast is utilized in the same manner.

Filter Aid

Much of the filter aid is used in the production of pectin, and, like the filter aid used in the brewing industry, it is not economic to reclaim or utilize it and it is disposed of by dumping.

OTHER FERMENTATION WASTES

Levels of wastes produced by other various industrial fermentations, including enzymes, vitamins, and steroid hormones, are virtually impossible to determine due to the high degree of secrecy in the industry. All that can be discussed here is which compounds are produced on an industrial scale by fermentation; it should be pointed out that the wastes which will be produced could have a market as feedstuffs or fertilizers if the security problems could be overcome.

A recent major publication in this field has been produced by Yamada (1977) and it gives production details of products produced by Japan's fermentation industry. Although this cannot be related to world production, it may be instructive to examine the production levels of these materials in Japan (Table 6.4).

Of the eight organic acids which are produced by fermentation in Japan, 51% of the production is gluconic acid, while of the sixteen amino acids produced by fermentation, 86% of the production, is L-glutamic acid.

Amylase is the enzyme produced in the largest quantities by fermentation and represents 87% of total enzyme production; together with protease, this represents 95% of total enzyme production. Of the seven nucleic acids produced by fermentation, the most important are 5-IMP and 5-GMP, which together account for 99% of nucleic acid production. Vitamins C and B1 account for 89% of the

TABLE 6.4. Some Products Produced by Fermentation in Japan.

	Numbers Produced	Production Level in Japan (tonnes/yr)
Amino acids	16	116,370
Organic acids	8	23,760
Enzymes	15	12,652
Vitamins	11	7,415
Nucleic acids	7	3,031
Steroid hormones	10	1,191.8 (kg)

Source: Yamada. 1977.

vitamins produced by fermentation. Very small amounts of steroid hormones are produced, and of the ten produced in 1974, prednisolone represented 79% of the total production.

Levels of wastes produced from these fermentations are not given, but if one considers that fermentation products in Japan in 1974 were worth $900,000,000, it suggests that large quantities of organic wastes from these processes are being produced. But due to the secrecy surrounding many of the fermentations it is unlikely that the materials will be marketed as feedstuffs or fertilizers in the near future. However, should the cost of feedstuffs and fertilizers continue to rise, the companies producing these materials may be induced to process their own wastes and market them. This would depend, of course, on the economics of the process at the time.

REFERENCES

Agostine, A. 1964. "Utilization of by-products of wines and vines." *Wynboer* **32**:13-16.

Amerine, M. A. and W. V. Cruess. 1960. *The Technology of Wine Making*. The Avi Pub. Co., Westport, Connecticut.

Amerine, M. A. and M. A. Joslyn. 1951. *Table Wines*. University of California Press.

Anon. 1973a. "Pfizer feeds mycelium-grown baked potatoes to employees." Norwich Bulletin, Oct. 18.

Anon. 1973b. "DEP grants approval for use of mycelium." Norwich Bulletin, Oct. 12.

Anon. 1974a. "Mycelium gets DEP green light." *Groton News*, Oct. 12.

Anon. 1974b. "State cites benefits of Pfizers waste." *The Hartford Courant*, Oct. 16.

Anon. 1977. "U.S. moves slowly on antibiotics in animal foods." *New Scientist* **75**:651.

Atkeson, F. W. and G. C. Anderson. 1927. "Apple pomace silage for milk production." *Idaho Agric. Exp. Sta. Bull.* **150**.

Barker, B. T. P. 1933. "A note on the uses of pomace." *J. Min. Agric.* **40**:710-715.

Barker, B. T. P. and C. T. Gimingham. 1915. "The use of pressed apple pomace." *J. Min. Agric.* **22**:851-858.

Barker, W. G., R. H. Otto, D. Schwarz, and G. F. Shipton. 1958. "Pharmaceutical waste disposal studies I. Turbo mixer aeration in an activated sludge pilot plant." *Proc. 13th Ind. Waste Conf. Purdue University*, pp. 1-7.

Berdy, J. 1974. "Recent developments of antibiotic research and classification of antibiotics according to chemical structure." *Adv. Appl. Micro.* **18**:309–406.

Bewick, M. W. M. 1978. "The effect of tylosin fermentation waste and pure tylosin on microbial activity in soil." *Soil Biol. Biochem.* **10**:403–407.

Bewick, M. W. M. 1979. "Adsorption and release of tylosin by clays and soils." *Plant and Soil*, **51**:363–372.

Bird, H. R. 1969. "Biological basis for the use of antibiotics in poultry feeds." *The Use of Drugs in Feeds*. National Academy of Science Publishers. **1679**:31–41.

Borrie, P. and J. Barrett. 1961. "Dermatitis caused by penicillin in bulked milk supplies." *Brit. Med. J.* **2**:1267.

Brian, P. W. 1957. "Effects of antibiotics on plants." *Ann. Rev. Plant Physiol.* **8**:413–426.

Brown, J. M. 1951. "Treatment of pharmaceutical wastes." *Sewage and Industrial Wastes* **23**:1017–1024.

Brown, J. M. and J. G. Niedercorn. 1952. "Antibiotics." *Ind. Eng. Chem.* **44**:468–472.

Casas-Campillo, C. 1951. "Rhizobacidin, an antibiotic especially active against the bacteria of the nodules of legumes." *Cienca (Mex.)* **11**:21–28.

Casida, L. E. 1968. *Industrial Microbiology*. John Wiley & Sons, New York.

Charley, V. L. S., A. W. Ling, and E. L. Smith. 1942. "Apple pomace silage. Report on experiments conducted in 1941." *Long Ashton Res. Sta. Ann. Rept.* pp. 101–106.

Colovos, G. C. and N. Tinklenberg. 1962. "Land disposal of pharmaceutical manufacturing wastes." *Biotech. Bioeng.* **4**:153–160.

Crosse, J. E. 1971. "Prospects for the use of bacteriocides for the control of bacterial diseases." *Proc. 6th Insectic. Fungic. Conf.* **3**:694–705.

Crowdy, S. H., D. Gardner, J. F. Grove, and D. Pramer. 1955. "Isolation of griseofulvin and chloramphenicol from plant tissue." *J. Expt. Bot.* **6**:371–383.

Cunha, T. J. 1955. "Antibiotics for swine, beef cattle, sheep and dairy cattle." *Proc. 1st Inter. Conf. on Use of Antibiotics in Agriculture*. National Academy of Science Publishers. **397**:9–17.

Edwards, H. M., R. Dam, L. C. Norris, and G. R. Heuser. 1953. "Probable identity of unidentified chick growth factors in fish solubles and penicillin mycelium residue." *Poult. Sci.* **32**:551–554.

FDA Task Force. 1972. *Report to the Commission of the Food and Drug Administration on the Use of Antibiotics in Animal Feed*. FDA 72-6008, U.S. Food and Drug Administration, Rockville, Maryland.

Folger, A. 1940. "The digestibility of ground prunes, winery pomace, avocado meal, asparagus butts and fenugreek meal." *Calif. Agric. Exp. Sta. Bull.* **635**:1–11.

Forbes, M. and J. T. Park. 1959. "Growth of germ free and conventional chicks—effect of diet, dietary penicillin and bacterial environment." *J. Nutr.* **67**:69–84.

Fuchs, C. S. 1911. "Uber traubenkernkuchen." *Chem. Ztg. Jakry.* **35**:30–31.

Ghosal, D. N. and S. K. Mukherjee. 1970. "Studies of sorption and desorption of two basic antibiotics by and from clays." *J. Ind. Soc. Soil Sci.* **18**:243–247.

Goodman, R. N. 1962. "Systemic effects of antibiotics." *Antibiotics in Agriculture.* M. Woodbine (Ed.). Butterworths, London, pp. 165–184.

Goodman, R. N. 1967. "Uses of antibiotics in plant pathology and production." *Proc. 6th Inter. Conf. on Antibiotics Amer. Soc. for Microbiol*, pp. 747–756.

Gottlieb, D. and P. Siminoff. 1952. "The production and role of antibiotics in the soil II. Chloromycetin." *Phytopath.* **42**:91–97.

Gottlieb, D., P. Siminoff, and M. Martin. 1952. "The production and role of antibiotics in the soil IV. Actidione and clavicin." *Phytopath.* **42**:493–496.

Gray, R. A. 1955. "Inhibition of root growth by streptomycin and reversal of the inhibition by manganese." *Amer. J. Bot.* **42**:327–331.

Hilgart, A. A. 1950. "Design and operation of a treatment plant for penicillin and streptomycin wastes." *Sewage and Ind. Wastes* **22**:207–211.

H.M.S.O. 1969. The Medicines Act, 1968, p. 1691. The Public General Acts and Church Assembly Measures, 1968.

Howe, R. H. L. 1960. "Handling wastes from the billion dollar pharmaceuticals industry." *Wastes Eng.* **31**:728–731.

Hurwitz, E., K. H. Edmondson, and S. M. Clarke. 1952. "Pharmaceutical waste treatment." *Water and Sewage Works* **99**:202–206.

Jacob, H. E. and E. L. Proebstring. 1937. "Grape pomace as a vineyard and orchard fertilizer." *Wines and Vines* **18**:22–23.

Jefferys, J. G. 1952. "The stability of antibiotics in soil." *J. Gen. Microbiol.* **7**:295–312.

Johnson, N. P. and G. H. Arscott. 1974. "Effect of a fermentation residue and an antibiotic on growth of chicks fed rations containing corn or wheat." *Poult. Sci.* **53**:1335–1341.

Jukes, T. H. 1973. "Public health significance of feeding low levels of antibiotics to animals." *Adv. Appl. Microbiol.* **16**:1–30.

Jukes, T. H., E. L. R. Stokstad, R. R. Taylor, T. J. Cunha, H. M. Edwards, and G. B. Meadows. 1950. "Growth promoting effect of aureomycin on pigs." *Arch. Biochem.* **26**:324–325.

Kellogg, T. F., V. W. Hays, D. V. Catron, L. Y. Quinn, and V. C. Spier. 1966. "Effects of dietary chemotheropeutics on the performance and faecal flora of baby pigs." *J. Anim. Sci.* 25:1102-1106.

Lawrence, K. 1974. Personal communication.

Lederman, P. B., H. S. Skovornek, N. J. Edison, and P. E. Des Rosiers. 1975. "Pollution abatement in the pharmaceutical industry." *Chem. Eng. Prog.* 71:93-99.

Lev, M. and M. Forbes. 1959. "Growth response to dietary penicillin of germ free chicks and of chicks with a defined intestinal flora." *Brit. J. Nutr.* 13:78-84.

Linton, A. H., K. Howe, and A. D. Osborne. 1975. "The effects of feeding tetracycline, nitrovin, and quindoxin on the drug resistance of *coli-aerogenes* bacteria from calves and pigs." *J. Appl. Bact.* 38:255-275.

Liontas, J. A. 1954. "High rate filters treat mixed wastes at Sharpe and Dohme." *Sewage and Ind. Wastes* 26:310-316.

Martin, M. and D. Gottlieb. 1952. "Production and role of antibiotics in the soil III. Terramycin and aureomycin." *Phytopath.* 42:294-296.

Martin, M. and D. Gottlieb. 1955. "The production and role of antibiotics in the soil V. Antibacterial activity of five antibiotics in the presence of soil." *Phytopath.* 45:407-418.

McGuire, J. M., R. L. Burch, R. C. Anderson, H. E. Boaz, E. H. Flynn, H. M. Powell, and J. W. Smith. 1960. "Ilotycin, a new antibiotic." *Antib. Chemother.* 2:281-284.

Melcher, R. R. 1962. "Pharmaceutical waste disposal by soil injection." *Biotech. Bioeng.* 4:147-151.

Miller, B. M. and W. Litsky. 1976. *Industrial Microbiology.* McGraw-Hill, New York.

Molof, A. H. 1962. "Pharmaceutical waste treatment by the trickling filter, activated sludge and compost process." *Biotech. Bioeng.* 4:197-209.

Moorehouse, E. C. 1969. "Transferable drug resistance in enterobacteria isolated from urban infants." *Brit. Med. J.* 2:405-407.

Mukherji, S. and A. Bag. 1974. "Mode of action of streptomycin in higher plants." *Ind. J. Exp. Biol.* 12:360-362.

Mukherji, S., A. Bag, and A. K. Paul. 1975. "Streptomycin induced changes in growth and metabolism of etiolated seedlings." *Biol. Plant.* 17:60-66.

Netherthorpe, Lord. 1962. *Report of the Joint Committee on Antibiotics in Animal Feeding.* H.M.S.O., London.

O.H.E. 1969. *Antibiotics in Animal Husbandry.* Office of Health and Economics, London.

Olive, T. H. 1949. "Chloromycetin by Parke Davis." *Chem. Eng.* **56**:107–110.

Perlman, D. 1968. "Are new antibiotics needed?" *Proc. Biochem.* **3**:54–58.

Perrone, A. F. 1958. "Composting of pomace piles." *Wine Inst. Tech. Advis. Comm.*, May 26, 1958.

Pinck, L. A., W. F. Holton, and F. E. Allison. 1969. "Antibiotics in soils. I. Physico-chemical studies of antibiotic-clay complexes," *Soil Sci.* **91**:22–28.

Pinck, L. A., D. A. Soulides, and F. E. Allison. 1961. "Antibiotics in soils II. Extent and mechanism of release." *Soil Sci.* **91**:94–99.

Pramer, D. 1953. "Observations on the uptake and translocation of five active antibiotics by cucumber seedlings." *Ann. Appl. Biol.* **40**:617–622.

Pramer, D. 1958. "Microbiological progress report. The persistance and biological effects of antibiotics in the soil." *Appl. Microbiol.* **6**:221–224.

Pramer, D. and R. L. Starkey. 1951. "Decomposition of streptomycin." *Science* **113**:127.

Pramer, D. and R. L. Starkey. 1962. "Influence of streptomycin on micro-biological development in soil." *Bact. Proc.* **52**:15–30.

Raback, F. 1921. "Commercial utilization of grape pomace and stems from the grape juice industry." *USDA Bull.* **952**.

Robinson, K. L. 1962. "The value of antibiotics for growth of pigs." *Antibiotics in Agriculture*. M. Woodbine (Ed.). Butterworths, London, pp. 185–202.

Rose, A. H. 1961. *Industrial Microbiology*. Butterworths, London.

Schreffler, C. 1954. "Recovery of sugar from grape stems." *Wine Inst. Tech. Advis. Comm.*, Dec. 3, 1954.

Semichon, L. 1907. "Dried grape marc as a feed for farm animals." *Compte Rendu Cong. Soc. Aliment Ration. Betail* **12**:144–150.

Siminoff, P. and D. Gottlieb. 1951. "The production and role of antibiotics in the soil I. The fate of streptomycin." *Phytopath.* **41**:420–430.

Smith, E. L. 1950. "Apple pomace silage." *Agriculture* **57**:328–332.

Smith, G. 1969. *An Introduction to Industrial Mycology*. Edward Arnold, London.

Soulides, D. A., L. A. Pinck, and F. E. Allison. 1961. "Antibiotics in soils III. Further studies on release of antibiotics from clay." *Soil Sci.* **92**:90–93.

Srovastra, S. N. S. 1966. "Uptake and translocation of sulphanilamide and griseofulvin by the rice plant." *Current Sci.* **35**:327–329.

Stokstad, E. L. R. 1954. "Antibiotics in animal nutrition." *Physiol. Rev.* **34**:25–51.

Stokstad, E. L. R., T. H. Jukes, J. Pierce, A. C. Page, and A. L. Franklin. 1949. "The multiple nature of the animal protein factor." *J. Biol. Chem.* **180**:647–654.

Struzeski, E. J. 1977. "Status of waste handling and waste treatment across the pharmaceuticals industry and 1977 effluent limitations." *Proc. 30th Ind. Waste Conf.* Purdue University, pp. 1095–1110.

Summers, J. D., W. F. Pepper, and S. J. Slinger. 1959. "Source of unidentified factors for practical poultry diets. 2. The value of fish solubles, dried whey, dried distillers solubles, and certain fermentation products for chick and broiler diets." *Poult. Sci.* **38**:846–854.

Swann, M. M. 1969. *Report of the Joint Committee on the Use of Antibiotics in Animal Husbandry and Veterinary Medicine.* H.M.S.O., London.

Tompkins, L. B. 1957. "Two stage filter operation at the Upjohn waste treatment plant." *Sewage and Ind. Wastes* **29**:1161–1166.

Trussell, P. C. and W. B. Sarles. 1943. "Effect of antibiotic substances on rhizobia." *Bact. Proc.* **43**:29–38.

Uhliar, J. and M. Bucko. 1974. "Moznost vyuzitia priemyslovych odpadov z vyroby antibiotik v rastlinnej vyrobe." *Rostlinna Vyroba* **20**:923–930.

Waksman, S. A. 1947. "What is an antibiotic or antibacterial substance?" *Mycologia* **39**:565–571.

Waksman, S. A. and H. B. Woodruff. 1941. "*Actinomyces antibioticus,* a new soil organism antagonistic to pathogenic and non-pathogenic bacteria." *J. Bact.* **42**:231–249.

Walter, R. H. and R. M. Sherman. 1976. "Fuel value of grape and apple processing waste." *J. Agric. Fd. Chem.* **24**:1224–1245.

Walton, G. P. and G. L. Bidwell, 1923. "Apple by-products as stock feeds." *U.S.D.A. Bull.* **1166**.

Warner, N. J. 1973. "Once dumped at sea, antibiotic by-product fattens potatoes." *The Providence Journal,* Oct. 22.

Whitehill, A. R., J. J. Oleson, and B. L. Hutchings. 1950. "Stimulatory effect of aureomycin on the growth of chicks." *Proc. Soc. Exptl. Biol. Med.* **74**:11–13.

Wicks, K. 1976. *Wine and Wine Making.* Macdonald Educational, London.

Wright, J. M. 1954. "The production of antibiotics in the soil. 1. Production of gliotoxin by *Trichoderma viride.*" *Ann. Appl. Biol.* **41**:280–289.

Wright, J. M. and J. F. Grove. 1957. "The production of antibiotics in the soil V. Breakdown of griseofulvin in soil. *Ann. Appl. Biol.* **45**:36–43.

Yamada, K. 1977. *Japan's Most Advanced Industrial Fermentation Technology and Industry*. International Technical Information Institute, Minato-Ku, Tokyo, Japan.

Young, R. J. 1969. "Unidentified factors in the growth of poultry." *Proc. 24th Distillers' Feed Conf.* Cincinnati, Ohio, pp. 23–35.

Zvara, J., P. Lacok, and J. Kolek. 1960. "Fysiologicke vlastnosti odpadu pri vyrobe penicilinu." *Biologia* 15:23–31.

7

Feedstuff potential of paper wastes and related cellulose residues

C. E. Dunlap, Ph.D.

Associate Professor, Department of Chemical Engineering, University of Missouri, Columbia, Missouri

INTRODUCTION

The systematic use of roughages in the diets of all ruminant animals is a precept of common nutritional practice. Morrison (1961) describes the importance of high fiber feeds in the diets of cattle, sheep, and goats, and also in non-ruminants such as horses, swine, and poultry. In many cases, roughages comprise the major part of ruminant rations, and are used in large volume wherever ruminant animal industry is concentrated. In the United States alone, between 114 and 123 million tonnes of hay is cut each year, making it one of the largest volume agricultural crops of the country (Agricultural Statistics, 1975).

The cost and availability of conventional roughage feeds are quite variable, with climatic upsets and planting practices yielding temporary or long-term shortages in many areas of the world. The pressure to put pasture and hay acreage into grain and other food crops promises an even more serious and longer-term reduction in forage crop production.

On the other hand, almost every nation is experiencing an increase in the total and per capita rates of municipal refuse generation. Much of the refuse is paper and paper products, and even more of these residues are generated by the pulp, paper, and paperboard industries. In all developed countries, where competition between animals and man for available land is most intense, these wastes are systematically collected, centrally gathered, and then usually disposed of by incineration or landfilling. Since the nature of refuse is so largely cellulosic, and because of the expense of its current collection and disposal, interest in the possibilities of its use as a roughage in ruminant feeds is enticing and of long standing.

Farber (1952) reported the use of wood residues in feeds as early as 1884, and the feed use of wood wastes during World War I (1911 to 1914) was widespread in Europe. For many years, it has been common practice to replace conventional roughages with lower quality residues whenever necessity dictates (Gohl, 1975). Generally, this practice has led to decreased production of meat, lower yields of milk, and poorer animal health; these results were, in fact, expected. The real challenge now is to develop and place in common practice techniques of treating, preparing, and feeding wastepaper and related materials so that acceptable animal performance is the normal consequence.

SOURCES OF PAPER WASTES

The United States produces about 46 million tonnes of wood pulp each year (Agricultural Statistics, 1975). With additional pulp imports, about 57 million tonnes/yr of paper and paperboard products are manufactured, and another 5 million tonnes of finished paper is imported. About 22% of this fiber is recycled back into paper and pulp use in the United States (Anon., 1974), while recycle rates in other nations range from 12 to 40% (Coombe and Briggs, 1974). Approximately 50% of the paper and board produced each year is not currently recoverable due to permanent use, exportation, or destruction. However, this leaves over 30 million tonnes/yr of recoverable wastepaper, of which only 14 million tonnes is recycled at present. Related wastes are produced by the forestry, pulp, lumber, and wood processing industries, and a number of these residues are of interest as possible roughages.

Municipal Refuse Generation

The United States produces about 209 million tonnes of municipal refuse per year—about one tonne per capita per year (Meller, 1969). On a dry basis, about 52% of this waste is composed of paper (48% on a fresh basis), the remainder being 7% leaves, 2% wood, 2% synthetics, 1% cloth, 10% garbage, 8% glass, 10% metals, and 8% ashes, stone, etc. The percentage of paper in refuse varies considerably from country to country, with German refuse averaging only 4.5% paper, England, 12.6%; Holland, 15.5%; and Switzerland, 22.6% (Meller, 1969).

Processes for segregating municipal refuse into its more or less homogeneous components have recently been developed which yield produce fractions which could possibly be used in animal feeding. These processes range from the production of a composted refuse from which glass, metals, and plastics have been removed (Fairfield Engineering Process; Schulze, 1965) to more rigorous processes which yield remarkably clean and homogeneous fiber streams—Hydrosposal/ Fiberclaim, Black-Clawson Co.; FPL Recovery System, U.S. Forest Products Laboratory; Solid Waste Separator, Franklin Institute Research Laboratories; Fedway Process, Strobel and Rongved Co. (Franklin, 1972); and the University of California Process, (Trezek and Golueke, 1976). Each of these schemes can claim as a strength the fact that the residue is collected and transported to the plant at zero or negative cost.

Wastepaper

Some paper components of refuse are collected and segregated on a regular basis. The wastepaper industry in the United States recycles about 22% of the annual paper and board production back into the primary fiber market. This volume has been about 13 to 15 million tonnes/yr for the past several years. Market volume of wastepaper responds to its market value, and this is dependent on the price and supply of primary wood pulp.

The Paper Stock Institute of America Circular PS-66 defines 45 grades of wastepaper suitable for marketing. Most of the volume, however, is comprised of such standard grades as: 1 mixed paper, 1 news, over-issue news, corrugated, white ledger, hard white envelope, IBM printout, IBM cards, and magazines. Each grade is more or less homogeneous, and each has a unique composition and value.

Wastepaper that enters the recycling industry has a number of advantages for possible use as a roughage feed; it has been segregated and cleaned, it is more homogeneous than refuse, and it is available through a defined and active market. On the other hand, prices fluctuate quite widely and have recently been relatively high.

Pulp and Paper Manufacturing Wastes

An estimated 41 kg fibrous wastes are generated per tonne of finished pulp produced (Millett et al, 1973). These residues are all produced at the pulping plant, and usually gathered for disposal or fuel use. Most of these wastes arise when fibers are rejected from the pulping process because of knots or insufficient length, and are collected in separate streams such as hog fuel, screen rejects, separator sludge, and bleached or unbleached fines.

Some of these wastes are used as boiler fuel in the pulp mill and some collect as residues. They are widely varying in composition and differ markedly in physical structure. Of primary interest in the evaluation of some pulping wastes as ruminant feeds is the fact that some of the wastes have been subjected to chemical pulping and have had some degree of lignin removal and physical structure disruption.

Although the total United States production rate of these residues, some 2.3 million tonnes annually, is not as high as several other residues mentioned, the production is continuous, non-seasonal, and occurs at a central point.

Summary of Paper Residue Generation

A summary of available information on the generation of paper wastes and related cellulose residues in the United States is shown in Table 7.1. The cellulose contained in these residues is the major component of interest in their digestion by ruminants. Cellulose content varies from under 50% to almost 90% in these materials and is dependent on the degree of processing or the amount of cleaning the material has received. It should be noted that only the materials entering the wastepaper market have well defined values. Although the processed refuse and pulping waste values are shown as undefined, it cannot be inferred that the costs

TABLE 7.1.   Wastepaper and Related Residues—Generation and Value in the United States.

Residue	Generation (10^6 tonnes/yr)	Fiber (%)	Ash (%)	Value Range ($/tonne)	Reference
Refuse:					
Municipal wastes	209	40–55	5–15	Negative	Meller (1969)
Fairfield Engineering composted product	–	51	11	–	Utley et al (1972)
University of California fiber product	–	76	2–3	–	Trezek and Golueke (1976)
Black-Clawson	–	76	1	–	Dunlap (1978)
Wastepaper:					
1 Mixed paper	2.9	60–75	1–3	5–16	
Newsprint	2.3	65–75	1	9–32	Dunlap (1975b);
Corrugated paper	4.7	65–85	1–3	16–36	Anon (1974);
IBM cards	–	88	1	127–164	Coombe and
IBM printout	–	78	2–4	27–46	Briggs (1974)
Writing papers	–	71–78	5–14	–	
Pulping Wastes:					
Hog fuel	–	57	2	5–14 as fuel	Ware (1976);
Separator sludge	2.3	62	21	–	Dunlap (1978)
Green liquor fines	–	60–70	1–3	–	Millett et al (1973)
Tailings	–	89	1	–	Dunlap (1978)

are negligible. Any use of these materials which demands processing, marketing, and the maintenance of an inventory will result in increasing the value and cost of the product.

COMPOSITION OF PAPER WASTES

Since all paper and related residues are initially produced from wood or grassy plants, they still retain some residual of the chemical and physical structure of their original form. The extent to which this structure is retained depends on the type and severity of treatment the fiber has received during its use history.

Refuse and wastepaper share similar processing histories and thus have similar composition and structure. The inclusion of extraneous materials in cruder refuse streams adds to ash, lipid, and protein levels, but does not change the essential structure of the fiber. A more important classification based on ruminant nutritional response is the difference between chemically pulped fiber and non-chemical or groundwood fiber. Later in this chapter it is shown that digestibilities of chemically pulped paper are often more than double those of mechanically pulped or groundwood papers.

Mechanical pulping involves only beating and washing prior to paper forming, and in some cases a bleaching operation is added to improve brightness. No lignin removal is accomplished, but almost all of the water soluble hemicelluloses are removed. Groundwood paper thus contains 55 to 65% cellulose, 15 to 25% lignin, 1 to 3% ash, and low levels of NFE (nitrogen-free extract), lipid, and protein. Groundwood pulp comprises 80 to 85% of newsprint, and is used in some magazine papers with additives such as clay, protein, and glues.

Chemical pulp has been subjected to a delignification by cooking in a sulfite, sulfate, or sodium hydroxide solution. The fibers are then beaten, washed, and may or may not be bleached prior to use. Chemical pulp contains 90 to 95% cellulose, and low levels of lignin and ash. When paper containing chemical pulp is manufactured, additions of groundwood pulp and other materials are made to yield the paper desired. Chemical pulp predominates in such papers as brown wrapping paper, paper bags, and fine writing papers.

All paper appearing in refuse or as wastepaper has been dried during paper manufacture. This drying step causes modification of fiber surface properties and pore structure (Stone et al, 1969) and results in a drastic decrease in water accessibility.

Other components present in some refuse and wastepapers in low concentrations are important for their toxicological properties. Polychlorinated biphenyls (PCB's), lead, and other metals are present in such treated papers as carbon paper, and in printing inks. Effects of these materials on animals fed wastepaper have been extensively investigated and are discussed later in this chapter.

The residues generated during paper manufacture vary from rough, knotty, hog fuel, which is essentially untreated wood chips, to treated and bleached pulp fines. Composition of hog fuel is the same as for native wood, while the composition of residues occuring after the pulping cook are essentially the same as for finished pulp, with the inclusion of more or less inorganic residue as a contaminant. Table 7.1 shows that separator sludge contains over 21% ash—mainly dirt, dust, and sand—while green liquor fines and tailings have ash contents under 3%.

It has been found that pulp residues which occur after delignification are much more digestible than those which have received no treatment. Also, it is important to note that most of these pulping residues have never been dried after being delignified. This results in higher digestibility and more efficient use of these materials by ruminants.

FEEDING UNTREATED PAPER RESIDUES

A number of controlled animal feeding studies have been carried out to determine the value of various wastepapers, paper pulps, and pulping residues in ruminant diets. Table 7.2 summarizes results from 29 feeding experiments with various paper residues on cattle, sheep, and goats. Many more residues have been analyzed by *in vitro* digestibility tests, and results of a number of these assays are given in Table 7.3.

Feeding of Municipal Refuse

The use of the fiber fraction of municipal refuse in ruminant diets is relatively new, since refuse segregation processes have only recently been developed. McClure *et al* (1970) reported the successful feeding of composted refuse to beef cattle and sheep, and Utley *et al* (1972) found composted refuse to be equal to peanut hulls at a level of 22% in the diet of steer calves and yearlings. Johnson and McCormick (1975) fed dairy heifers a ration containing 18.4% composted and ensiled refuse and found that although feed intake and efficiency were lower than for the control ration, they were better than the values obtained from the same refuse which had not been ensiled.

Feeding of Wastepaper

Because of the ready availability and more or less homogeneous nature of wastepaper, a number of feeding trials have been carried out to evaluate its use in the ruminant diet.

Newspaper was used by Furr *et al* (1974) at 30% in the diet of lactating dairy cattle with an *in vivo* DMD of 59.8%. The study monitored levels of PCB's and metals in milk and organs. Sherrod and Hanson (1973) fed newspaper at 3.3, 6.7, and 10% levels to heifers and noted successively lower feed intake, feed

TABLE 7.2. Animal Feeding Results—Wastepaper and Residues.

Residue	Animal (Number In test)	Level in Diet (%)	Duration of: Conditioning (days)	Trial (days)	In vivo DMD of Diet (%)	Feed Efficiency (Kg feed / Kg gain)	Comments	Reference
Municipal refuse:								
1. Municipal refuse, composted (Fairfield)	cattle, steer, calves (6)	0 / 22	12 / 12	7 / 7	— / 80.8	— / —	Refuse better than peanut hulls at same level	Utley et al (1972)
2. As above	cattle, steer, yearlings (27)	0 / 22	— / —	140 / 140	— / —	— / 7.35	Refuse resulted in faster gain, better efficiency than peanut hulls	Utley et al (1972)
3. As above, except ensiled	cattle, heifers, dairy (16)	0 / 18.4	21 / 21	91 / 91	47.4 / 54.2	11.8 dry / 13.9 dry	Intake and feed efficiency significantly lower than control, better than unensiled refuse	Johnson and McCormick (1975)
Wastepaper:								
4. Newspaper (used)	cattle, dairy, lactating (4)	30	8	38	59.8	—	Levels of PCB's and metals in milk were monitored	Furr et al (1974)
5. Brown cardboard	As above	30	8	38	51.6	—	As above	As above
6. Grey cardboard	As above	30	8	38	52.7	—	As above	As above
7. Computer printout	As above	30	8	38	77.9	—	As above	As above
8. Newspaper (used)	cattle, heifers (8)	0 / 3.3 / 6.7 / 10.0	—	110 / 110 / 110 / 110	—	9.18 / 9.51 / 10.24 / 10.89	Intake lower at increased paper levels, conformation score lower successively	Sherrod and Hanson (1973)

TABLE 7.2. (*Continued*).

Residue	Animal (Number In test)	Level in Diet (%)	Conditioning (days)	Trial (days)	In vivo DMD of Diet (%)	Feed Efficiency ($\frac{Kg\ feed}{Kg\ gain}$)	Comments	Reference
9. Newspaper (used)	Sheep, wethers (24)	0	7	7	74.7	—	Intake approximately same all levels, TDN lower successively	As above
		7.5	7	7	72.9	—		
		15	7	7	69.0	—		
		30	7	7	61.4	—		
10. As above	cattle, steer (3)	8	—	98	82	—	Steers sorted feed without molasses, rumen protozoa successively lower	Dinius and Oltjen (1972a)
		16	—	98	76.5	—		
		24	—	98	69.7	—		
11. As above	cattle, steer (48)	0	14	98	—	9.48	Intake and gain declined significantly at higher paper levels	As above
		8	14	98	—	9.39		
		16	14	98	—	11.20		
		24	14	98	—	17.63		
12. As above	cattle, steer, finishing (40)	0 (hay)	14	70	—	6.97	Same dressing % and carcass grade, daily gain same, intake lower at 10% paper	As above
		0 (grain)	14	70	—	7.07		
		5	14	70	—	6.85		
		10	14	70	—	7.95		
13. Newspaper	cattle, steer (4)	0	—	14	65.5	—	No significant difference between used or unused paper, intake lower than with hay	Dinius and Oltjen (1972b)
		24	—	14	60.9	—		
14. Newspaper (used)	cattle, dairy, lactating (15)	0	14	28	—	—	No difference in milk flavor at any level	Cloninger et al (1972)
		10	14	28	—	—		
		20	14	28	—	—		

Pulp, pulp residues, and other:

No.	Feed	Animal (no.)						Remarks	Reference
15.	As above	cattle, dairy, lactating (15)	0	14	28	77.4	—	Intake lower with paper, milk yield lower at 20% paper, butterfat increased with paper	Mertens *et al* (1971a)
			10	14	28	77.4	—		
			20	14	28	77.5	—		
16.	Writing paper, waste	Sheep, wethers (9)	0	21	49	48.9	—	Intake decreased as paper levels increased	Coombe and Briggs (1974)
			59.6	21	49	–	—		
			65.4	21	49	56.2	—		
			84.2	21	49	–	—		
17.	Newspaper (used)	Sheep, ewes (18)	0	14	49	–	—	Poor results with paper diets	As above
			42.2	14	49	–	—		
			84.2	14	–	Rejected	—		
18.	Pulping screen rejects, aspen (air dry)	Goats (4)	0	7	14	59.9	—	Dry matter digestion at 100% dilution was 66% for rejects	Millett *et al* (1973)
			20	7	14	57.7	—		
			35	7	14	62.7	—		
			50	7	14	64.2	—		
19.	Pulp fines, aspen, unbleached (dried)	Goats (4)	0	7	14	53	—	Dry matter digestion at 100% dilution was 50% for fines	As above
			20	7	14	55.7	—		
			35	7	14	54.4	—		
			50	7	14	50.0	—		
20.	Pulp fines, hardwood, bleached (air dry)	Goats (4)	0	7	14	58.7	—	Dry matter digestion at 100% dilution was 80% for fines	As above
			20	7	14	71.9	—		
			35	7	14	77.8	—		
			50	7	14	78.0	—		
21.	Pulp fines, pine, Kraft, unbleached	Cattle, steer (20)	0	13	7	68	—	Diluted dry matter disappearance of fines was 43%, fine particles may pass through rumen quickly	As above
			20	13	7	66.1	—		
			35	13	7	61.9	—		
			50	13	7	60	—		
			65	13	7	63.9	—		
22.	As above	Cattle, steer (10)	0	–	58	–	9.44	Tended to reject fines, feed intake lower than control	As above
			50	–	58	–	11.69		

TABLE 7.2. (*Continued*).

Residue	Animal (Number In test)	Level in Diet (%)	Duration of: Conditioning (days)	Trial (days)	*In vivo* DMD of Diet (%)	Feed Efficiency ($\frac{\text{Kg feed}}{\text{Kg gain}}$)	Comments	Reference
23. Pulp fines, aspen, unbleached	Sheep, ewes (30)	77	—	77	—	—	Ration accepted well, 21 gained, 9 lost weight, dustiness a problem	As above
24. Sawdust, aspen	Cattle, dairy, lactating (20)	0 10 20 30	—	42 42 42 42	—	—	Milk production unchanged, fat content decreased, feed intake tended to increase	Satter et al (1973)
25. As above	Cattle, dairy, lactating (12)	0 12	—	49 49	—	—	Milk production unchanged, fat higher with sawdust	As above
26. Sawdust, hardwood	Cattle, dairy, steer (24)	25	—	84	—	7.4	Gain slow, no control reported	Keith and Daniels (1976)

TABLE 7.3. *In Vitro* Digestibility Tests—Wastepaper and Residues.

Residue	*In Vitro* Test (Method)[1]	Digestibility of Residue (% DMD)	Reference
Wastepaper:			
Newspaper (with black ink)	IVRF-DMD (Mellenberger)	33.2	Mertens *et al* (1971b)
Newspaper (unused)	As above	32.6	As above
Newspaper (with colored ink)	As above	26.5	As above
Newspaper (used)	IVRF-DMD (Sayre)	39.3	Becker *et al* (1975)
Newspaper (used)	IVRF-DMD (Christian)	21.3	Coombe and Briggs (1974)
Newspaper (unused)	As above	24.5	As above
Brown cardboard	IVRF-DMD (Mellenberger)	77.8	Mertens *et al* (1971b)
Cardboard	IVRF-DMD (Sayre)	52.2	Becker *et al* (1975)
Magazine, glossy	As above	38.7	As above
Magazine, glossy I	IVRF-DMD (Mellenberger)	46.1	Mertens *et al* (1971b)
Magazine, glossy II	As above	45.1	As above
Magazine, glossy III	As above	41.0	As above
Magazine, coarse I	As above	24.0	As above
Magazine, coarse II	As above	20.1	As above
Brown wrapping paper	As above	90.8	As above
Brown wrapping paper	IVRF-DMD (Christian)	62.7	Coombe and Briggs (1974)
Magazine, glossy	As above	52.6	As above
Cardboard box	As above	45.5	As above
Computer cards	As above	42.7	As above
Computer printout	As above	64.1	As above
Photo and multilith copy paper	As above	49.7	As above
Government waste writing paper	As above	52.5	As above
Writing paper	As above	51.3	As above
Typing paper	As above	65.8	As above
White pages (phone book)	IVRF-DMD (Sayre)	37.2	Becker *et al* (1975)
Yellow pages (phone book)	As above	41.7	As above
Brown bags	As above	47.9	As above
Feed sacks	As above	68.3	As above
Telephone book covers	As above	73.3	As above
Computer printout	As above	77.8	As above
Computer cards	As above	81.7	As above

TABLE 7.3. (Continued).

Residue	*In Vitro* Test (Method)[1]	Digestibility of Residue (% DMD)	Reference
Pulp, pulping residues, and other:			
Sawdust, hardwood	IVRF-DMD (Barnes)	7.8	Keith *et al* (1976)
Pulp, hardwood, sulfite, unbleached	IVRF-DMD (Mellenberger)	72	Baker *et al* (1973)
As above (fines)	As above	67	As above
Pulp, hardwood, sulfite, bleached	As above	94	As above
As above (fines)	As above	89	As above
Pulp, hardwood, Kraft, unbleached	As above	92	As above
As above (fines)	As above	71	As above
Pulp, hardwood, Kraft, bleached	As above	96	As above
As above (fines)	As above	93	As above
Pulp, softwood, Kraft, unbleached	As above	90	As above
As above (fines)	As above	79	As above
Pulp, softwood, Kraft, bleached	As above	91	As above
As above (fines)	As above	98	As above
Pulp, softwood, ground-wood, unbleached)	As above	0	As above
As above (fines)	As above	7	As above
Pulp, pine, groundwood, unbleached	As above	0	As above
As above (fines)	As above	0	As above
Pulp fines, groundwood, aspen	As above	37	Millett *et at* (1973)
Pulp fines, groundwood, pine	As above	0	As above
Pulp fines, groundwood, spruce	As above	0	As above
Pulp screen rejects, sulfite, aspen	As above	66	As above
Pulp screen rejects, sulfite, hardwood	As above	54	As above
Pulp screen rejects, Kraft, hardwood	As above	44	As above
Pulp fines, hardwood, Kraft, bleached	As above	95	As above
Pulp fines, hardwood, sulfite, unbleached	As above	90	As above
Pulp fines, pine, Kraft, unbleached	As above	53	As above

[1] IVRF-DMD refers to *in vitro* rumen fluid dry matter disappearance; the name in parentheses refers to the technique used.

efficiency, and conformation scores as the level of paper was increased. The same work cited a test with sheep wethers fed 7.5, 15, and 30% newspaper with the result of lower *in vivo* diet digestion and successively lower TDN values, although the feed intake was about the same at all levels. Dinius and Oltjen (1972a, b) fed newspaper to steers at 5, 8, 10, 16, and 24% of a balanced ration and again noted declines in intake and feed efficiency at paper levels over 8%. Carcasses dressed at the same percentages, however, and daily gain with the 5% paper ration was the same as for the control diet. It was noted that the cattle sorted the paper out of the diet unless molasses was added. Cloninger *et al* (1972) reported that milk flavor was not affected in dairy cattle fed up to 20% newspaper, but Mertens *et al* (1971a), in the same test, found lower intake at 10 and 20% paper, and lower milk yields at the higher paper level. Butterfat content, however, increased as the level of paper in the diet was raised. Coombe and Briggs (1974) fed newspaper to ewes at very high levels—42 and 84%—in pellet form and found poor results at both levels, with complete rejection of the 84% paper diet.

Wastepapers other than newsprint have been used with somewhat better results. Furr *et al* (1974) used brown and grey cardboard, and computer printout paper at 30% of the diet of cattle, and Coombe and Briggs (1974) used waste writing papers with sheep at 65.4% of the diet.

In vitro digestibilities of wastepapers vary over quite a wide range, depending on the type of paper tested. Inspection of Table 7.3 indicates a DMD of newsprint ranging from 21 to 39%. Other papers which also contain high levels of groundwood pulp display similarly low DMD values; coarse magazine papers, for example, result in DMD's of 20.1 and 24%.

As the percentage of chemical pulp in the paper is increased, the DMD values also increase up to levels of from 70 to 90% for wrapping paper, sacks, typing paper, and computer cards. It would be expected that these high chemical pulp papers would result in better animal performance, but it must be remembered that they are, in fact, more expensive commodities in the wastepaper market.

Feeding of Pulp and Pulping Residues

A number of different pulping residues weie fed to goats and cattle by Millett *et al* (1973) with much better results than those provided by newsprint or other wastepapers. Goats fed 20, 35, and 50% levels of three different pulping wastes in digestibility trials showed good acceptance and high *in vivo* DMD values at all levels of pulp in the diet. Unbleached Kraft pulp fines from pine were fed to cattle at levels 20, 35, 50 and 65%, and showed a decrease in dietary *in vivo* DMD as pulp levels were increased. Ewes fed the same pulp at 77% of the diet accepted it with 70% gaining and 30% losing weight during a 77 day trial. Problems with dustiness and feed sorting were pointed out.

Extensive *in vitro* DMD studies on pulp and pulping residues by Baker *et al* (1973) and Millett *et al* (1973) are shown in Table 7.3 and graphically demonstrate the differences in digestion of the groundwood pulps (0 to 7%) and the sulfite and Kraft pulps (70 to 95%). Other residues, such as screen rejects (44 to 66%) and pulp fines (53 to 95%) demonstrate various DMD values, but all chemical pulps and pulp residues have higher DMD values than non-chemical pulps, groundwood pulps, and most wastepaper. These chemical pulping wastes are a promising source of replacement roughage for ruminants.

Problems With Feeding Paper Wastes

Problems with feeding refuse, wastepaper, and pulping residues can be generally divided into two areas: problems associated with animal acceptance and utilization efficiencies; and toxic components in these materials.

Numerous trials with wastepaper have found decreased voluntary feed intake as paper levels are increased, with concommitant decreases in rates of gain and feed efficiencies. Dinius and Oltjen (1972b) examined the tendency of newsprint to fill the rumen with bulk while providing low digestible energy. They concluded that at 24% of the diet newsprint did impair feed intake because of ruminoreticular fill.

The problem of animals sorting out wastepaper from a ration has also been of concern (Dinius and Oltjen, 1972a; Millett *et al* 1973), but the addition of molasses (Dinius and Oltjen 1972a) or the use of pelleted feeds (Coombe and Briggs, 1974) seems to alleviate this problem if particle size is small enough. However, Millet *et al* (1973) pointed out the tendency of very small particles— pulp fines in this case—to pass through the rumen very quickly, thus decreasing utilization efficiency.

The fate of PCB's and metals (such as lead, mercury, and copper) which occur in wastepaper and refuse has been of major concern to animal feeders. PCB's occur in printing inks and dyes, and are at very high levels in carbon paper and carbonless transfer paper. Lead is a component in printing inks and is at a high enough level in used newspaper that it has been suggested as a source of dietary lead for dogs (Hankin *et al* 1974).

Furr *et al* (1974) monitored PCB's and 44 metals in the milk of cows fed 30% newspaper, brown or grey cardboard, or computer printout paper. Differential metabolism and storage of certain PCB's was indicated with excretion of up to 76 parts per billion in the milk and storage of up to 1540 parts per billion in renal fat over 38 days.

Accumulation of various heavy metals in muscle, organs, bones, and milk has been discussed by Sherrod and Hanson (1973), Dinius and Oltjen, (1972a), Millett *et al* (1973), and Utley *et al* (1972). General agreement has been that no dangerous accumulation or concentration of metals has occurred in feeding trials with wastepaper, but monitoring should continue with longer-term feeding studies.

TREATED PAPER AND PULPING RESIDUES

The literature on the use of various treatment processes to increase the digestibility of a number of low-grade forages is voluminous and of long standing. Treatment of such materials as straws, corn cobs, sawdusts, wood chips, sugarcane bagasse, cotton linters, leaves, and grasses has been thoroughly covered. Processes using sodium hydroxide, ammonia, urea, sulfur dioxide, nitrogen dioxide, gamma and electron irradiation, ball milling, hammer milling, shear milling, steaming, boiling, and a number of combinations of the above techniques have been applied to these wastes. Review articles by Cowling and Brown (1969), Pidgen and Heaney (1969), Dunlap et al (1976), Tarkow and Feist (1969), Donefer et al (1969), and Millett et al (1976) provide an excellent perspective of the application of these treatment processes to low-grade, native holocelluloses. However, very few references can be found that deal with the application of a treatment process to wastepapers, refuse fibers, or pulping residues.

A great many (if not the majority) of the treatments listed previously serve to delignify native fibers and increase their accessibility to hydrolytic enzymes. Under relatively mild conditions, a treatment with sodium hydroxide can often more than double the digestibility of a straw or of bagasse, while treatment of refuse fiber or wastepaper under the same conditions results in a DMD increase of only a few percent. Certainly, the gain is not worth the expense of the treatment. All papers have, of course, already been through either a mechanical or chemical pulping and drying operation. Stone et al (1969) have shown that drying these pulped fibers essentially cuts in half the fiber saturation point of these materials. Fibers after drying, therefore, absorb only half as much water upon rewetting as those which have never been dried. This certainly limits the accessibility of these papers to enzymes, and probably also to solutes such as the alkali and other chemicals used for treatment. Additional chemical treatments have, therefore, proven to be of limited value when applied to refuse fiber and wastepaper. Pulp and pulping wastes, with the exception of groundwood, have already received a more severe treatment than can be afforded for an animal feed; thus, their further treatment by chemical means is also impractical.

Treatment of wastepaper by physical means such as intensive milling has been investigated by a few researchers with successful, if not economically promising, results. Andren et al (1976) used ball milling to increase the reactivity of a number of cellulosics with cellulase from *Tricoderma viride*. The following materials were subjected to a 48 hour enzyme saccharification and resulted in the stated percent saccharification when run as received or after ball milling: hydropulped government documents, 65 to 75%; Bureau of Mines municipal refuse fiber (MRF), 30 to 60%; Black-Clawson MRF, 35 to 55%; ADL Process MRF, 15 to 25%; and newspaper, 30 to 45%.

A differential speed, two-roll mill has been used by Tassinari and Macy (1977) to improve the hydrolysis of cotton, newspaper, and wood chips by *T. viride* cellulase. Milling for three minutes resulted in a 12-fold increase in the digestibility of raw cotton and a 17-fold increase for maple chips. Milling pine chips for 8 minutes caused a 7-fold increase in hydrolysis. Newspaper which was roll-milled for 8 minutes yielded about 27 mg/ml sugar in a 24 hr enzyme digest while newspaper ball milled for 24 hr yielded about 18 mg/ml, and untreated newspaper yielded about 10 mg/ml. The test showed that milling time and mill spacing were the two main variables controlling reactivity increase. At this time, neither ball milled nor roll milled paper has been fed to animals. One of the major reasons is the cost of processing, which has been estimated for ball milling as 9.9 c/kg of material ground (Nystrom, 1975).

Another problem with milled papers which contain very fine particles may be significantly decreased ruminal retention time, as discussed by Pidgen and Heaney (1969).

Although a number of different treatment processes look promising for enhancing the value of low-grade roughages as ruminant feeds, the same cannot be said for refuse fiber and wastepaper. Chemical treatments thus far tested have little effect at economically feasible reaction conditions, and the intensive milling techniques may also be too costly, although their expense has yet to be verified.

NEW POSSIBILITIES AND ECONOMIC ASPECTS

The search for suitable treatment techniques to increase the digestibility of wastepaper is continuing. A better understanding of the mechanism of hydrolytic enzyme attack and those unique properties of wastepaper that hinder hydrolysis should allow the development of treatment techniques which can successfully upgrade these materials at reasonable costs.

Two-roll shear milling has shown good initial promise in increasing paper digestibility, and the possibility of enhancing its effectiveness with a simultaneous or subsequent chemical treatment is most interesting.

Another new prospect for the use of paper and pulping residues as feeds involves the production of microbial single-cell protein by fermentation. This concept has been carried out with aerobic, submerged bacterial cultures by Dunlap (1975a) and Callihan and Dunlap (1971), and with thermophilic actinomyces by Bellamy (1975) and Armiger et al (1976). A promising semi-solid fermentation with either yeast or bacteria has been developed by Han and Anderson (1975), and its application to grass straw has been discussed by Grant et al (1977). The process has not yet been applied to paper wastes, but the prospects are interesting for such an application.

The success of these processes is dependent on the use of a highly digestible cellulose, and wastepaper has not been in favor as a substrate. Residues from chemical pulping operations, on the other hand, have proven to be excellent substrate materials.

The most immediate improvement in the application of wastepaper and pulping wastes as feeds can probably be gained by the dissemination of information on better feeding practice. Earlier in the chapter, a number of feeding trials were reviewed in which refuse and wastepaper were used successfully at certain levels in rations for beef and dairy cattle and sheep. Most trials showed suitable animal performance when these materials were used at levels up to 10% of the ration. Chemical pulping wastes were found suitable at levels up to 50%. Also noted, however, were problems associated with animal acceptance, preferential feed sorting, small particle short-circuiting of the rumen, and dustiness. These are problems which can be solved by improved feed formulations and feeding practices. It is apparent that a clear economic incentive must exist before the feeding

TABLE 7.4. Economics of Substituting Waste Newspaper or Pulping Fines for Alfalfa Meal.

Basis: • Waste newspaper costs $27.30/tonne
  • Pulp fines must be dried
  • Newspaper must be ground
  • Both rations must be pelleted
  • Newspaper can be used as 10% of ration
  • Pulp fines can be used as 25% of ration

Cost Component	Cost (US$/tonne of ration)	Cost Component	Cost (US$/tonne of ration)
Newspaper purchase (100 kg at 27.30/tonne	2.73	Pulp fines purchased (250 kg obtained free)	0
Hammer milling (100 kg at 3.64/tonne)	0.36	Drying, from 80 to 5% moisture content	3.24
Molasses, pellet binder (5% of ration at 36.40/tonne)	1.82	Transportation (50 miles at 6 cents/tonne-mile)	0.8
Pelleting	3.50	Molasses, pellet binder	1.82
Cost of using 10% news per tonne of ration	8.42	Cost of using 25% pulp fines per tonne of ration	9.36
Breaks even with alfalfa meal costing: 8.42 0.10 = 84.20/tonne		Breaks even with alfalfa meal costing: 9.36 0.25 = 37.46/tonne	

of paper wastes—and the systematic use of good waste feeding practice—is common.

The definition of when and where the economic incentive becomes clear for the substitution of wastepaper or pulping residues for a conventional fodder (usually hay) in a ration is not at all straightforward. To illustrate this, the following scenario will be considered. Alfalfa meal prices are rising; at what meal price will it be economically attractive to use 10% waste newspaper or 25% unbleached Kraft fines as meal replacement in a dairy ration originally containing 40% meal? The brief calculations shown in Table 7.4 indicate that newsprint can be used at meal prices above US$84.20/tonne, and the pulping fines at meal prices above US$37.46/tonne. These calculations indicate the additional processing, drying, and transportation costs which would be incurred whenever the proposed substitutions are made. The current price of alfalfa meal in the United States is US$57.33/tonne, and was US$70.48/tonne a year ago. This would indicate that, in areas where pulping fines could be obtained, their use in some rations would be economical at the present time. Obviously, the use of newspaper would be uneconomical. Since fuel, wastepaper, molasses, and transportation costs vary from country to country and fluctuate with time, these calculations must be done on a local basis with current prices of all components to arrive at a reasonable economic picture. The cost and availability of other roughages such as straws and low-grade hay would also affect the decision.

CONCLUSION

The large-scale use of paper and pulping wastes has been practiced only in regions where conventional forages are chronically unavailable, or during times of emergency when usual cultivation practices are disrupted. In the longer term, however, the competition between food and forage crops for available arable land may introduce chronic forage shortages in areas heretofore very productive of those crops. In this instance, the more efficient use of paper and pulping wastes will be mandated by necessity rather than by less compelling economic incentives.

Studies have demonstrated the practices needed for effectively using some of these wastes in feeds at significant levels; in fact in some cases it is already economically attractive to do so. Better techniques for upgrading the digestibility of wastepaper are needed. Their implementation would certainly enhance the attractiveness of these materials a great deal.

REFERENCES

Agricultural Statistics. 1975. U. S. Government Printing Office, Washington, D. C.

Andren, R. K., R. J. Erickson, and J. E. Medeiros. 1976. "Cellulosic substrates for enzymatic saccharification." *Enzymatic Commersion of Cellulosic Mate-*

rials: Technology and Applications. E. L. Gaden *et al* (Eds.). Interscience/John Wiley & Sons, New York, pp. 177–204.

Anon. 1974. *Chem. Engr.* June 10, pp. 44–45.

Armiger, W. B., D. W. Zabriskie, A. E. Humphrey, S. E. Lee, A. Horeira, and G. Joly. 1976. "Computer control of cellulose fermentations: SCP production by thermophilic actinomyces." *Biochemical Engineering: Energy, Renewable Resources, and New Foods.* AIChE Symposium Series, No. **158**, Vol. **72**. Barnett *et al* (Eds.). American Institute of Chemical Engineers, New York, pp. 77–85.

Baker, A. J., A. K. Mohaupt, and D. F. Spino. 1973. "Evaluating wood pulp as feedstuff for ruminants and substrate for *Aspergillus fumigatus." J. Anim. Sci.* **37**:179–182.

Becker, B. A., T. R. Campbell, and F. A. Martz. 1975. "Paper and whey as a feedstuff for ruminants." *J. Dairy Sci.* **58**:1677–1681.

Bellamy, W. D. 1975. "Conversion of insoluble agricultural wastes to SCP by thermophilic micro-organisms." *Single Cell Protein II.* Tannenbaum and Wang (Eds.). MIT Press, Cambridge, Massachusetts, pp. 263–272.

Callihan, C. D. and C. E. Dunlap. 1971. *Construction of a Chemical-Microbial Pilot Plant for Production of Single-Cell Protein from Cellulosic Wastes.* Report SW-25c to U.S. Environmental Protection Agency, Cincinnatti, Ohio.

Christian, K. R. 1971. "Detergent method for total lignin in herbage." *Field Station Record* **10**:29. C.S.I.R.O. Division of Plant Industry.

Cloninger, M. R., R. E. Baldwin, and R. T. Marshall. 1972. "Effects of rations containing newsprint on flavor of milk." *J. Dairy Sci.* **55**:1018–1019.

Coombe, J. B. and A. L. Briggs. 1974. "Use of wastepaper as a feedstuff for ruminants." *Austr. J. Exp. Agric.* **14**:292–301.

Cowling, E. B. and W. Brown. 1969. "Structural features of cellulosic materials in relation to enzyme hydrolysis." *Cellulases and Their Applications.* Advances in Chemistry Series, No. **95**. American Chemical Society, Washington, D.C., pp. 152–187.

Dinius, D. A. and R. R. Oltjen. 1972a. "Newsprint as a feedstuff for beef cattle." *J. Anim. Sci.* **33**:1344–1350.

Dinius, D. A., and R. R. Oltjen. 1972b. "Effect of newsprint on ration palatability and ruminoreticular parameters of beef steers." *J. Anim. Sci.* **34**:137–141.

Donefer, E., I. O. A. Adeleye, and T. A. O. C. Jones. 1969. "Effect of urea supplementation on the nutritive value of NaOH treated oat straw." *Cellulases and Their Applications.* Advances in Chemistry Series, No. **95**. American Chemical Society, Washington, D.C., pp. 328–342.

Dunlap, C. E. 1975a. "Production of single-cell protein from insoluble agricultural wastes by mesophiles." *Single Cell Protein II.* Tannenbaum and Wang (Eds.). MIT Press, Cambridge, Massachusetts, pp. 244–262.

Dunlap, C. E. 1975b. "A note on the value of cellulose." In *Cellulose as a Chemical Energy Source*, Biotech. Bioeng. Symposium Series, No. 5, C. R. Wilke (Ed.). John Wiley & Sons, New York, pp. 73–75.

Dunlap, C. E. 1978. Unpublished results.

Dunlap, C. E., J. Thompson, and L. C. Chiang. 1976. "Treatment processes to increase cellulose microbial digestibility." *Biochemical Engineering: Energy, Renewable Resources, and New Foods.* AIChE Symposium Series, No. **158**, Vol. **72**. Barnett *et al* (Eds.). American Institute of Chemical Engineers, New York, pp. 58–63.

Farber, E. 1952. "Vegetable aliment from wood. *J. Forestry* **50**:446.

Franklin, W. E. 1972. *Resource Recovery Processes for Mixed Municipal Solid Waste, Part II.* Final report project No. **3634-D** by Midwest Research Institute, Kansas City, Missouri.

Furr, A. K., D. R. Mertens, W. H. Gutenmann, C. A. Bache, and D. J. Lisle. 1974. "Fate of polychlorinated biphenyls, metals and other elements in papers fed to lactating cows." *J. Agr. Food Chem.* **22**: 954–959.

Gohl, B. 1975. *Tropical Feeds.* Food and Agriculture Organization of the United Nations, Rome.

Grant, G. A., A. W. Anderson, and Y. W. Han. 1977. "Preliminary cost estimates for commercial fermentation of straw as animal feed." *Biotech. Bioeng.* **19**: 1817–1830.

Han, Y. W. and A. W. Anderson. 1975. "Semisolid fermentation of ryegrass straw." *Appl. Microbiol.* **30**:930.

Hankin, L., G. H. Heichel, and R. A. Botsford. 1974. "Newspapers and magazines as a source of dietary lead for dogs." *J. Amer. Vet. Med. Assn.* **164**:490.

Johnson, J. C. and W. C. McCormick. 1975. "Ensiled diet containing processed municipal garbage and sorghum forage for heifers." *J. Dairy Sci.* **58**:1672–1676.

Keith, E. A. and L. B. Daniels. 1976. "Acid or alkali-treated hardwood sawdust as a feed for cattle." *J. Anim. Sci.* **42**:888–892.

Kruse, C. G. 1969. *The Value of Wood By-Products in Ruminant Feeding.* M.S. thesis, University of Missouri, Columbia.

McClure, K. E., E. W. Klosterman, and R. R. Johnson. 1970. "Palatability and digestibility of processed garbage fed to ruminants." *J. Anim. Sci.* **31**:249.

Mellenberger, R. W., L. D. Satter, M. H. Millott, and A. J. Baker. 1970. "An *in vitro* technique for estimating the digestibility of treated and untreated wood." *J. Anim. Sci.* **30**:1005.

Meller, F. H. 1969. *Conversion of Organic Solid Wastes into Yeast.* Public Health Service Publication No. **1909**. Superintendent of Documents, Washington, D.C.

Mertens, D. R., J. R. Campbell, E. A. Martz, and E. S. Hilderbrand. 1971a. "Lactational and ruminal response of dairy cows to ten and twenty percent dietary newspaper." *J. Dairy Sci.* **54**:667–672.

Mertens, D. R., E. A. Martz, and J. R. Campbell. 1971b. "*In vitro* ruminal dry matter disappearance of selected waste papers." *J. Dairy Sci.* **54**:931–933.

Millett, M. A., A. J. Baker, L. D. Satter, J. N. McGovern, and R. A. Dinius. 1973. "Pulp and papermaking residues as feedstuffs for ruminants." *J. Anim. Sci.* **37**: 599–607.

Millett, M. A., A. J. Baker, and L. D. Satter. 1976. "Physical and Chemical pretreatments for enhancing cellulose saccharification." *Enzymatic Conversion of Cellulosic Materials: Technology and Applications.* E. L. Gaden *et al* (Eds.). John Wiley and Sons, New York, pp. 125–154.

Morrison, F. B. 1961. *Feeds and Feeding, Abridged,* 9th ed. The Morrison Publishing Co., Claremont, Ontario, Canada.

Nystrom, J. 1975. "Discussion of pretreatments to enhance enzymatic and microbiological attack of cellulose," In *Cellulose as a Chemical and Energy Resource.* Biotech. Bioeng. Symposium Series, No. 5, C. R. Wilke (Ed.). Wiley Interscience, New York, pp. 221–224.

Pidgen, W. J. and D. P. Heaney. 1969. "Lignocellulose in ruminant nutrition." *Cellulases and Their Applications.* Advances in Chemistry Series, No. **95**. American Chemical Society, Washington, D.C., pp. 245–261.

Satter, L. D., R. L. Lang, A. J. Baker, and M. A. Millett. 1973. "Value of aspen sawdust as a roughage replacement in high-concentrate dairy rations." *J. Dairy Sci.* **56**:1291–1297.

Sayre, K. D. and P. J. Van Soest. 1971. "Comparison of types of fermentation vessels for an *in vitro* artificial rumen procedure." *J. Dairy Sci.* **55**:1496.

Schulze, K. L., 1965. "The Fairfield-Hardy composting pilot plant at Altoona, Pa." *Compost Sci.* **5(3)**:5–10.

Sherrod, L. B. and K. R. Hansen. 1973. "Newspaper levels as roughage in ruminant rations." *J. Anim. Sci.* **36**:592–596.

Stone, J. E., A. M. Scallen, E. Donefer, and E. Ahlgreen. 1969. "Digestibility as a simple function of a molecule of a similar size to a cellulose enzyme." *Cellulases and Their Applications.* Advances in Chemistry Series, No. **95**. American Chemical Society, Washington, D.C., pp. 219–241.

Tarkow, H. and W. C. Feist. 1969. "A method for improving the digestibility of lignocellulosic materials with dilute alkali and liquid ammonia." *Celluloses and*

Their Applications. Advances in Chemistry Series, No. **95**. American Chemical Society, Washington, D.C., pp. 197–218.

Tassinari, T. and C. Macy. 1977. "Differential speed two-roll mill pretreatment of cellulosic materials for enzymatic hydrolysis." *Biotech. Bioeng.*, **19**:1259–1268.

Trezek, G. J. and C. G. Golueke. 1976. "Availability of cellulosic wastes for chemical or biochemical processing." *Biochemical Engineering: Energy, Renewable Resources, and New Foods*. AIChE Symposium Series, No. **158**, Vol. **72**. Barnett *et al* (Eds.). American Institute of Chemical Engineers, New York, pp. 52–57.

Utley, P. R., O. H. Jones, and W. C. McCormick. 1972. "Processed municipal solid waste as a roughage and supplemental protein source in beef cattle finishing diets." *J. Anim. Sci.* **35**:139–143.

Ware, S. A. 1976. *Single Cell Protein Production and Other Food and Feed Recovery Technologies From Waste Materials*. Report on EPA Contract No. CI-76-0088, by Ebon Research Systems, Silver Spring, Maryland.

8

Wood and bark wastes as feedstuffs and fertilizers

J. E. Smith

Professor, Department of Applied Microbiology, University of Strathclyde, Glasgow, Scotland

INTRODUCTION

Single cell protein (SCP), or protein biomass produced by microorganisms, is widely being considered as an important means of alleviating world protein shortage. Recently, the European Association of Single Cell Protein Producers (UNICELPE) have estimated that the EEC market for animal grade SCP will be 500,000 to 600,000 tonnes/yr in 1980 and the Japanese market may well equal this (Maclennan, 1976). The production of beef in America is estimated to use between 0.5 and 0.75% of the national energy, with most of this being consumed in the production of animal feed, (Lemieux and Wilson, 1978). Feed can represent up to 75% of total beef production costs. Many animal feeding systems already make use of crop residues and waste products as a means of reducing the energy and costs factors. However, it is essential that SCP be shown to be a safe, reliable, and good quality feed for animals and perhaps even as a food for direct consumption by man. The success of

SCP will be judged by society on economic considerations and on the acceptability of the final product.

Much of the early interest in SCP was concerned with the use of *n*-alkane components of petroleum, either as gas oil or purified *n*-paraffin fractions (Fig. 8.1). However, the massive increase in crude oil prices and the problems of contamination have been a major concern. Methanol derived from natural gas is now considered an important feed stock for SCP production, largely on economic grounds and because it is free from contamination with potentially carcinogenic polycyclic aromatics.

Alternative feed stocks are now being sought from more traditional waste organic materials, and this chapter considers, in particular, the use of lignocellulosic wastes, namely wood and bark. When considering the suitability of an organic waste for SCP production, the following criteria have been set out by Maclennan (1976) and are well worth examining before embarking on a waste utilization program:

1. It must be cheap.
2. It must be readily degradable microbiologically.
3. On economic grounds, it is only possible to produce animal feed grade SCP in large-scale plants, typically of 100,000 to 300,000 tonnes/yr/plant.
4. The feed stock should be available on a year round basis and be capable of storage safely and economically.
5. Transport to site of plant must be at low cost.
6. It must be available on a reliable basis and be of constant and predictable quality.

There can be little doubt that cellulose, both from agricultural and forestry sources and from waste material, must constitute a future major source for the production of feed, energy, and chemicals. Cellulose is the most abundant and renewable natural resource available to man. It has been realistically calculated that c 3.3×10^{14} kg/yr of carbon dioxide are fixed on the surfaces of the earth and that approximately 6% of this (i.e., 22 billion tonnes/yr) will be cellulose (Humphrey, 1975). Four billion tonnes/yr are readily available for processing, with 2.6 billion tonnes existing as wood and pulp. On a worldwide basis, land plants produce 24 tonnes cellulose/person/yr (Bellamy, 1974). The three main naturally occurring sources of cellulose from agriculture and forestry are wood, straw from grains, and sugarcane after removal of sugar.

Although cellulose can be readily attacked microbiologically, it invariably occurs in combination with lignin and in a complicated interwoven state. Dry wood has an approximate composition of 40 to 45% cellulose, 20 to 30% hemicellulose, and the rest mostly lignin (Koch, 1972). Since lignin is only sparingly attacked by microorganisms, it has caused a serious impediment to the wide acceptance of such materials for SCP production. However, new technologies

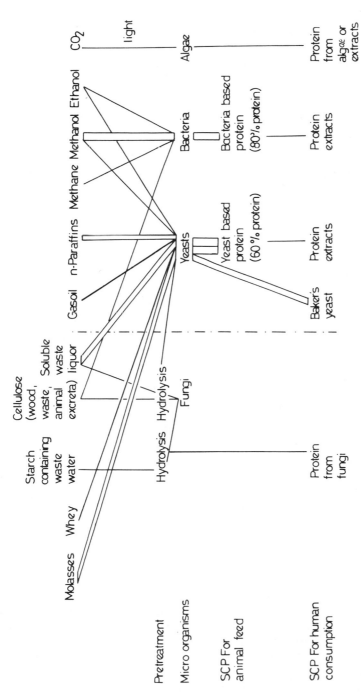

Fig. 8.1. Microorganisms and substrates for SCP production. *Source:* Dimmling and Seipenbusch. 1978. (Reprinted by permission of *Process Biochemistry*.)

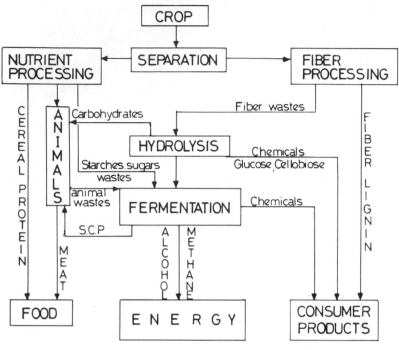

Fig. 8.2. General scheme for cellulose utilization. *Source:* Humphrey *et al.* 1976. (Reproduced with permission from *Continuous Culture* 6: *Applications and New Fields.* Published by Ellis Horwood, Ltd., Chichester, England, for Society of Chemical Industry, London.)

are being developed which will allow this abundant and renewable source of organic carbon to be used for fermentation purposes (Fig. 8.2; Humphrey *et al*, 1976). The following sections consider the nature of the availability of lignocellulose substrate derived from timber, the enzymology of cellulose and lignin degradation, and some of the worthwhile fermentation processes that are using lignocellulose substrate for proteinaceous biomass and fertilizer production.

SOURCES OF LIGNOCELLULOSE

Associated with the timber industry worldwide, large tonnages of logging and milling wood residues are annually being generated. Limited markets exist for most of these cellulose-rich by-products. Approximately one-third of the timber tonnage that is harvested will eventually become a residue or potential waste product (Stone, 1976).

TABLE 8.1. Estimates of Wood and Bark Residues and By-Products from Timber Harvesting, Primary Manufacturing, and Secondary Manufacture.

Source of Residues	Total	Wood	Bark
	(million tonnes oven dry)		
Timber harvesting	130	110	20
Primary manufacturing	76	59	17
Secondary manufacturing	14	14	—
Total	220	183	37

Source: Stone. 1976.

There are three main categories of residue: logging residues (80%—tree tops, branches, etc.), primary manufacturing residues (17%—saw mills, veneer mills, and pulp mills), and secondary manufacturing residues (3%) (Table 8.1). Logging residues constitute the greatest wastage and are also the most unlikely to be used in some other process because of the inconvenience and economic impracticality of collecting and transporting to a central area. Primary manufacturing residues have much more potential for worthwhile use because they occur at centralized positions. For example, in America, in 1973, 59 million tonnes of wood residues and by-products (such as sawdust, sides, edgings, and chips) and 17 million tonnes of bark were available at mills throughout the country. Although much of the milling residues are being profitably used in the pulp and paper industries, and particle board manufacture is expanding each year, it has been calculated that, in America, there are still 10 million tonnes of wood residue and 7 million tonnes of bark that are not being used. Such materials are mainly discarded into water courses or burnt, either way contributing to environmental pollution (Ellis, 1975; Phelps and Hair, 1974).

ENZYMOLOGY OF LIGNOCELLULOSE DEGRADATION

A vast literature has been amassed concerning the nature of cellulose degradation by microorganisms (see Szajer and Targonski (1977) and Eriksson (1978). The studies involving two organisms, *Trichoderma viride* and *Sporotrichum pulverulentum*, have dominated the research efforts.

Studies with *T. viride* have been primarily concerned with relatively pure sources of cellulose and have led to the development of sophisticated fermentation technologies involving enzyme hydrolysis of cellulose (e.g., waste paper) to dextrose sugar (Mandels *et al*, 1974; Spano *et al*, 1975). However, current work in America indicates that the problems of recovering crystalline dextrose from enzymic hydrolysis of cellulose materials are at present insurmountable (Elder, 1976). Thus, it is more probable that the dextrose in solution should be considered as a raw material for the production of ethyl alcohol or as a carbon source for growing filamentous fungi or yeasts. The technological implications of this work have been examined by Brown and Fitzpatrick (1976).

Eriksson's group in Sweden has realistically considered lignocellulosic materials as potential substrates for fungi; particularly the white rot fungi, which can degrade lignin as well as the other major components of wood. Such fungi (e.g., *Sporotrichum pulverulentum*) will normally attack cellulose and lignin simultaneously when degrading wood. It seems a general factor that when lignin is being microbiologically degraded there must be a parallel degradation of one of the other polysaccharides (Eriksson, 1978). Lignin is a three-dimensional polymer compound of one or more phenylpropane monomers. The simplest monomers are hydroxycinnamyl alcohol, coniferyl alcohol, and sinapyl alcohol. Coniferyl alcohol is the major component of most conifer lignins and is a significant proportion of the lignins of other species.

Clearly, a knowledge of the mechanisms of enzyme degradation of both cellulose and lignin will be of considerable value for the future development of applied research to use these complex substrates for food purposes. Such an understanding can then be coupled to fermentation technology.

The celluloses of wood are linear polymers of D-anhydrogluco-pyranose units linked by β-1,4-glucosidic bonds, and are variously folded, bundled, and stabilized by hydrogen bonding. The key enzyme in all cellulolytic systems is *endo*-1,- 4-β-glucanase, which attacks at random the 1,4-β linkages along the cellulose chain. Different amounts of *endo*-glucanases have been found in different enzyme preparations. Almin *et al* (1975) and Eriksson and Pettersson (1975) distinguished five *endo*-glucanases in *Sporotrichum pulverulentum*, while only two *endo*-glucanases could be demonstrated in culture liquids from *Trichoderma viride* (Pettersson, 1975). These enzymes are normally associated with an *exo*-1,- 4-β-glucanase, which splits off cellobiose or glucose units from the non-reducing end of the cellulose, and one or several 1,4-β-glucosidases which hydrolyze cellobiose to glucose and cellobionic acid to glucose and glucolactone. In *S. pulverulentum*, an additional oxidative enzyme, cellobiose oxidase, is necessary for *in vitro* cellulose degradation. However, no such enzyme has yet been found to be involved in the extracellular enzymic action of *T. viride*. The present understanding of the enzymology of cellulose and lignin breakdown is given in Fig. 8.3 (Eriksson, 1978).

The enzymology of lignin degradation is less well understood but may involve a laccase and an oxidoreductase, cellobiose:quinone oxidoreductase. The cellobiose:quinone oxidoreductase is important in both the degradation of cellulose and lignin. Studies with cellulase-less mutants have shown that lignin degradation is more favored by the presence of cellulose and it has been suggested that this is due to the action of the cellobiose:quinone oxidoreductase (Eriksson, 1978).

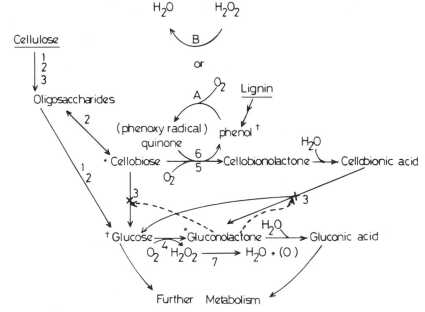

Fig. 8.3. Enzyme mechanisms for cellulose degradation and their extracellular regulation in *S. pulverulentum. Source:* Eriksson. 1978.

Enzymes involved in cellulose degradation: 1) endo-1, 4-β-glucanases; 2) exo-1, 4-β-glucanase; 3) β-glucosidase; 4) glucose oxidase; 5) cellobiose oxidase; 6) cellobiose:quinone oxidoreductase; 7) catalase. Enzymes involved in lignin degradation: A) laccase; B) peroxidase. An asterisk denotes products regulating enzyme activity; gluconolactone inhibits 3; cellobiose increases transglycosylations. A dagger denotes products regulating enzyme synthesis: glucose, gluconic acid → catabolite repression; phenols → repression of glucanases.

Cellobiose:quinone oxidoreductase reduces quinones or phenoxy radicals in the presence of cellobiose, which is then oxidized to cellobiono-δ-lactone.

The importance of the phenol oxidase, laccase, for lignin degradation is now well accepted. A laccase-less mutant of *S. pulverulentum* could not degrade lignin (Eriksson, 1978).

The studies on the basic enzymology of cellulose and lignin degradation will lay the foundation on which fermentation bioconversion processes can be developed.

PRETREATMENT OF LIGNOCELLULOSE MATERIALS

Chemical and Physical Treatments

In nature, cellulose invariably is present in the crystalline form and combined with various proportions of lignin. This subtle combination of two of the most common naturally occurring organic materials has been both a great biological blessing and an effective barrier to useful conversion techniques. Lignification restricts enzymological and microbiological access to the cellulose and other polysaccharides, while the crystallinity of the cellulose restricts the rate of all modes of enzymological attack on the cellulose. Without this unique natural barrier to microbiological attack, the durability and longevity of nature's trees would be greatly reduced.

Therefore, with a view to harnessing the cellulose part of lignocellulosic materials, much effort has been given to pretreatment techniques which will effectively and economically reduce both the lignification and the crystallinity of these natural polymers. The physical and chemical pretreatments of lignocellulose materials have been reviewed by Millett et al (1976).

Acid hydrolysis has been increasingly proposed as a major approach to saccharification of cellulose. In this way, cellulose is completely broken down to glucose units which may then form the feedstock of further chemical synthesis or for SCP production. Such techniques are being widely considered for straw and wastepaper, but so far not for wooden materials. Dilute alkali pretreatment (5 to 10% NaOH) may well have a role to play in upgrading the feeding value of forage and certain hardwood wastes. The saponification of intermolecular bonds can cause increased enzymological and microbiological access to the carbohydrate components of cell walls and enhance conversion to more utilizable end-products. Electron irradiation and vibratory ball milling have greatly altered the crystallinity of cellulose.

Delignified wood pulp of all species provides an excellent source of dietary energy for ruminant animals and is an excellent substrate for further microbiological conversion. Removal of only one-third of the lignin of hardwood and two-thirds of softwoods yields a product equivalent to hay in ruminant digestibility. Wood fines, a by-product remaining after hardwood is pulped to produce tissue paper, have been shown to be suitable partial replacement for forage and grains in beef and sheep rations as a source of dietary energy (Lemieux and Wilson, 1978).

Enzymatic Delignification of Lignocellulose

A novel approach to the delignification of wood has been the subject of a recent patent (Eriksson et al, 1976) which makes use of the ability of certain filamentous fungi to form lignin decomposing enzymes that can at least partially delignify wood products. The end-product is not unlike semi-chemical pulp and can then be subjected to mechanical de-fibration or further chemical digestion, including a possible utilization for biomass production.

The fungi involved in this process are white rot basidiomycete fungi capable of completely digesting all the wood components. This destructive ability is well known in forest product management and protective steps are routinely taken to prevent their growth. Such fungi, therefore, possess a complement of enzymes capable of completely degrading the complex molecular structure of lignocellulose. Eriksson *et al* (1976) have demonstrated that it is possible to regulate the enzymic ability of selected fungi in such a way that the fungi specifically attack the lignin components of the wood with little effect on the cellulose or other carbohydrate materials. This control can be engineered by genetic changes, by using special additives, or by the use of cellulase inhibitors.

As a first step, fungi were selected which had a natural ability to form lignin decomposing enzymes and a lesser potential to produce cellulose decomposing enzymes. Such fungi occur quite generally in the environment and include *Peniophora sanguinea, Phellinus isabellinus, Polyporus resinosus, Trametes pini*, and, in particular, *Peniophora cremea PB1*. The latter fungus could digest at least 50% of the lignin content of birch, while only decomposing 2% of the cellulose.

Mutants were also developed for their ability to produce lignin decomposing enzymes but with a greatly reduced ability to form cellulose decomposing enzymes. Such mutants were obtained by irradiating the spores of several white rot fungi—*Polyporus adustus* and *Sporotrichum pulverulentum* in particular, but also *Pheniophora gigantea, Trametes cinnabarina, Polyporus hirsutus, Pycnoporus sanguineus, Polyporus versicolor*, and *Polyporus zonatus*. The selective abilities of several cellulase-less mutants of *Polyporus adustus* are shown in Table 8.2,

TABLE 8.2. Decomposition of Cellulose, Glucomannan, Xylan, Laminarin, Pectin, and Lignin by Natural *Polyporus Adustus* and by Cellulase-less Mutants.

	Substrate (%)					
Strain	Cellulose	Xylan	Glucomannan	Laminarin	Pectin	Lignin
Natural	12	24	18	28	21	38
Cel-3	0	0	0	28	21	35
Cel-5	0	0	0	26	21	43
Cel-6	0	0	0	28	21	52
Cel-7	0	12	0	30	23	43
Cel-21	0	0	0	28	23	36
Cel-22	0	0	0	28	27	38
Cel-25	0	0	0	27	23	42
Cel-27	0	5	0	33	23	30
Cel-28	0	14	0	29	21	43
Cel-30	0	12	0	29	23	42
Cel-101	0	0	0	28	22	38
Cel-102	0	0	0	28	21	6
Cel-103	0	0	0	24	21	
Cel-108	0	0	0	28	22	30

Source: Eriksson *et al.* 1976.

which shows that these mutants, while retaining their ability to decompose lignin, have a greatly reduced ability to decompose cellulose and other carbohydrate components.

Yet another embodiment in this patent makes use of fungi with both lignin and cellulose decomposing enzymes; by the addition of selective inhibitors, the ability of the cellulose enzymes can be reduced without affecting the lignin decomposing enzymes. Compounds that have been used include glucose, arabinose, cellobiose, zylose, mannose, saccharose, ammonium salts, nitrate, asparagine, and casein hydrolysate. Many fungi have been used, including *Pheniophora sanguinea*, *Phellinus isabellinus*, *Trametes pini*, *Polyporus abietinus*, *P. hirsutus*, and *Stereum hirsutus*.

These methods of delignifying wooden materials offer a new and appealing approach to utilizing waste lignocellulosic materials. The cellulose product that results from these biological treatments could then readily be further upgraded to serve as animal feed.

THE UTILIZATION OF WASTE LIQUORS OF SULFITE PULPING INDUSTRY

The residual monosaccharides in the waste liquors of the sulfite pulping industry have long been utilized for the growth of yeast cells, *Candida utilis*. The yeast cells can utilize hexoses and pentoses and produce a nutritious but bitter proteinaceous biomass. There are a number of factories throughout the world, mostly concentrated in Eastern Europe, which produce this product.

A new process, the Pekilo process, in which a filamentous fungus *Paecilomyces varioti* is cultivated continuously on the waste liquors of the acid sodium-calcium-magnesium or ammonium bisulfite cooking process is now in full operation at the Jämsänkoski pulp mill in central Finland. This is the only industrial scale plant for continuous cultivation of fungal mycelium which produces a high-grade protein from a natural renewable carbohydrate source.

At present, two 360 m^3 fermenters are in operation, and although they are not yet working to full capacity because of lack of raw material, they have each been able to produce 15 to 16.5 tonnes/24hr of dry biomass corresponding to a growth rate of 2.7 to 2.8/m^3/hr (Romantschuk and Lehtomäki, 1978). The dried biomass has a crude protein level of 52 to 57% and is sold to feed compounding mills. Feeding and toxicological trials have been successfully carried out in several countries and the Pekilo protein is now recognized officially in Finland as a food ingredient. A simple flow sheet of the Pekilo process is shown in Fig. 8.4, and the chemical composition is given in Table 8.3.

The construction and operation of the Pekilo fermentation system has presented many technical difficulties, in particular the problem of oxygen limita-

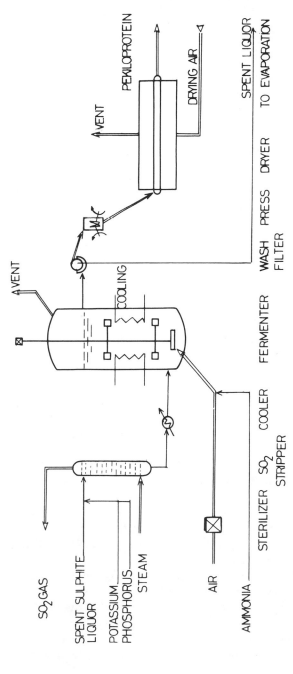

Fig. 8.4. A simplified flow sheet of the Pekilo Process. *Source:* Romantschuk and Lehtomäki. 1978. (Reprinted by permission of *Process Biochemistry*.)

TABLE 8.3.   Average Chemical Composition of Pekilo Product

	(g/16g N)
Threonine	4.8
Valine	5.0
Methionine	1.6
Isoleucine	4.6
Leucine	7.1
Tyrosine	4.0
Phenylalanine	4.2
Lysine	6.5
	(%)
Moisture, max.	6.0
Crude fat	1.3
Crude protein	55
Crude fiber	7
N-free extracts	25
Ash	6
Nucleic acids	10
	(ppm)
Thiamine	7
Riboflavin	70
Pyridoxin	20
Niacin	450
Pantothenic acid	60
Biotin	2
Folic acid	15

Source: Romantschuk and Lehtomäki. 1978. (Reprinted by permission of *Process Biochemistry*.)

tion due to the special viscosity characteristics associated with the cultivation of fungal mycelia.

A major advantage of the Pekilo process is the very considerable reduction in pollution resulting from the pulp mill. In addition to the monosaccharides, the fungus also utilizes the acetic acid in the spent liquor. The BOD of the entire pulp mill has been reduced by at least 50%.

It is difficult to compare this process with conventional yeast and bacterial fermentations because, for commercial reasons, relevant information is not available. However, the Finnish workers consider that the Pekilo protein can be economically produced in small units producing 5,000 to 10,000 tonnes annually, whereas *n*-alkane and methanol based processes are generally considered to require production of 50,000 to 100,000 tonnes annually to be viable (Romäntschuk 1976).

Neither *Candida utilis* or *Paecilomyces varioti* can use water soluble oligomers and insoluble polysaccharides, which are present originally in spent sulfite liquor. For this reason, hydrolysis must be used before fermentation. However, Ek and Eriksson (1975) and Ek *et al* (1976) have developed a symbiotic process whereby *Sporotrichum pulverulentum* degrades the complex polymers to monomers which are then utilized for growth by *Candida utilis*.

TREE BARK AS FERTILIZER AND ANIMAL FEED

Bark is that part of the tree that is external to the vascular cambium and is composed of different types of cells, mostly rich in organic molecules. On the average, bark comprises about 10% of the fresh weight of most coniferous trees, and in most countries it has been considered a waste material to be disposed of by dumping or incineration. Disposal by destruction is wasteful and must not be accepted without careful scientific consideration. Methods should be explored and developed to turn this otherwise useless material into something of benefit to man and the ecosystem.

Historically, bark has been used by the Romans for floats to assist swimmers, by the Chinese to prepare paper, and in more recent times, on the Continent as bottle casks, insulation boards, and linoleum (Smith, 1976). The use of bark for heating purposes has had limited value. Some advances have been made in the use of bark as one of the ingredients in hardboard and insulating board manufacture. Chopped or pulverized bark can be used as a deep litter for animal bedding, having good water and ammonia absorbing properties. Chemical extraction of bark has been considered because of the wide range of potentially useful compounds to be found in various species. However, the economical aspects of this approach have yet to be proved (Smith, 1976).

Chemically, bark is normally more complex than wood. As with wood, the structural or cell wall components are composed primarily of polysaccharides such as cellulose and hemicellulose and the phenolic compound lignin. However, a unique group of compounds (the phenolic acids) are also present in bark (Jensen *et al*, 1967) and these may be separated from lignin by dilute alkali. Present also in bark are waxes, fats, and numerous organic molecules which are readily soluble in neutral solvents.

From a horticultural point of view, various grades of chipped, ground, or pulverized bark have had some use as a mulching cover to suppress weeds and to reduce water loss by evaporation. There are, however, many reports which imply a toxic effect of bark to plants. Indeed, this is not surprising when it is shown how many truly toxic molecules are present in bark. Much of the longevity of trees resides in the potency of the bark deposits to microbial and insect life, and many of the compounds are powerful uncouplers of oxidative phosphorylation. The possibility of detoxifying bark by broad spectrum microbial composting has at last been realized and numerous commercial products are now on the market

TABLE 8.4. Horticultural Uses of Bark.

Species	Mulching and Landscaping	Plunge Beds	Potting Composts	Bulb Forcing	Mushroom Cultivation	Orchid Culture	Pollution Control	Tannin Extraction
Scots pine (*Pinus silvestris*)	Well suited (ornamental)	Suitable	Well suited	All species suitable if lime is added	Well suited if lime is added	Well suited	Suitable	Not recommended
Corsican pine (*Pinus nigra*)	Suitable	Suitable	Well suited	added	Well suited if lime is added	Well suited	Suitable	Not recommended
European larch (*Larix decidua*)	Well suited (ornamental)	Suitable				Suitable	Not tested	Suitable
Japanese larch (*Larix leptolepsis*)	Well suited (ornamental)	Suitable				Suitable	Not tested	Suitable
Norway spruce (*Picea abies*)	Suitable	Suitable				Not recommended	Suitable	Suitable
Sitka spruce (*Picea stitchensis*)	Suitable	Suitable				Not recommended	Suitable	Well suited
Douglas fir (*Pseudotsuga menziesii*)	Suitable	Suitable				Not recommended	Not tested	Suitable

Source: Aaron. 1973. (Reprinted from the *Journal of the Institute of Wood Science.*)

which claim to be the end-product of a composting process. Details of such processes are not given, but the product no longer exhibits toxicity to plant life. The success of these processes is due in part to the more efficient pulverization of the bark, reducing it to relatively small sizes and thus allowing more efficient microbial penetration and utilization. A fuller appreciation of composting processes can be found in most applied microbiological texts. In modern fermentation terminology, the term composting is synonymous with solid state fermentation.

Thus, given the correct experimental parameters, it now seems that bark may find an expanding horticultural market as a soil conditioner and fertilizer (Table 8.4). A fuller consideration of the horticultural uses of bark has been considered by Smith (1976).

The possibility of using bark as a means of supporting the growth of microbes for biomass production has been considered by several workers. Such studies have, in particular, used solvent extracted tree bark and tree bark extracts. The early studies of Kuhlman (1970) using *Pinus taeda* and *Lenzites saepierie* were not very successful. However, Updegraff and Grant (1975), with *Pinus radiata*, and Nordstrom (1973), with *Picea abies*, using simple flask cultures, produced respectively 13 mg/g and 37.5 mg/g mycelial protein/g bark. Recently, Daugulis and Bone (1977) have shown that alkaline extracted bark can be degraded in submerged culture by *Polyporus anceps* and *Phamerochaete chrysoprium*, giving yields of 116 and 136 mg/g bark, representing a substantial improvement over all the previous workers. A non-linear relationship between bark concentration and protein yield was found. This process is now being stepped up to pilot fermentation stage.

Thus, at the present, bark is finding an increasing horticultural market throughout the world, and in the future, bark may well enter the more lucrative market of animal feed. Time and man's inventive genius will tell.

Undoubtedly, lignocellulose complexes of wood and bark are difficult feedstocks for SCP, and for a successful outcome of this problem, "there is a strong need for co-operation between biochemists, microbiologists, geneticists and engineers if the ultimate goal—the development of an economically feasible large-scale bioconversion process based on ligno-cellulosic waste material shall be realised." (Eriksson, 1978).

REFERENCES

Aaron, J. R. 1973. "Horticultural uses of bark." *J. Inst. Wood Sci.* **6**:22–30.

Almin, K. E., K. E. Eriksson, and B. Pettersson. 1975. "Activation of the 5 endo-1,4-glucanases towards carboxymethyl cellulose." *Eur. J. Biochem.* **51**:207–217.

Bellamy, W. D. 1974. "Single cell proteins from cellulosic wastes." *Biotech. Bioeng.* **16**:869–880.

Brown, D. E. and S. W. Fitzpatrick. 1976. "Food from waste paper." *Food from Waste*. G. G. Birch, K. J. Parker, and J. T. Worgan (Eds.). Applied Science Publishers, London, pp. 139–155.

Daugulis, A. J. and D. H. Bone. 1977. "Submerged cultivation of edible white-rot fungi on tree bark." *Eur. J. Appl. Microbiol.* **4**:159–166.

Dimmling, W. and R. Seipenbusch. 1978. "Raw materials for the production of SCP." *Proc. Biochem.* **13**(3):9–15.

Ek, M. and K. E. Eriksson. 1975. "Conversion of cellulosic waste into protein." *Appl. Polymer Symp.* **28**:197–203.

Ek, M., K. E. Eriksson, S. Hamp, and E. Bach. 1976. Swedish Forest Products Research Laboratory, Serial No. **B381**.

Elder, A. L. 1976. "Cellulose saccharification as an alternate source of glucose for commercial and food use." *Enzymatic Conversion of Cellulosic Materials: Technology and Applications, Biotech. Bioeng. Symp.* **6**: E. L. Gaden, M. H. Mandels, E. T. Reese, and L. A. Spano (Eds.). Wiley Interscience, New York, pp. 275–288.

Ellis, T. H. 1975. Paper presented at Forest Products Research Society meeting on Wood Residues as an Energy Source, Denver, Colorado, Sept. 3–5.

Eriksson, K. E. 1978. "Enzyme mechanisms involved in cellulose hydrolysis by the rot fungus *Sporotrichum pulverulentum*." *Biotech. Bioeng.* **20**:317–332.

Eriksson, K. E. and B. Pettersson. 1975. "Extracellular systems utilised for the breakdown of cellulose. 1. Separation, purification and physical-chemical characteristics of 5 endo-1, 4-β-glucanases. *Eur. J. Biochem.* **51**:193–201.

Eriksson, K. E., P. Ander, B. Henningsson, T. Nilsson, and B. Goodell. United States Patent 3,962,033, 1976.

Humphrey, A. E. 1975. "Economics and utilisation of enzymatically hydrolysed cellulose." Paper presented at SITRA *Symp. on Enzymatic Hydrolysis of Cellulose*, Hameenlinna, Finland.

Humphrey, A. E., W. B. Armiger, D. W. Zebriskie, E. Lee, A. Moreira, and G. Joly. 1976. "Utilisation of waste cellulose for the production of single cell protein. *Continuous Culture* **6**: *Applications and New Fields*. A. C. R. Dean, D. C. Ellwood, C. G. T. Evans, and J. Melling (Eds.). Ellis Horwood, Chichester, England, pp. 85–99.

Jensen, W., K. E. Fremer, P. Sierilaz, and V. Wartiovaara. 1967. "The chemistry of bark." *The Chemistry of Wood*. B. L. Browning (Ed.). New York, John Wiley & Sons.

Koch, K. 1972. *USDA Forest Service Agricultural Handbook No.* **420**, p. 1663.

Kuhlman, E. G. 1970. "Decomposition of loblolly pine bark by soil- and root-inhabiting fungi." *Can. J. Bot.* **48**:1787–1793.

Lemieux, P. G. and L. L. Wilson. 1978. "Energy savings using waste wood pulp as feed." *Science in Agriculture* **26**, in press.

Maclennan, D. G. 1976. "Single cell protein from starch." *Continuous Culture* **6**: *Applications and New Fields.* A. C. R. Dean, D. C. Ellwood, C. G. T. Evans, and J. Melling (Eds.). Ellis Horwood, Chichester, England, pp. 68–84.

Mandels, M., L. Hontz, and J. Majstrom. 1974. "Enzymatic hydrolysis of waste cellulose." *Biotech. Bioeng.* **16**:1471–1482.

McDowell, R. E., R. G. Jones, H. C. Pant, A. Roy, E. J. Siegenthaler, and J. R. Stouffer, 1972. *Improvement of Livestock Production in Warm Climates.* W. H. Freeman and Co., San Francisco.

Millett, M. A., A. J. Baker, and L. D. Satter, 1976. "Physical and chemical pre-treatments for enhancing cellulose saccharification." *Enzymatic Conversion of Cellulosic Materials: Technology and Applications, Biotech. Bioeng. Symposium* **6**: E. L. Gaden, M. H. Mandels, E. T. Reese, and L. A. Spano (Eds). Wiley Interscience, New York, pp. 125–154.

Nordstrom, U. M. 1973. "Bark degradation by *Aspergillus fumigatus*. Growth studies." *Can. J. Microbiol.* **20**:283–298.

Pettersson, L. S. 1975. "The mechanism of enzymatic hydrolysis of cellulose by *Trichoderma viride*." *Symp. Enzymatic Hydrolysis.* M. Baily, T. M. Enaris, and M. Linko (Eds.). SITRA, Helsinki, p. 261.

Phelps, R. B. and D. Hair. 1974. *USDA Forest Services Misc. Pub. No.* **1292**, p. 85.

Romantschuk, H. 1976. "The Pekilo Process: a development project." *Continuous Culture* **6**: *Applications and New Fields.* A. C. R. Dean, D. C. Ellwood, C. G. T. Evans, and J. Melling (Eds.). Ellis Horwood, Chichester, England, pp. 116–121.

Romantschuk, H. and M. Lehtomäki. 1978. "Operational experiences of first full scale Pekilo SCP-Mill application." *Proc. Biochem.* **13**(3):16–17.

Smith, J. E. 1976. "Tree bark—a usable commodity." *Proc. Biochem.* **11**(6): 41–48.

Spano, L. A., J. Medeiros, and M. Mandels. 1975. *Enzymatic Hydrolysis of Cellulosic Wastes and Glucose.* Pollution Abatement Division, U.S. Army.

Stone, R. N. 1976. "Timber, wood residues, and wood pulps as sources of cellulose." *Enzymatic Conversion of Cellulosic Materials: Technology and Appli-*

cations, *Biotech. Bioeng. Symp.* **6**: E. L. Gaden, M. H. Mandels, E. T. Reese, and L. A. Spano (Eds). Wiley Interscience, New York, pp. 223–234.

Szajer, C. and Z. Targonski. 1977. "Lignocellulose fermentations." *Biotech. and Fungal Ferment.*, *FEMS Symp. No.* **4**. J. Meynath and J. D. Bullock (Eds.). Academic Press, London, New York, and San Francisco.

Updegraff, D. M. and W. D. Grant. 1975. "Microbial utilisation of *Pinus radiata* bark." *Appl. Microbiol.* **30**:722–725.

9

Recovery of protein and fat from slaughterhouse effluents by physicochemical means

R. A. Grant, D.Sc.

Aquapure Services, Piper Works, Poole, Dorset, England.

INTRODUCTION

Waste effluents from meat, poultry, and fish processing plants contain large amounts of protein and fat and usually have much higher values of biochemical oxygen demand (B.O.D.) than does town sewage. Such effluents are highly polluting and can impose heavy loads on public sewage treatment works.

It has been estimated that between 2 and 5% of the total carcass protein is lost in the effluents from abattoirs and poultry processing plants. In the United Kingdom, this amounts to tens of thousands of tonnes per annum, with a potential value at present in the region of £200/tonne. For the world as a whole, this loss needs to be multiplied by a factor of about 100. At the present time, when world population is increasing rapidly and outstripping food production in many such areas, such a wastage is hard to justify if the means of preventing it are available.

Conventional biological effluent treatment plants suffer from the inherent disadvantage that potentially valuable materials such as protein and fat are degraded to useless sludge which in itself presents a disposal problem. On the other hand, in the case of physicochemical treatment plants designed to recover fat and protein, the revenue from the sale or recycling of by-products can be used to defray, either in whole or part, the capital and running costs of the plant.

Effluent Treatment

The process consists essentially of a flocculation reaction whereby soluble protein, together with insoluble suspended protein, is removed from the effluent. The floc entraps fat globules, which are removed simultaneously with the protein. The flocculated protein plus fat is separated from the effluent by air flotation and skimming in the form of a sludge containing up to about 15% total solids. The relative amounts of protein and fat in the separated sludge are dependent on the composition of the effluent. In general, the protein/fat ratio is higher in the case of slaughterhouse effluents than for poultry processing effluents. Where the effluent contains exceptionally large quantities of fat, as in the case of cooking and bone degreasing effluents, the bulk of the fat may be recovered in a separate air flotation stage. It has been found from analytical studies on a large number of effluents that the B.O.D. and C.O.D. (chemical oxygen demand) levels can be reduced by about 70 to 90% of the initial value with virtually complete removal of fat and suspended solids.

By-products

The protein content of a typical slaughterhouse effluent by-product is given in Table 9.1. Other samples of recovered meat works effluent were analyzed for amino acids and the results (Table 9.2) are compared with various reference proteins. The amino acids were found to be quite evenly distributed, with no major deficiencies. The contents of essential amino acids in two specimens are shown in Table 9.3 compared with the F.A.O. recommendation for human nutrition. Apart from tryptophan, which was not determined, the essential amino acid content appeared adequate. A similar analysis for amino acids was carried out on a sample of protein recovered from poultry processing effluent (Table 9.4). In certain instances, as in the case of effluents from cooking and rendering operations, the effluent may contain fat which is firmly complexed with protein, in addition to free and emulsified fat. Even with a separate fat recovery stage, the by-products from such effluents usually have a low protein/fat ratio. However, the presence of fat does not appear to impair the value of the protein as an animal feed.

The nutritional value of recovered meat works effluent protein was deter-

TABLE 9.1.   Composition of By-Product Recovered from Slaughterhouse Effluent.

Batch No.	1	2	3	4	5	6	Means
Nitrogen (%)	10.5	11.0	11.3	11.5	11.5	10.6	11.1
Protein (%)	65.5	68.0	70.5	72.0	72.0	65.5	68.9
Total organics (%)	74.5	78.3	77.6	77.4	76.2	72.7	76.1
Ash (%)	21.8	17.7	18.5	19.1	20.0	23.6	20.1
Moisture (%)	3.7	4.0	3.9	3.5	3.8	3.7	3.8

mined in a standard feeding trial on chicks. The basal ration consisted of wheatmeal, maizemeal, barley, lucerne, and salts, and contained 13% protein. This comprised 50% of the diet and contributed 6.5% protein. The remainder of the diet consisted of sugar and sufficient casein, meatmeal, fishmeal, or effluent protein to contribute a further 6.5% of protein. The relative growth rates, feed consumption, and feed efficiencies are shown in Table 9.5. The 280 chicks used

TABLE 9.2.   Amino Acid Composition of Recovered Solids from Meat Works Effluent (g amino acid/16 g nitrogen).

Amino Acid	Recovered Solids Fraction			Reference Proteins		
	A	B	Fibrin	Hemo-globins	Serum Proteins	Casein
Lysine	8.8	8.5	9.1	9.1	10.0	8.5
Histidine	3.9	5.9	2.9	8.0	3.3	3.2
Arginine	4.4	4.7	7.8	3.9	5.8	4.2
Aspartic acid	14.3	9.4	11.9	9.8	10.3	7.0
Threonine	7.9	4.5	7.3	5.6	12.6	4.5
Serine	7.7	5.7	12.5	5.5	18.2	6.8
Glutamic acid	19.3	10.0	15.0	8.1	14.2	23.0
Proline	6.6	3.2	5.3	4.7	5.5	13.1
Glycine	5.5	3.5	5.4	5.3	2.0	2.1
Alanine	8.8	6.8	4.0	9.8	–	3.3
Cystine (half)	trace	trace	3.8	1.0–2.2	7.0	0.8
Valine	11.0	8.2	5.6	9.0	7.5	7.7
Methionine	2.8	3.2	2.6	1.0	4.0	3.5
Isoleucine	5.5	4.1	5.6	0.2	3.4	7.5
Leucine	17.1	15.0	7.1	14.4	10.1	10.0
Tyrosine	5.5	3.2	6.0	2.9	5.5	6.4
Phenylalanine	9.9	7.9	4.5	7.8	5.2	6.3

TABLE 9.3. Essential Amino Acids (g amino acid/16 g nitrogen).

Amino Acid	F.A.O.*	Egg	Fraction A	Fraction B	Casein
Isoleucine	4.2	6.8	5.5	4.1	7.5
Leucine	4.8	9.0	17.1	15.0	10.0
Lysine	4.2	6.3	8.8	8.5	8.5
Phenylalanine	2.8	6.0	9.9	7.9	6.3
Tyrosine	2.8	4.4	5.5	3.2	6.4
Threonine	2.8	5.0	7.9	4.5	4.5
Tryptophan	1.4	1.7	—	—	—
Valine	4.2	7.4	11.0	8.2	7.7
Sulphur containing:					
Total	4.2	5.4	2.8	3.2	4.3
Methionine	2.2	3.1	2.8	3.2	3.5

*Food and Agricultural Organization "provisional pattern" of essential amino acids for human nutrition. Rome, 1957.

were crossbred WL/AO cockerels randomized in groups of 14. The ratio of feed consumption to weight gain shown in Table 9.5 is a measure of the efficiency of the feed. The dried effluent protein was approximately equal to the meatmeals and casein in nutritional value, slightly inferior to fishmeal but superior to the meat and bone meal and protein extracted from grass, which were tested at the same time. It is evident that the effluent protein could be used as a concentrate for poultry production. There was no evidence of toxic side-effects which could be attributed to the recovery process. Samples tested bacteriologically had low total amounts and were negative for coliform organisms. A further feeding trial

TABLE 9.4. Amino Acid Analysis of Protein Recovered from Poultry Processing Plant Effluent (μmoles/100 μmoles).

Aspartic acid	10.8	Threonine	5.1
Serine	7.7	Glutamic acid	11.5
Proline	4.5	Glycine	6.9
Alanine	8.9	Cystine (half)	1.2
Valine	7.4	Methionine	1.5
Isoleucine	5.3	Leucine	9.1
Tyrosine	2.3	Phenylalanine	3.8
Lysine	6.6	Histidine	2.3
Arginine	4.9		

Note: Tryptophan not estimated.

TABLE 9.5. Chick Growth, Feed Consumption, and Feed Efficiency of Experiment Rations.

	Body Weight Gain/Chick (g) 1–4 Weeks	Food Consumption/ Chick (g) 1–4 Weeks	Feed Consumption/ Weight Gain (g)
Reference ration (casein)	137	359	2.62
Mm 41	147	400	2.74
Mb 43	103	347	3.38
Mb 44	100	302	3.02
Mm 45	166	413	2.49
Mb 46	83	282	3.42
Mm 47	141	386	2.73
Gross protein	104	338	3.16
Fishmeal 6	169	418	2.48
Effluent protein	125	345	2.75
MSD 5%	25	91	0.54

on the effluent protein was carried out using pigs. The composition of the experimental rations and the results obtained are shown in Table 9.6. A satisfactory growth rate was obtained over 68 days, when the diet included effluent protein at the 5% level.

TABLE 9.6. Pig Trial Rations and Results.

	Ecotech Effluent By-product (%)	Whey (%)	Control (%)
Barley	42.0	31.2	43.0
Maize	41.75	31.2	43.0
Meatmeal (60%)	11.0	10.0	13.75
Whey mix (13.4%)	—	26.75	—
Trace nutrients	0.25	0.25	0.25
Steamed bone flour	0.0	0.6	—
Effluent by-product	5.0	—	—
Total (%)	100.0	100.0	100.0
Estimated total protein (%)	17.0	16.9	17.0
Results:			
Original liveweight (kg)	14.0	14.4	14.3
68 trial days	—	—	—
Liveweight (kg)	39.5	33.4	41.6
68 day gain (kg)	25.5	19.0	27.3
Average daily gain (kg)	0.38	0.28	0.40

ION EXCHANGE

In the past, the great majority of ion exchange applications have been in processes involving small inorganic ions such as the softening and demineralization
of water supplies. The development of suitable resins and ion exchange processes
for the removal of large organic molecules from solution has been relatively
neglected, although there has been an upsurge of interest in this area in recent
years. The conventional synthetic polymer or condensation types of resin have
gel or micro-reticular structures which allow the diffusion of small inorganic
ions but preclude significant absorption of all but the smallest organic ions.
In the case of raw water purification and effluent treatment, a large variety of
organic compounds are encountered, ranging from relatively simple vegetable
breakdown products to macromolecules such as proteins and lipo-protein complexes, which occur in food processing effluents and having molecular weights
in the range 10^4 to 10^6.

A significant breakthrough in macromolecular ion exchange occurred when
Peterson and Sober (1956) published their work on the use of ion exchange

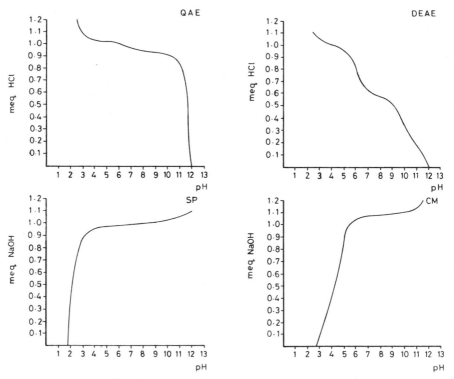

Fig. 9.1.   Resin ionization characteristics.

cellulose for the uptake and separation of proteins. Although these fibrous cellulose ion exchangers possessed high capacities for proteins and other biological macromolecules, their inherently poor hydraulic properties presented a severe obstacle to their application other than for analytical and small scale preparative work. Attempts were made by Lockwood and Rafer (1959) to overcome this disadvantage by substituting granular regenerated cellulose for natural fibrous cellulose and it was found that the ability to exchange large organic molecules was retained by the cellulose in the regenerated state. A major defect of simple substituted ion exchangers derived from regenerated cellulose was their ready solubility in all but dilute solutions of acids or alkalis, which severely limited the range of applications. It was not until the successful development of procedures for the introduction of a controlled degree of crosslinking into the matrix that ion exchangers based on regenerated cellulose could be produced with physical, chemical, and hydraulic properties comparable with conventional ion exchange resins (Tasman Vaccine Laboratories, 1971).

Crosslinked regenerated cellulose ion exchangers are now available in strongly and weakly acidic or basic forms, corresponding to the main types of conventional ion exchange resins. The principle types so far developed are the carboxymethyl, diethylaminoethyl, sulfopropyl, and quaternary base derivatives. The ionization characteristics for these resins determined by titration using a glass electrode in the presence of salt are shown in Fig. 9.1. The effective working pH range covered by these types of resin is approximately 2 to 12.

Preparation and Properties

Generally, the ion exchangers are prepared by reacting granular regenerated cellulose having the desired particle size range with the appropriate chloro derivative of the group which it is desired to introduce under alkaline conditions. For example:

$$R \cdot ONa + Cl \cdot CH_2 \cdot CH_2 \cdot N (Et)_2 \xrightarrow{\text{NaOH}} R \cdot O \cdot CH_2 \cdot CH_2 \cdot N (Et)_2$$

$$R \cdot ONa + Cl \cdot CH_2 \cdot COONa \longrightarrow R \cdot O \cdot CH_2 \cdot COONa.$$

The sulfopropyl derivative is prepared by reaction with propane sultone and the quaternary base resin is made by treating the diethylaminoethyl derivative with propylene oxide. Crosslinking is effected by incorporating a suitable amount of a crosslinking compound, such as epichlorhydrin, into the reaction mixture. The degree of crosslinking should not exceed a defined limit for each type of resin, otherwise the effective capacity for large molecules is decreased (Fig. 9.2). The preferred degree of crosslinking for most applications has been found to lie in the range of 1 to 10%. Ion exchangers of this type have been made using regenerated cellulose from a variety of sources, and although some variation in

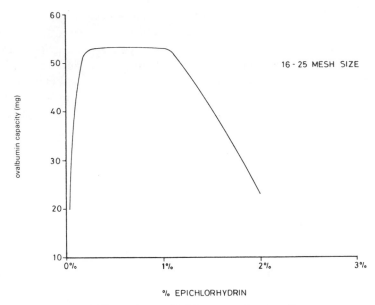

Fig. 9.2. Effect of cross-linking on resin capacity.

the products was noted, it has been found that satisfactory resins can be made from most types of currently available regenerated cellulose.

The capacities of these resins for macromolecular ion exchange may be conveniently expressed in terms of the uptake of pure proteins of known molecular weight under specified conditions. Generally, this is determined by stirring a known weight of freshly regenerated and washed resin with an aqueous solution of protein for a period of 30 minutes. The amount of protein removed from the solution is determined by ultraviolet spectroscopy or Kjeldahl analysis. Useful proteins for standardization are bovine serum albumin, ovalbumin, and lysozyme. In general, it has been found that maximum capacity for macromolecules is obtained with resins having small ion capacities in the region of 1 meq/g. High small ion capacities have been found to result in excessive swelling in acid and alkali with consequent poor hydraulic properties.

In order to study the effective porosity of the regenerated cellulose resins and the degree of penetration of large molecules into the grains, recourse was made to microscopical techniques. The distribution of diethylaminoethyl groups was studied by embedding in epoxy resin and section after staining with Ponceau S. It was found that resins with high capacities for proteins were deeply stained into the centers of the grains. However, large grains (14 to 25 mesh, dry) were not always stained in the central regions, indicating that the usual preparative methods did not fully utilize the raw material. In other experiments, stains were used which distinguished protein from the amino groups of the resin and this work showed that protein was absorbed to the full depth of penetration of the ion exchange groups. This work clearly demonstrated the highly porous nature of the resins and that protein absorption did not occur only on the surface of the resin grains.

The highly porous nature of the resin was also demonstrated by examination of freeze dried resin grains with a scanning electron microscope which showed a sponge-like structure.

Desorption of organic material (regeneration) has been found to be a relatively fast process, bearing in mind the low diffusion rates of macromolecules. Efficient regeneration has been achieved with dilute solutions (0.1 to 0.5M) of salts, acid, or alkalis. Usually 30 minutes exposure was sufficient to allow complete desorption of protein, even with coarse resin grains. In large scale applications involving food process effluents, the preferred regenerant has been alkaline brine (1% Na OH + 2 to 5% NaCl) in the case of the most commonly used DEAE resin. When selective regeneration is required to isolate a particuarly valuable compound, such as an enzyme, stepwise or gradient elution may be employed.

Although sodium hydroxide solution (1 to 2%, W/V) is an effective regenerant for proteins, the addition of sodium chloride has been found advantageous for resin regeneration after effluent treatment. The higher ionic strength of the alkaline brine enables certain non-protein organic compounds found in effluents to be desorbed more effectively than with alkali alone.

It has been found, in the case of cellulose based resins used for treatment of slaughterhouse effluents, that a form of fouling occurs, resulting in a progressive fall in protein capacity. This has been attributed to the deposition of insoluble denatured protein material on the surfaces of and within the resin grains. Such fouling can be reversed and the resins restored to full capacity by treatment with warm sodium hydroxide solution (5% W/V) for a period of 1 to 5 days.

With regard to the plant, most of the work up to the present has been carried out using columns or beds of the conventional type and provided with means for fluidization and backwashing of the resin.

Adsorption and Desorption of Pure Proteins

A typical absorption/desorption cycle is shown in Fig. 9.3 for the DEAE anion exchange regenerated cellulose resin and 0.1% aqueous bovine serum albumin solution. Protein concentration was monitored in the column effluent by ultraviolet absorption measurement at a wavelength of 280 nm. The completely flat baseline indicates total uptake of protein up to the breakthrough point. This was confirmed by the absence of turbidity when phosphotungstic acid was added to the column effluent. The uptake curve may be compared with a similar curve obtained with an identical column of a macroporous condensation type

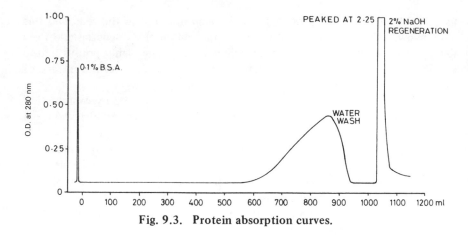

Fig. 9.3.   Protein absorption curves.

resin (Fig. 9.4). It may be seen that breakthrough occurred very rapidly in this case, after the solution was applied to the column indicating a negligible capacity of this resin for protein.

The ultraviolet absorbance measured at 280 nm is directly proportional to the protein concentration for solutions of pure proteins. In Figs. 9.3 and 9.4, the base lines were initially adjusted using distilled water, and the height of the first peak in each case indicates the absorption due to the bovine serum albumin test solution. The volume during which the ultraviolet absorption falls to the blank (H_2O) level indicates the volume for which complete protein uptake was obtained in each case.

Following breakthrough, the cellulose resin column was washed with water applied by downflow until the ultraviolet absorbance fell to zero. Regenerant solution consisting of 0.5M sodium hydroxide was then applied. This stripped the protein from the bed in a concentrated solution, as shown by the second peak. It may be noted that the elution curve is tall and narrow and shows little

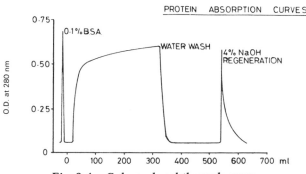

Fig. 9.4.   Column breakthrough curve.

evidence of tailing. Usually, complete regeneration is achieved with one bed volume of regenerant followed by rinsing with two bed volumes of water. In contrast, the regeneration peak for the macroporous condensation type resin was small and asymmetric, indicating minimal uptake of protein.

Repeated cycling of regenerated cellulose resins with pure proteins over many cycles has shown no significant changes in absorption efficiency, indicating that there is little irreversible binding of protein under these conditions.

In general, the capacities for proteins have been found to lie within the range of 300 to 500 mg protein (bovine serum albumin or lysozyme) per gram of dry resin. Exceptionally, capacities up to 1g protein per gram of resin have been recorded.

SLAUGHTERHOUSE EFFLUENT TREATMENT

Most of the work to date on the application of regenerated cellulose ion exchangers to effluent treatment has been carried out on waste waters from abattoirs (Tasman Vaccine Laboratories, 1971). Such effluents are very polluting, with high B.O.D. values, the principal contaminants being fats and protein. In general, such effluents contain large amounts of suspended insoluble matter as well as fat, and if applied directly to the resin bed, they will cause rapid clogging of the bed with a large fall off in flow rate. Therefore, such effluents should be subjected to a pretreatment, preferably consisting of flocculation and air flotation to remove particulate matter and fat. Typical results obtained with this type of effluent are shown in Tables 9.7 and 9.8, where ion exchange has been combined with a pretreatment of the above type. Combined treatment of this type, employing the DEAE cellulose anion exchanger in the second state, has given reductions in the B.O.D. level in excess of 95%. The general forms of the absorption and breakthrough curves are shown in Figs. 9.5 and 9.6. The resin columns were regenerated with a solution of 1% NaOH + 3.5% NaCl, followed by a water backwash and downward rinse. Since the DEAE resin

TABLE 9.7. Residual B.O.D. (%) in Treated Slaughterhouse Effluent.

Treatment	Sample					Means
	1	2	3	4	5	
Flocculation	38	41	31	23	27	32
Ion exchange	10	13	23	19	18	17
Flocculation plus ion exchange	TL*	5	12	7	TL*	5

*Value too low for measurement by standard method.

TABLE 9.8.   Residual C.O.D. (%) in Slaughterhouse Effluent After Treatment.

Treatment	Sample							Means
	1	2	3	4	5	6	7	
Flocculation	19	40	31	37	35	31	31	32
Ion exchange	18	31	31	28	25	27	15	25
Flocculation plus ion exchange	TL*	14	9	22	TL*	11	TL*	11

*Value too low for estimation by standard method.

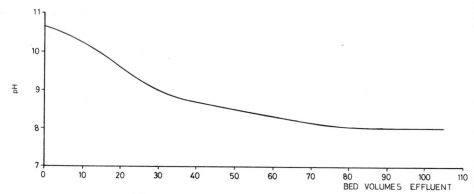

Fig. 9.5.   Column breakthrough curve.

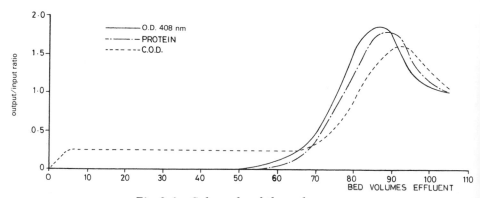

Fig. 9.6.   Column breakthrough curve.

has a certain degree of strong base function and since the effluent contains salts, a high initial pH is produced which falls steadily thereafter. Usually the pH remains at about 10 for the first 10 to 15 bed volumes (BV) after which it decreases to about 9 during the next 30 to 50 BV, and eventually falls to about 8. Under normal conditions, both the color (optical density of 408nm) and the protein concentration of the effluent are minimal until breakthrough occurs, usually at 40 to 100 BV, depending on the strength of the effluent.

Following breakthrough, the value for protein and color in the effluent often exceed the input values for a time before decreasing to a level equal to the input. In the case of slaughterhouse effluent, this overshoot appears to be due to hemoglobin accumulating in a band which travels down the column to emerge at the breakthrough point. C.O.D. measurements result in a different type of curve, in which a fairly constant residual value equal to 5 to 20% of the initial C.O.D. value is maintained up to the breakthrough point. This behavior is explained by the presence in the effluent of uncharged organic compounds, such as carbohydrates and urea, which are not absorbed by the resin, and also to low molecular weight protein breakdown components produced by the action of bacteria in the effluent. The behavior tends to become more pronounced the longer the effluent is kept before treatment. It is generally found that following breakthrough, the protein curve tends to lag behind the optical density curve, and the C.O.D. lags behind the protein curve with this type of effluent. Generally, the absorption cycle is terminated at the breakthrough point and the resin bed regenerated with alkaline brine solution (1 bed volume), which is allowed to stand in the bed for 30 minutes to 1 hour before rinsing off with water. In cases where a small amount of color and protein can be tolerated in the effluent, the absorption cycle may be continued beyond the breakthrough point.

Since the spent regenerant from the ion exchange bed consists of concentrated protein solution, this should be treated to remove as much as possible of the organic matter and reduce the B.O.D. level before discharge. It is possible to recover virtually all the desorbed protein by adjustment of the pH to about 4.5, followed by heat coagulation. The insoluble heat coagulated protein is readily de-watered by screening or filtration and may be dried to yield a by-product suitable for inclusion in animal feeds. The nutritional value of effluent protein recovered by ion exchange has been evaluated by amino acid analysis and animal feeding trials, and there was no indication of toxic effects arising from the treatment process.

Applications

Reduction in the B.O.D. and C.O.D. levels obtained with slaughterhouse effluents are given in Tables 9.7 and 9.8. The best results, from the point of view of

TABLE 9.9.   Reduction of C.O.D. and By-product Yields for Poultry
Processing Effluents.

Specimen No.	1	2	3
C.O.D. (ppm)			
Before treatment	6050	5920	7400
After treatment	1150	1040	945
Reduction in C.O.D. (%)	81.0	82.4	87.0
By-product yield (g/1)	3.4	3.2	4.0

effluent quality, were obtained using the combined process. However, there is great variability in the strength and composition of effluents from different plants and also in the consent conditions laid down by local water authorities for discharge to sewers, so that depending on circumstances, either the first stage only or both stages may be employed. In some cases, it may be preferable to recover fat and protein separately. Table 9.9 shows the C.O.D. reductions obtained with poultry processing effluents together with the yields of dried by-product for different effluent strengths. In the case of a fish processing effluent, the C.O.D. level was reduced by 90% using the flocculation process only, with a dry product yield of 2 g/l. Although, in general, the concentrations of protein found in effluents are relatively low—ranging from 1000 to 2000 parts per million on the average—since large volumes are involved, the ultimate yield of by-product may be considerable. For example, at 1000 parts per million, a yield of about 1 tonne per million liters (4.5. tonnes per million gallons) would be obtained.

Vegetable Wastes

Vegetable processing effluents may also contain significant amounts of protein which can be recovered by suitable *chemicophysical* treatment for use in animal or human nutrition. It has been found possible to recover the soluble protein from rice starch factory effluent by flocculation and air flotation in a yield amounting to about 5% of the weight of the rice processed. Since the possibility of contamination with excreta or pathogenic organisms does not arise in this case, the product will be suitable for human consumption and have a relatively high value.

In the case of the palm oil industry, vast quantities of centrifugation sludge and sterilizer condensate are produced having extremely high B.O.D. values in the region of 30,000 parts per million. By means of flocculation followed by centrifuging or air flotation and drying, a product containing about 12% protein and 80% low grade carbohydrate was obtained in a yield of about 4% W/V, the product being possibly useful as a ruminant feed. This treatment reduced the B.O.D. level by about 75%, with virtually complete removal of residual oil and suspended solids.

Economics

From the economic point of view, rapidly escalating costs of water, effluent disposal, and animal feed protein are making the case for protein recovery much more attractive. In terms of plant size, the new system is considerably more compact than conventional sewage type or biological treatment plants and compares favorably in terms of capital and running costs. The costs of biological treatments have been evaluated by Chipperfield (1970), who concluded that the evidence casts considerable doubt upon the traditional view that biodegradable industrial wastes are most effectively and cheaply treated at local sewage works. The evidence suggests that liquid wastes should be treated at source. The possibility of re-use of water should also be considered carefully, in view of rapidly increasing water costs.

The question of disposal of industrial effluents with domestic sewage and the related costs have also been reviewed, by Calvert (1970). He estimated the cost at 6–14c/5000 liters in 1970. However, this figure has increased considerably since then, and at present one may quote a cost per 5000 liters for a poultry processing effluent with a C.O.D. value of 4,000 of 48c and for a meat works effluent with a C.O.D. of 2000 of 36c/5000 liters. Calvert (1970) also states that biological treatment on site is not always desirable, and the fact that it may appear economical may be a fallacy of the Mogden formula.

The financial benefit is best illustrated by a typical example for an abattoir where C.O.D. is reduced by 75% and suspended solids reduced to 100 mg/1.

Assumed daily flow = 400 cubic meters.
Assumed raw effluent C.O.D. = 4000 mg/1.
Assumed raw effluent suspended solids = 750 mg/1.
Typical water authority charging formula for 1977/79 in U.K. = $R + V + B \times Ot/Os + S \times St/Ss$ pence per cubic meter.

$$R = 1.39, \quad V = 2.31, \quad B = 2.31, \quad S = 3.24.$$

Ot and St are the C.O.D. and suspended solids of the effluent and Os and Ss are the values for standard sewage in the area in question. In this case, Os = 400 mg/1, and Ss = 350 mg/1.
Untreated effluent charge = 1.39 + 2.31 + 2.31 × 4000/400 + 3.24
                                         × 750/350 = 33.74 pence per cubic meter.
Treated effluent charge = 1.39 + 2.31 + 2.31 × 1000/400 + 3.24 × 100/350
                                  = 10.39 pence per cubic meter.
Saving per cubic meter = 23.35 pence.

Annual saving based on 250 working days = 400 × 23.35 × 250/100 = £23,350.

Chemical and electrical costs for treatment are 5 pence per cubic meter.

Yearly operating cost = 400 × 5 × 250/100 = £5,000.

Net benefit yearly = £18,550.

Yield of protein with a C.O.D. of 4000 mg/1 can be expected to be about 3 g/1.

Daily yield = 400 × 3 = 1200 kg dry solids.

This originates as 15% solids air float = 8000 kg.

Water to be evaporated = 6800 kg.

Steam cost at £5 per tonne and 60% efficiency = 6.8 × 5 × 100/60 = £56.

Value of product at 50% protein content and market value of £140 per tonne = £168.

Daily benefit = £112.

Annual benefit = £28,000.

Total benefit from discharge savings and protein sales = £28,000 + £18,550 = £46,550.

In the case of ion exchange units, these can only be operated after a pretreatment such as the above. Even then the economics do not make them attractive for normal effluent. Where they are significantly useful is in the preparation of high value end products such as bovine serum albumin or various enzymes. In cases such as these, over extended periods there has been no tendency towards biological fouling.

In view of the foregoing, it now appears that a very strong case can be made for physicochemical effluent treatment on site in the food industry with by-product recovery to offset capital and running costs. This appears to apply particularly in the case of meat and poultry works effluents.

REFERENCES

Calvert, J. T. 1970. "Disposal of industrial effluents with domestic sewage." *Chem. and Ind.* June 1970, pp. 733–734.

Chipperfield, P. N. J. 1970. "The cost of biological treatment for industrial wastes." *Chem. and Ind.* June 1970, pp. 735–737.

Lockwood, A. R. and A. H. Rafer. 1959. British Patent 911,223.

Peterson, E. A. and H. A. Sober. 1956. "Chromatography of proteins I. Cellulose ion-exchange adsorbents." *J. Amer. Chem. Soc.* 78:751–756.

Tasman Vaccine Laboratories, Ltd. 1971. British Patent 1,226,448.

10

Factory canning and food processing wastes as feedstuffs and fertilizers

L. L. Wilson and P. G. Lemieux

Professor and Research Graduate Assistant, Department of Dairy and Animal Science, Pennsylvania State University, University Park, Pennsylvania

INTRODUCTION

Feeding or "recycling" of waste materials from various agricultural and non-agricultural industries is not new. Although most of the volume of different types of cannery wastes have been recycled directly through the soil as fertilizers, a large proportion of the total wastes from various types of food and fiber processing procedures have become important livestock feeds. Among the most common are soybean meal, beet pulp, linseed meal, and dried citrus pulp. The primary reasons these waste materials have become widely used in the animal industries are the large, centralized processors, the palatibility of the feedstuffs when dried, and the nutritional value of the waste material. However, there are still large amounts of wastes that are not being completely utilized. In most cases, these wastes are produced from isolated sources and not by large, centralized companies. In practically all cases, horticultural and vegetable products are processed in relatively small canning plants and distributed for use generally in the same geographical areas as the produce is grown. This is true throughout the world, since horticultural and vegetable wastes cannot usually be transported long distances because of their bulk, high moisture, and low to moderate feeding values relative to conventional feeds.

As with other phases of livestock and dairy production, efforts must be made to adapt available feed supplies, including waste materials, to individual animal operations. In many areas of the world, there are specialized crops available that are processed for direct human consumption resulting in a wide variety of waste materials. Since these possible waste feeds are diverse, and frequently available in large volumes, it is difficult to make specific recommendations for all types of livestock operations. In addition, each type of animal has different nutritional requirements, and there are many different standard feedstuffs available throughout the world which must be integrated into waste-containing rations. All of these factors determine the methods by which wastes and by-products can contribute to animal production and society.

Within the past several years, animal agriculture, particularly the beef cattle industry, has been accused of being wasteful of grain that might be consumed directly by humans. This criticism has been so pointed that even those closely associated with agriculture have been examining more carefully the use of the ruminant animal as a waste recycling device. However, in some ways, the ruminant is already being used to recycle waste materials, even though these materials are now considered as standard feeds because of their frequency of use (soybean meal, beet pulp, citrus pulp, etc.). In addition, considering all the land used by United States beef production, approximately 90% is land which can only produce forage crops and cannot produce grain crops for either feed or food purposes. In fact, only about 10% of the world's surface can be tilled for intensive crop production (Trenkle and Willham, 1977). In the United States, approximately 77% of the agricultural land was used for grazing in 1969 (Nix, 1975). In many areas, particularly in South America, Australia, and New Zealand, essentially no feed grains are used to produce beef cattle, and there is a similar trend toward maximizing grass utilization in the United States. Therefore, the ruminant is capable of utilizing land which cannot economically be used for grain production and at the same time has the ability to utilize a wide variety of waste materials.

Although the primary subject here is methods (and opportunities) of utilizing waste materials from plant material processed by canneries, there are many other types of waste materials which can be utilized. As an example, approximately one-half of the food energy of plants raised on existing croplands remains after harvesting (N.R.C., 1977). Another study was carried out by Fontenot (1977), in which it was calculated that 1.5 billion tonnes of animal wastes were excreted in the United States per year. Approximately 50% of this waste output is produced under intensive rearing conditions and is collectable, amounting to 725 million tonnes, or approximately 218 million tonnes of dry matter. Yeck *et al* (1975) presented a comprehensive and detailed estimate of the amount of nitrogen excretion by various types of American livestock. These calculations indicated that about 2.27 billion kg nitrogen is collectable in the United States each year, which is more total nitrogen than in the 1972 soybean crop. The uses of these materials as animal feed are dealt with in more detail in Chapter 2.

GENERAL WASTE HANDLING CONSIDERATIONS

There are several considerations in the utilization of any waste material from a canning or food processing plant, and many are applicable not only to horticultural and vegetable products, but also to several other types of waste materials, regardless of origin.

Economic Considerations

The amount of waste material generated by a processing plant may seem large to the individual processor and even to the individual contemplating utilizing the waste in his livestock operation. However, the amount of dry matter in the waste produced by a processing plant may not be sufficient to justify the inconvenience and expense of collecting, transporting, and perhaps making changes in the methods of feeding and feed storage. Before considering the use of certain waste materials in livestock rations, the animal producer should determine the relative costs of the different components of the existing ration, the actual dry matter in the waste, the total dry matter available, and whether or not the waste material might substitute partially or entirely for a standard, costly ration component.

Variability and Frequency of Waste Collection

Most canneries produce waste which is predictable in volume and relatively uniform in composition. However, prior to utilizing a waste material which has not previously been used as an animal feed, the animal producer should obtain samples over a few weeks and submit these to a laboratory for analysis. Samples should be subject to a proximate analysis (A.O.A.C., 1975), mineral analysis, and pH and dry matter determinations. Predictions of total digestible nutrients (TDN, %) and digestible, metabolizable and net energy (kcal/kg) can be calculated based upon the results of the proximate analysis (N.A.S., 1971). If the waste material is to be used for ruminants, additional analyses can be conducted to determine rumen *in vitro* digestible dry matter (Tilley and Terry, 1963) and fiber composition (Goering and Van Soest, 1970). Calculations of digestible protein can be obtained from equations reported by N.A.S. (1971) and Goering *et al* (1972). The uniformity of the waste material is important if it is to be fed directly from the processing plant on a day to day basis because of the necessity to maintain generally consistent rations for most livestock. If the processor is truly interested in making available a material which is of consistent quality, there are ways of ensuring this. These may include separation of various types of waste material (from different vegetable types), or the removal of more water from the waste material.

Metal, Glass, and Similar Contaminants

In many canneries, the organic waste from the processing system is combined with the other types of nonusable wastes, such as hardware, glass, cans, and nails. This is true not only in canneries but also in other types of processing plants, such as bakeries. It must be remembered that most cannery wastes have been disposed of in landfills, and it is usually convenient for a cannery to have one composite waste stream from the plant if the wastes are not being used for livestock feeds or further processing. Naturally, metal or glass materials must be removed before the material is fed to livestock, or "hardware" disease may result from puncturing of the linings of the gastrointestinal tract or the diaphragm. As a precaution against metal contaminants, the processor or livestock producer may use magnets at strategic points in a mechanized handling system, removing hardware from the waste material. Glass and aluminium present a special problem, and the most desirable way of making certain that these contaminants are not contained in the waste is to divert these materials from the organic waste stream.

Method and Total Cost of Transporting and Using the Waste

Practically every type of waste material used in a livestock operation requires some adjustment in feed storage and handling systems on the farm. This may range from a relatively small change, such as the addition of a concrete apron and a roof for relatively dry material which can be piled for short periods of time after delivery to the farm, to a specialized storage facility for a semi-liquid waste material, such as potato slurry. The latter waste material is a good example of the additional investment which may be required on the part of the livestock producer to transport and utilize cannery wastes. In the United States producers utilizing potato slurry (average composition given in Table 10.1) usually invest in large, tank-type, completely closed semi-trailer trucks for transporting the waste material. It should be emphasized that the responsibility for hauling the waste material in a safe manner is usually vested with the livestock operator or a specialized contractor. Vehicles which allow excess moisture to drain onto roads, or uncovered trucks which allow materials to be lost onto the roadside, may cause problems with regulatory agencies.

TABLE 10.1. Composition of Certain Waste Products and Conventional Reference Feeds.

Feed Name	Reference No.	Dry Matter (%)	Crude Fiber (%)	Digestibility Coefficient	Crude Protein (%)	Digestible Protein (%)			Digestible Energy (Mcal/kg)		(kcal/kg)	Ash (%)	Ca (%)	P (%)	Vitamin A Equivalent (international units per gram)
						Cattle	Sheep	Swine	Cattle	Sheep	Swine				
Alfalfa, aerial part, fresh, mid-bloom	200185	24.2	27.7	46	20.4	15.9	15.2	—	2.68	2.63	—	8.8	2.01	0.28	—
Almond, hulls	400359	91.0	15.2	—	4.2	0.6	0.4	—	3.42	3.18	—	6.5	—	—	—
Almond, shells	107754	92.5	45.2	—	—	—	—	—	—	—	—	3.7	—	—	—
Apples, fruit, ensiled	300419	12.4	14.6	—	5.5	1.2	1.2	—	3.16	3.03	—	4.8	—	—	—
Apples, pulp, ensiled	300420	21.4	20.6	—	7.8	3.3	3.3	—	3.26	2.94	—	4.9	0.10	0.10	—
Artichoke, tubers, ensiled	300430	30.2	27.3	65	7.2	2.8	1.2	—	2.13	2.29	—	18.5	—	—	—
Asparagus, stem butts, dehydrated ground	100437	91.0	31.9	41	15.6	10.4	9.7	—	2.17	2.07	—	16.0	—	—	—
Avocado, seeds, extender unspecified ground	500451	91.4	19.3	86	20.4	—	9.0	—	2.22	2.52	—	19.8	—	—	—
Banana, aerial part, fresh	200483	16.0	23.7	54	6.4	3.3	3.5	—	2.44	2.86	—	13.1	—	—	—
Banana, skins, dehydrated ground	400486	88.0	8.6	22	7.7	3.1	2.6	—	2.61	2.82	24.60	10.5	—	—	—
Barley, flour by-products without hulls	400547	89.3	10.4	39	16.3	10.9	13.3	—	3.01	3.49	29.76	4.8	—	—	—
Barley, grain cooked	400524	85.3	7.9	—	11.2	6.3	7.5	7.5	3.50	3.57	34.38	2.6	—	—	—
Barley, pearl by-product	400548	89.6	11.9	18	14.7	11.0	11.9	11.6	3.33	3.05	32.80	5.7	0.05	0.46	0.8
Barley, brewers' grains, dehydrated	500516	91.1	19.9	—	22.0	—	—	—	2.86	2.94	30.60	4.1	—	—	—
Barley, distillers' grains, dehydrated	500518	92.0	11.0	—	30.1	—	—	—	3.05	3.12	32.45	2.0	—	—	—
Bean, cannery residue, fresh	200587	9.4	13.5	—	23.5	17.9	18.9	—	2.74	2.90	—	—	—	—	41.5
Bean, pods, fresh	200589	28.2	38.3	—	7.4	4.2	3.9	—	2.81	3.22	—	4.2	—	—	—
Beet, sugar, pulp, wet	200671	11.3	30.1	82	11.7	5.3	6.4	—	—	—	—	4.7	0.86	0.10	—
Broccoli, leaves, fresh	200882	13.5	13.5	—	28.8	22.5	23.6	—	—	—	—	—	—	—	559.9
Cabbage, aerial part, fresh	201046	9.6	10.5	91	20.8	15.5	17.9	—	3.80	3.75	—	10.3	0.64	0.35	17.1
Cacao, hulls	101051	90.0	13.5	—	15.8	10.6	10.7	—	—	—	—	6.5	—	—	—
Carob bean, gluten, fresh etiolated	501129	90.0	—	—	—	—	—	—	—	—	—	—	—	—	—
Carrot, aerial part, fresh	208371	16.0	18.1	—	13.1	9.0	9.2	—	3.28	3.37	—	15.0	1.94	0.19	—
Carrots, roots, fresh	201146	12.9	9.1	108	10.3	5.1	7.8	7.3	3.62	3.85	36.38	9.7	0.37	0.32	932.2
Cassava, roots	201150	32.4	4.6	53	3.6	-0.5	-0.0	—	3.47	3.60	34.19	3.9	—	0.12	—
Cassava, meal	401152	90.9	4.9	—	2.9	-1.3	0.0	—	3.50	3.81	35.44	2.3	—	0.03	—
Cassava, seeds, extender unspecified	401151	87.2	2.5	32	1.6	-2.4	-1.2	—	3.72	3.82	37.87	2.2	—	—	—
Cassava, tapioca flour	401154	88.3	2.8	—	2.6	-1.5	-0.3	1.8	3.63	3.87	41.86	2.4	—	—	—

247

TABLE 10.1. (Continued).

Feed Name	Reference No.	Dry Matter (%)	Crude Fiber (%)	Digestibility Coefficient	Crude Protein (%)	Digestible Protein (%) Cattle	Sheep	Swine	Digestible Energy (Mcal/kg) Cattle	Sheep	(kcal/kg) Swine	Ash (%)	Ca (%)	P (%)	Vitamin A Equivalent (international units per gram)
Castorbean, seeds with hulls, toxicity extender, ground	501155	86.7	41.0	—	30.0	23.1	—	—	1.16	—	—	8.6	—	0.22	—
Chicory, aerial part, fresh	201220	17.4	15.3	—	14.4	10.1	—	10.4	—	—	—	18.0	1.89	0.22	—
Citrus, pulp, fresh	208376	18.3	12.6	—	6.6	2.0	3.2	—	3.67	3.40	22.61	7.7	—	—	—
Citrus, orange, fruit, fresh, cull	201252	12.8	9.4	82	7.8	3.2	5.0	—	4.17	3.97	—	4.7	—	—	—
Coffee, grounds	101576	73.7	29.2	—	13.8	8.9	9.0	—	—	—	—	1.6	0.12	0.08	—
Corn, aerial part, ensiled	302824	26.0	23.9	—	7.9	3.4	3.4	—	2.94	2.99	—	5.0	0.41	0.18	—
Corn, dent, aerial part, ensiled	302910	27.3	25.8	—	8.2	3.7	3.7	—	2.90	2.94	—	6.2	0.35	0.23	120.9
Corn, dent, grain, ground	402914	86.3	2.3	30	10.2	7.7	8.0	8.2	4.02	4.33	40.65	1.5	0.02	0.32	—
Corn, sweet, aerial part, fresh	202968	20.2	22.0	64	10.1	6.4	6.4	—	2.88	3.12	—	6.5	—	—	—
Corn, sweet, cannery residue, fresh	202975	76.6	22.5	—	8.9	5.0	5.3	—	3.09	2.72	—	3.3	—	0.90	22.5
Cotton, bolls process residue, S-C	101605	88.5	42.5	—	9.3	0.6	4.9	—	1.90	1.98	—	5.9	0.65	0.12	—
Cotton, gin by-product	108413	90.3	36.7	—	7.4	3.4	3.2	—	1.98	2.16	—	8.5	—	—	—
Cottonseed meal	501617	92.7	11.8	60	44.7	36.2	36.2	—	3.47	3.63	36.23	6.6	0.20	1.18	—
Crab, shells, ground	601664	94.8	13.8	—	13.0	—	—	—	—	—	—	53.2	23.74	2.25	—
Fish, residue meal	501966	94.3	—	—	73.2	—	54.9	—	—	2.70	—	24.0	—	—	—
Fish, condensed solubles	501969	50.1	0.2	—	60.2	—	—	58.1	3.64	3.35	39.22	18.8	0.34	1.64	—
Flax, fiber by-product	102035	90.9	41.6	19	10.8	6.9	5.2	—	2.12	1.38	—	6.9	—	—	—
Flax, seed screenings	402056	91.5	14.2	—	17.3	11.9	13.1	—	2.92	2.83	23.45	7.8	0.40	0.47	—
Garbage, cooked, dehydrated, high fat	407863	95.9	20.9	—	18.2	12.8	14.0	—	3.91	3.53	39.93	13.5	—	0.34	—
Garbage, municipal, cooked, dehydrated, ground	407876	52.7	7.7	—	19.3	13.7	14.9	—	3.53	3.55	39.11	9.2	1.60	0.45	—
Grains, distillers' grains with potato flakes added	402145	88.7	7.0	—	18.1	12.6	13.8	10.9	3.02	3.25	32.17	9.7	—	—	—
Grains, brewers' dried grains	502141	91.0	16.1	47	28.3	21.0	21.0	22.4	2.92	3.06	19.56	4.2	0.30	0.53	—
Grains, brewers' grains, wet	502142	23.8	16.1	—	23.0	16.8	—	—	2.96	3.13	40.26	4.8	0.30	0.51	—
Grains, distillers' grains, dehydrated	502144	92.5	13.8	70	29.6	—	20.4	—	3.64	3.62	42.31	1.7	0.17	0.15	14.0
Grains, distillers' dried grains with solubles	507987	91.2	8.6	—	31.6	—	—	—	3.79	3.89	43.66	4.8	0.29	0.86	—
Grains, distillers' stillage, wet	502148	6.2	8.1	—	30.6	—	—	—	2.96	3.05	33.24	4.8	—	—	—

Feed name	Ref. No.														
Grapes, pulp, dehydrated, ground	102208	90.7	30.3	20	12.7	7.9	1.7	—	1.34	1.20	—	11.7	—	—	—
Grapes, fruit, dehydrated cull	408427	84.8	5.2	—	4.0	-0.2	0.9	—	2.24	1.98	20.81	3.5	—	—	—
Hay-potato, aerial and tubers, ensiled	303792	33.7	17.8	—	11.3	6.5	6.5	—	2.83	2.79	—	5.9	—	—	—
Helianthella, oneflower, aerial part, fresh, mature	208839	21.0	13.6	—	13.0	8.9	9.1	—	2.91	3.02	—	15.8	1.61	0.51	—
Kale, aerial part, fresh	202446	11.6	—	—	20.8	16.8	16.3	—	—	—	—	—	—	—	—
Leek, seeds with hulls	509271	—	—	—	28.0	—	—	—	—	—	—	—	—	—	—
Lentil, common, husks, S-C	102507	87.0	29.1	67	12.6	7.9	1.5	—	2.59	2.67	—	3.6	—	—	226.2
Lettuce, aerial part, fresh	202624	5.4	11.2	—	22.0	16.6	17.5	—	2.26	2.40	—	15.9	0.86	0.46	—
Lettuce, refuse, fresh	202625	—	19.1	—	12.8	8.8	8.9	—	—	—	—	20.8	—	—	—
Oak, acorns	407755	70.7	13.9	—	4.8	0.5	1.7	—	2.09	2.06	—	2.5	—	—	—
Oats, groats by-product	103332	92.2	29.4	38	6.1	3.3	3.6	—	1.38	1.83	11.05	6.4	0.16	0.22	—
Oats, meal	403302	90.8	3.1	80	17.7	12.3	16.0	—	4.40	4.46	40.21	2.5	0.08	0.50	—
Oats, grain screenings	403329	90.2	16.2	18	12.1	7.1	10.6	—	3.89	3.23	26.54	5.5	—	—	—
Okra, seed hulls	103411	91.3	—	—	4.2	0.6	0.3	3.6	—	—	—	—	—	—	—
Olives, fruit without pits, extender unspecified ground	103413	92.0	39.6	-19	6.4	2.5	-0.6	—	1.84	1.37	—	2.7	—	—	—
Onion, aerial part, fresh	203417	91.4	22.6	—	12.6	8.6	8.7	—	—	—	—	8.0	1.80	0.21	20.1
Oysters, meat, raw	503480	18.2	—	—	52.6	—	—	10.6	3.93	3.77	41.97	9.9	0.51	0.83	—
Oysters, shell flour	603481	99.6	—	—	1.0	—	—	—	—	—	—	90.6	38.10	0.07	—
Parsnip, stems, fresh	203535	—	17.2	—	6.0	3.0	2.6	—	—	—	—	3.5	—	—	6.6
Pea, bran	103602	89.8	41.3	83	11.6	7.0	4.8	—	3.36	3.29	36.58	3.9	0.75	0.14	—
Pea, meal	108478	90.0	26.3	—	19.7	14.0	14.2	—	3.72	3.77	—	14.1	—	—	—
Peanut, hulls with skins, ground	103630	94.2	14.1	—	13.6	8.7	8.8	—	2.48	2.34	—	7.6	—	—	—
Peanut, seed screenings	108485	90.6	26.3	—	15.2	10.1	10.2	—	1.93	2.08	—	5.6	—	—	—
Peanut, meal and hulls	503655	92.3	14.0	—	46.9	—	—	—	3.31	3.38	36.20	2.0	0.11	0.65	—
Pears, cannery, residue, fresh	208486	15.2	17.1	—	3.9	1.2	0.7	—	3.06	3.28	—	—	—	—	—
Pineapple, cannery, residue, dehydrated	403722	87.1	19.6	70	4.6	0.7	1.0	—	3.26	3.13	—	3.4	—	—	—
Potato, aerial part, S-C	103762	87.2	25.9	61	12.4	7.7	6.4	—	2.34	2.28	—	18.1	0.24	0.12	—
Potato, tubers, ensiled	303768	24.4	3.0	15	7.1	2.7	3.8	5.3	3.74	3.64	38.86	5.5	—	—	—
Potato, tubers with alfalfa hay, ensiled added	303770	33.7	20.2	-129	13.5	8.5	8.5	—	2.49	2.64	—	6.1	—	—	—
Potato, process residue, dehydrated	403775	88.6	6.9	—	8.7	4.0	2.3	2.3	3.93	3.08	39.46	4.7	—	—	—
Potato, process residue with lime added, dehydrated	408491	88.0	11.8	—	4.1	-0.1	1.0	—	3.86	3.33	31.78	10.8	4.19	0.18	—
Potato, distillers residue, dehydrated	503773	95.7	21.5	61	23.9	—	—	11.7	2.81	2.85	34.39	7.0	—	—	—
Pumpkins, fruit, fresh	203815	8.7	14.2	—	16.2	11.3	12.5	—	3.75	3.78	35.68	8.9	0.24	0.43	—
Rhubarb, leaves, fresh	203918	11.4	8.0	—	27.6	21.4	22.7	—	—	—	—	11.9	—	—	—
Rice bran	403929	90.8	13.7	30	13.8	9.0	9.4	9.8	2.85	3.37	32.28	—	0.08	1.77	190.5
Rice, hulls	108075	92.4	44.5	12	3.1	-0.3	0.2	—	0.48	0.68	—	19.9	0.09	0.08	671.8

TABLE 10.1. (Continued).

Feed Name	Reference No.	Dry Matter (%)	Crude Fiber (%)	Crude Digestibility Coefficient	Crude Protein (%)	Digestible Protein (%) Cattle	Sheep	Swine	Digestible Energy Cattle (Mcal/kg)	Sheep (Mcal/kg)	Swine (kcal/kg)	Ash (%)	Ca (%)	P (%)	Vitamin A Equivalent (international units per gram)
Rice, straw	103925	90.5	35.1	60	4.5	0.2	1.7	–	1.99	1.82	–	17.0	0.21	0.08	–
Rutabaga, aerial part with crown, dehydrated	103995	87.3	9.8	96	11.2	6.6	8.7	–	2.43	2.34	–	39.4	–	–	–
Rutabaga, aerial part, ensiled	303999	14.7	12.3	85	16.4	11.1	13.0	–	2.61	2.66	–	30.8	–	–	–
Rye, brown, dry milled, dehydrated	404022	90.7	7.6	23	17.5	14.7	11.7	12.2	3.76	2.70	30.51	5.2	–	–	–
Rye, distillers' stillage, wet	208498	5.9	8.5	–	32.2	–	–	–	2.66	2.78	29.10	5.1	–	–	–
Scrghum, distillers' grains, dehydrated	508512	94.0	14.8	–	29.8	–	–	–	3.42	3.62	40.27	5.0	0.16	0.82	–
Scrghum, gluten feed	508086	90.7	7.2	–	27.5	–	–	–	3.48	3.62	40.07	8.0	0.10	0.65	–
Soybean, hulls	104560	91.6	36.1	72	12.4	7.6	6.0	–	2.83	3.09	–	4.2	0.59	0.17	–
Soybean, pods	104564	–	–	–	–	–	–	–	–	–	–	–	0.99	0.20	–
Soybean, flour by-product, ground	504594	89.4	36.9	–	14.9	17.0	18.0	–	2.14	2.19	–	5.0	0.42	0.21	–
Spinach, stems, fresh	204665	6.5	9.3	–	22.5	-1.5	-1.8	–	–	–	–	3.1	–	–	–
Sugarcane, cane bagasse, dried	104686	91.5	48.6	–	1.7	3.8	3.7	–	1.24	2.06	–	6.0	–	–	–
Sugarcane, aerial part, fresh, mature	204687	27.2	32.0	–	7.3	5.9	7.1	–	2.56	2.66	–	–	–	–	–
Sugarcane, molasses	404695	93.5	2.9	–	10.7	–	–	–	3.76	3.37	32.05	14.4	1.23	0.15	–
Tomato, leaves with stems, fresh	205039	–	15.4	–	26.4	20.3	21.6	–	–	–	–	26.6	–	–	–
Tomato, pulp, dehydrated	505041	91.9	26.3	–	23.9	13.5	–	–	3.22	2.83	–	6.5	0.44	0.59	–
Tucum, oil meal	405057	91.3	11.5	–	12.8	7.8	9.0	–	–	–	–	3.0	–	–	–
Vanilla, hulls	105071	26.3	–	–	10.0	5.6	5.5	–	–	–	–	6.0	1.90	0.07	–
Watermelon, seeds with hulls	509131	–	–	–	38.0	–	–	–	–	–	–	–	–	–	–
Wheat, bran	405191	88.1	11.3	36	17.2	13.1	12.9	–	3.09	2.92	26.70	6.7	0.15	1.38	–
Wheat, endosperm	405197	87.9	0.3	–	12.6	7.6	8.8	–	3.01	3.80	42.85	1.4	–	–	–

Source: Atlas of Nutritional Data on U.S. and Canadian Feeds. National Academy of Science. 1971.

ECONOMIC VALUE AS DETERMINED BY NUTRITIVE VALUE

A livestock feeder may often overestimate the economic value of a cannery waste material which is available without a direct cost. To emphasize the decrease in the actual economic value (nutritional value in a ration) of waste materials necessitating handling and transportation for different distances, some information from a study by Wilson and Borger (1969) is presented in Table 10.2. The costs of transporting several different types of waste materials available in Pennsylvania were calculated, and finishing rations for steer—containing each type of waste and requiring different amounts of conventional feeds—were formulated. Monetary values of the waste materials were determined from the values of the conventional feeds which were replaced by the waste materials.

It should be emphasized that costs of transportation and handling were determined with farm-sized trucks and farm labor for loading and unloading. The economic values of each waste at the plant and at distances of 5 and 50 miles from the plant are presented in comparison with whole plant, field harvested, corn silage, which has been assigned a relative value of 100. The costs obtained by Wilson and Borger (1969) were increased in Table 10.2 by 20% to compensate for some of the increased costs of handling and transportation which has occurred since their study was conducted.

The primary determinants of the relative values of the waste materials are their energy and protein contents. The values of even the standard feedstuffs (corn silage) will vary widely, depending on corn grain or standard forage price, crop

TABLE 10.2.   Relative Value of Certain Waste Materials and Standard Feeds at Different Distances from the Processing Plant.

Feed	Dry Matter (%)	Relative Value—Distance from Plant		
		At Plant	5 Miles	50 Miles
Corn silage	27.6	100	66	51
Citrus pulp	22.8	127	87	72
Citrus pulp	90.0	501	385	371
Corn cannery wastes	22.4	93	61	45
Grape wastes	30.0	41	19	4
Apple wastes	30.0	115	78	63
Tomato wastes	11.1	45	22	7
Pea cannery wastes	24.5	86	55	40
Potato wastes	20.0	72	66	29

Source: Wilson and Borger. 1969.
NOTE: Value of field harvested whole plant corn silage equals 100%. Trucking and handling by conventional farm trucks and systems. Costs of transporting and handling assumed 20% greater than in original study.

growing conditions, and other factors. However, it is interesting to note that according to the calculations of Wilson and Borger (1969), corn silage can be transported for 50 miles and still have approximately one-half the value that it had at the harvest site.

Two different cost comparisons are given for citrus pulp: one at 22.8% dry matter, and the dry pulp containing 90% dry matter. The cost for transporting citrus pulp would be much less in large commercial operations than the costs presented in Table 10.2, in which farm-scale collection and transportation methods were used. The relative economic values of wastes with lower nutritive values, such as grape and tomato wastes, decline quite rapidly as distance from the plant is increased. (The feeding characteristics of tomato, apple, and potato wastes are discussed individually in more detail later.) It should be emphasized that any change from the percentage dry matter of wastes given in Table 10.2 necessitates re-calculation of the relative values. As an example, the dry matter content of tomato pomace has recently been increased from 30 to 35% in some of the United States processing plants, by improved water separation methods. This would essentially double the relative economic value of tomato wastes on an as-received basis. Therefore, although the nutritive value of the dry matter contained in the waste material is very important, the percentage dry matter content is of great importance in determining values at different distances from the plant.

INDIVIDUAL WASTE MATERIALS

It is not possible to discuss in detail all the research detailing feeding trials which have been conducted to determine forage and grain replacement values for different types of cannery wastes. However, some of the wastes which have proved to be most efficient in livestock production are discussed below. The composition of several different types of waste materials from food processing or related operations are presented in Table 10.1 (N.A.S., 1971). The composition of several conventional feeds are also presented for comparison. Additional analyses of other conventional or waste products can be found in the *Atlas of Nutritional Data on U.S. and Canadian Feeds* (N.A.S., 1971).

Citrus Pulp

Probably the food processing waste which has been most successfully recycled and merchandized for livestock is dried citrus pulp. Much of the original research with citrus pulp was conducted at the Florida Agricultural Experiment Station, Gainesville, Florida, U.S.A., in the late 1950's to mid-1960's. Peacock and Kirk (1959) conducted three drylot feeding trials comparing citrus pulp, corn meal, and ground ear corn (140 day trial) that indicated there was no significant difference in weight gain, energy requirement per 100 pounds of gain, or carcass characteristics between steer fed citrus pulp or other rations composed of standard feeds. The different rations consisted of 70 parts of either citrus pulp or the two types of corn, plus 25 parts cotton seed meal and 5 parts alfalfa. Several other workers (Baker, 1953, 1955; Chapman *et al*, 1953) have also reported essentially equal replacement values of citrus pulp for corn grain, or for ground ear corn with a comparable corn grain content. In Florida, as well as in several other parts of the United States and other countries, tropical pastures produce relatively large amounts of forage, but these pastures become rather low in energy density with advanced maturity. Therefore, since corn grain production in these areas is not usually highly successful, the availability of a high energy feed such as citrus pulp is quite important in the total animal production scheme.

Citrus pulp is dehydrated and shipped throughout the United States, particularly in the midwestern and eastern states, and is utilized primarily in dairy rations. Citrus molasses is also a valuable feed in cattle finishing and dairy rations in many areas. It is not only a high energy feed, but also usually increases feed consumption, thereby resulting generally in a higher performance rate and increased feed efficiency. With molasses, as with citrus pulp, a high energy feed can be made available for inclusion in livestock and dairy rations to complement a stable, predictable, but not usually high quality tropical forage.

Ammerman *et al* (1965) evaluated different processing methods, including dehydration temperatures, which varied from plant to plant with regard to the effect on the nutritional value of citrus pulp. Increases in drying temperature from 104 to 127°C generally decreased digestion coefficients for both protein and energy. However, even at the higher temperature, digestion coefficients for protein and energy were 77.6 and 80.1%, respectively. The effect of high temperature on the proportion of heat damaged protein, or that protein which is less available even to ruminant animals, was summarized by Goering *et al* (1972). The research of Ammerman *et al* (1965) and Goering *et al* (1972) emphasized the need for periodic analysis of waste materials processed by different plants or by different methods.

Potato Processing Wastes

Potato processing wastes are becoming more important as livestock feeds for several reasons, including the fact that a larger proportion of potatoes used for human consumption is being processed prior to retail sale, and the wastes from potato peeling and processing, particularly with caustic peeling methods, have a high B.O.D. (which results in relatively high sewage costs for the producer). Therefore, potato processors are making increased efforts to recycle potato wastes through livestock.

Many potato processors are changing from the "wet" caustic processing method, which usually results in wastes of pH 9.0 to 13.0 when large amounts of caustic are used late in the potato processing season, to a dry peeling process, which requires less alkali and water. This is advantageous even if the processor does not use the waste for livestock feed. The dry peeling process generally has a lower sewage cost, and, if utilized, the potato wastes can be transported more economically to agricultural land for use as feed or fertilizer. The high pH of potato processing wastes makes it a desirable fertilizer, particularly in areas of acid soils.

Because of the many different methods of processing potatoes, different potential feeds may result. Only a few of these, including ensiled whole tubers, vines, and stems are included in Table 10.1. In some of the potato producing areas, where excess or spoiled potatoes are frequently available, potatoes are mixed with alfalfa and other types of hay and ensiled. Most of the resulting waste products are in protein, particularly the tuber portion, without added protein sources such as legume hays or non-protein nitrogen. However, potato wastes generally have quite a high feeding value, even if there have been added alkalis. The dry peeling waste material usually has a similar or slightly lower pH. The use of heat and concentrated alkali in the peeling process causes the starch to gelatinize after several days of storage. The high pH and dryness of the waste peel inhibits bacterial growth and fermentation, a situation not found in the wet peeling wastes, since there is ordinarily enough solid potato particles in the wet peel wastes to permit bacterial growth and organic acid production to neutralize the alkali. However, if the dry peeling waste is mixed with cull potatoes or other fresh materials, the formation of the starch gel is prevented and the total mixed waste can be handled in a normal fashion. Dry peeling wastes may be stored as obtained without expensive storage facilities, since the lack of bacterial growth allows preservation of starches for long periods of time (Howes and Sauter, 1975).

Potato screenings and filter cake are usually higher in energy and protein than are other types of potato wastes. Screenings consist of portions of peeling, discarded potatoes, and other potato rejects. The use of screens allows processing plants to reduce the amount of dissolved and suspended solids which must be removed from the effluent water. Generally, most plants pump waste water to a central location where large particles are screened, and the resulting waste water is transferred to a central settling or clarifier tank where much of the remaining suspended solids settle out as a sludge. Subsequently, this sludge is augered or pumped to a vacuum drum filter, where water is removed and filter cake produced. Filter cake has a dry matter content of 12 to 16%, depending on the pH. Outside the pH range of 6 to 8, the filter cake will have a higher amount of adsorbed water. A typical analysis of filter cake from potato wastes is 7.6% crude protein, 4.9% crude fiber; it is similar to ensiled potato tubers with respect to energy content and mineral profile.

Research reported by Hinman and Sauter (1978) on the effect of ensiling filter cake in open pits is interesting, because their results demonstrate the decrease in starch concentration with fermentation. As mentioned previously, ensiling is usually an effective, low-cost method of handling wastes, particularly if conventional feedstuffs are to be added to improve the nutritional and storage characteristics. In the ensiling of filter cake, which is high in starch (593 mg/g), starches were broken down by enzymic and microbial action. From an initial pH of 6.5, after three weeks of storage, the pH decreased to 3.5 and approximately 53% of the initial starch had disappeared as a result of fermentation. In addition, the microflora of fresh filter cake was found to be highly variable, with bacterial counts ranging from 10 to 100 million per gram, generally divided about equally between aerobic and anaerobic forms. Certain species of bacteria and fungi produce toxins during growth which can cause problems in cattle fed filter cake. Acidosis in cattle can also result from the acidity caused by bacterial growth during storage of filter cake. For these reasons, filter cake can be used most efficiently on a fresh basis with dry ingredients of the ration, ensiling with other feedstuffs being a less desirable feeding and storage method.

Apple Processing Wastes

It is interesting to trace the history of apple processing waste utilization through beef cattle in the United States. In the early 1960's (Bovard *et al*, 1961), DDT and other chlorinated hydrocarbons were used extensively for the control of many pests in most orchard crops. These persistent pesticides were also deposited in the soil of the orchards and were detectable several years after the use of DDT had been suspended. Because of these sprayings, DDT residues appeared in the processing wastes. According to Wilson *et al* (1970), residues of chlorinated hydrocarbons remained in the soil, even five years after orchards had last been exposed to these materials, and residues were deposited in the fat of animals eating the resulting processing wastes. However, Wilson *et al* (1971) found that finishing steer receiving up to 60% of their total ration dry matter from apple pomace averaged only slightly greater levels of chlorinated hydrocarbons in fat tissues than did the control group during a 112 day trial. Although chlorinated hydrocarbons are generally no longer a concern in recycling horticultural and vegetable waste materials, there are continuing introductions of other types of spray materials and chemicals. These materials include drop preventatives and pesticides, which, although they are less persistent in the environment and are biodegradable, still need to be tested for safety in waste recycling systems. In Pennsylvania and Virginia, essentially all the apple processing wastes are presently used in ruminant rations, almost entirely in beef brood cow and steer growing and finishing rations.

Bovard *et al* (1977) reported that ensiled apple processing waste used together with a non-protein nitrogen source, such as biuret or urea, apparently causes abortions and birth abnormalities in calves. The exact mechanisms for these abnormalities have not been discovered, although the same observations have been made in private commercial herds exposed to a combination of NPN and apple waste feeds. Some probable causes are mycotoxins or similar materials, which may interfere with trace mineral utilization, or which act directly as a poisonous substance within the body.

As indicated in Table 10.1, the energy value of apple processing wastes is quite high, although the protein content suggests that protein supplementation is required for most types of animals.

Apple pomace is seldom dehydrated because of the cost of fossil fuels; it is generally fed fresh from the factory or is ensiled. In most cases, rice hulls are added on an approximate 1:13 proportion of rice hulls to apple processing wastes (on a volume basis). The rice hulls are added to increase the amount of liquids pressed from the apple waste before the waste leaves the processing plant. Addition of rice hulls does not appear to reduce the feeding value, although rice hulls are essentially an indigestible portion of the waste material. There were indications of digestive impaction in a feeding trial conducted by Wilson *et al* (1971) in which 70% of the total dry matter of the ration consisted of the rice hull/apple waste product. However, this impaction decreased weight gain and feed efficiency only slightly compared to the gain and efficiency calculated from the energy values of the materials.

Cassava

Cassava (manioc) is widely grown in tropical areas, with the principle food value contained in the tuber. Cassava is not only used widely for human food, but also for livestock, particularly swine. Since the area in which cassava is grown and processed is increasing, it will be discussed briefly. Cassava contains a cyanogenic glucoside and accompanying enzymes which can liberate hydrocyanic acid, which is toxic to animals and man. The concentration of these toxic substances is higher in bitter (0.27%) than in sweet cassava (0.007%). Cassava should be dried or ensiled prior to feeding swine to reduce the toxicity. However, for both types of cassava, the concentration of cyanogenic glucoside varies with growing conditions, moisture, temperature, and plant age (Castillo *et al*, 1964).

Cassava meal has been used in swine rations to replace 20 to 40% of the corn grain in the ration. Cassava silage has also been used to replace 25 to 40% of the corn in the ration, producing similar growth and feed efficiency responses. The nutritive values of several different types of cassava products and waste materials are presented in Table 10.1.

Pineapple Wastes

Pineapple wastes include the dried residue from the canning of pineapple: slips, suckers, immature fruits, stumps, and some dead leaves. It is generally a bulky material and as a result has been pelleted for improved handling. It is usually considered as a roughage source for ruminants. Pineapple bran was found to be highly digestible and palatable in Hawaiian feed trials (Otagaki *et al*, 1961). When included in soybean meal/barley-coconut meal rations at the rate of 30%, it had no significant effect on milk or total energy production.

Dronawat *et al* (1966) concluded that pelleting of pineapple hay may reduce its feeding value, although the addition of loose or pelleted pineapple hay increased the percentage of milk fat in the Hawaiian studies—probably a reflection of the increased roughage content of the ration, which reportedly increases the percentage of milk fat in high-concentrate rations.

Rice Wastes

A wide variety of by-products are available from the production and milling of rice. Rice straw is probably the least palatable of all the small grain straws, but can be used in relatively low proportions in a ration as a roughage source (Lansburg, 1958). Rice bran is the pericarp or outer covering of the rice seed, and has a relatively high fat content. Because of the high fat content, it may become rancid quite quickly. According to Arnott and Lim (1966) the development of rancidity can be negated by heating rice brans to destroy the lipolytic enzymes responsible for the breakdown of the fat. Rice bran oils produce soft pork fat when fed at moderate levels to hogs. However, both rice bran and rice polish are relatively high in B vitamins and niacin. Rice hulls, as mentioned in the discussion of the feeding of apple wastes, have very low feeding value and they irritate the digestive tract or cause impaction if fed at moderate to high levels.

Cotton By-products

One tonne of cotton seed yields about 48% cotton seed cake, 26% hulls, and 15% oil, and the remaining 11% is composed of fiber and shrinks during ginning. Although usually high in phosphorus, cotton seed products are low in lysine and vitamins A and D. Dairy cattle fed cotton seed products produce milk with a higher fat content, since cotton seed oil consists of saturated fats. Raw cotton seed contains gossypol, a compound which is toxic to swine, poultry, and young cattle (Morrison, 1959). However, adult cattle have been fed as much as 4.5 kg cotton seed meal daily for three years without injury. Fortunately, free toxic gossypol is changed to a bound form (*d*-gossypol) during the heating process of oil manufacture, and therefore the primary concern is in the feeding of unprocessed cotton seed. Because of the high fat content, quantities which can be consumed, particularly by the ruminant, are usually 20% of the total dry matter content of the ration. For dairy cattle, 77 to 91 kg of cotton seed meal are equivalent to 45 kg of cotton seed. In beef cattle feeding trials, 47 to 64 kg of cotton seed was found to be equivalent to 45 kg of cotton seed meal, when the animals were fed 2.7 kg per day of whole cotton seed. Although cotton seed hulls are rather palatable for this type of product, they are low in most of the required minerals and contain only about 43% TDN. Generally, cotton seed hulls are worth slightly less in feeding value than is sudan grass or sorghum hay.

Sugarcane By-products

There are several different wastes associated with the production of cane sugar, including cane tops, molasses, and bagasse. Sugar cane bagasse is the residue remaining after as much juice as possible has been pressed from the crushed stalks. Bagasse is usually used as a fuel source for the operation of the sugar extraction plants, and with the increased cost of fossil fuels it is becoming even more valuable as a fuel. Bagasse is low in protein; in fact, the digestion coefficients for cattle and sheep have a negative value for protein. It is also high in fiber, as indicated in Table 10.1, and therefore is primarily used as a roughage replacement or extender in conventional rations. With the reduced availability of corn grains and the increased use of forage in livestock rations, roughage replacements such as bagasse are not as important as they were several years ago in most areas. Randel (1966) has suggested that it takes more energy to digest bagasse than is obtained from it by the animal. In addition, weight gain and milk production in cattle is usually reduced when bagasse is used, and it has been found to be inferior to rice straw and cotton seed hulls when fed at 62% of the ration (Brown, 1954).

Harvesting sugarcane (strip cane) produces waste materials that include green cane tops, leaf sheaths, and miscellaneous dry cane trash. Percentage crude fiber in the strip cane is still rather high (44.3%), but it has been found to be a satisfactory substitute for Napier grass in Hawaiian trials with dairy cows.

Bagasse has been incorporated into chicken litter to increase the protein equivalent of the bagasse. This results in a slightly more palatable product. Molasses is available worldwide and serves as a readily available source of energy as well as increasing the palatability of many different types of feed. It is low in digestible protein, but is particularly useful in short-term seasonal supplementation in areas where sugarcane is processed and the molasses is available at a relatively reasonable rate compared to other concentrate sources. In most areas, continuous supplementation with molasses is not economical, although one of the most convenient methods of providing protein supplements is by using liquid molasses as a carrier for the protein supplement, and feeding on a free choice basis or as a mechanized method of adding liquid to a dry feed during mixing.

Tomato Processing Wastes

Tomato waste (frequently called pomace) is a mixture of tomato skins, pulp, and crushed seeds resulting from tomato juice, catsup, or the processing of other tomato products. Tomato pomace is high in thiamine, contains moderate levels of riboflavin, and has a considerable carotene content. Morrison (1959) indicated that tomato pomace was chiefly used in special feeds such as dog food in which it acts as a paragoric (tending to constipate), resulting in drier feces. As indicated in Table 10.1, tomato pomace has a rather high crude protein content (23.9%). Chapman and Haines (1957) reported on a trial in which dry tomato pulp was included in concentrate rations containing 0, 10, 20, 30, and 40% dehydrated tomato pulp fed *ad libitum* to animals on pasture. Average daily weight gains of the five groups were 1.26, 1.39, 1.06, 1.24 and 1.16 kg/day, respectively. The feeding of tomato pulp did not affect carcass grades, or lean and fat tissue characteristics. Tomato pomace is ordinarily not palatable when fed fresh, although palatability increases with drying. In a study by Hampden *et al* (1977), tomato processing wastes ensiled favorably when combined with hay and corn grain in a ratio of 65 : 17.5 : 17.5 of waste to hay to grain. It was also of a consistency which could be handled mechanically and was fairly acceptable in a lamb metabolism trial. Ordinarily, tomato waste is high in water content, and in these cases may be handled successfully as a slurry (much like potato slurry) but is most successfully used by ensiling with conventional feedstuffs.

DISPOSAL OF WASTES THROUGH THE SOIL

Many solid cannery wastes have traditionally been disposed of via soil application, which adds valuable organic matter and inorganic elements to the soil. In many areas where soils lack organic matter, fertility can be increased dramatically through the application of wastes to soil. However, recently there has been more attention given to disposal of effluents through the soil, which not only provides moisture needed for growing crops, but also adds to the soil dissolved nutrients or solids contained in the effluent. It should be emphasized that if a processor wishes to improve the value and usefulness of solid wastes for recycling through livestock, one of the first priorities is to reduce the water content in the waste material by as much as is practical and economic. Although this usually results in a higher quality of solid waste for recycling, it can also markedly increase the amount of liquid effluent which must be disposed of from the plant. The maximizing of the in-plant de-watering of wastes has the additional advantage that more water is available for recycling within the plant, thereby decreasing water and sewage costs.

In comparing the economic value of most of the solid waste materials from cannery processing as fertilizers or livestock feeds, most of the wastes have more value as feeds. The wastes are rather low in nitrogen and therefore do not contribute significant amounts of nitrogen to the soil. Although most of the wastes are also low in protein when considered as feedstuffs, the protein is usually of relatively high quality, particularly important in the feeding of non-ruminant animals. Probably the greatest value of wastes which have low nitrogen and feed energy content is in adding organic material or 'mulch' to soils low in organic matter. Some of these waste materials are sugarcane bagasse, rice hulls, fibrous flax by-products, and other similar wastes. A further consideration is that many of these same fibrous materials have an extremely high ash content (either from the material itself or from in-plant addition of soil, dirt, etc.). These may have significant value as soil additives but may be too high in ash to be used successfully in livestock rations.

Liquid Wastes

Ratios of biological oxygen demand to chemical oxygen demand (B.O.D./C.O.D.) vary widely between plants and between different processed products. In 1965, liquid waste from cannery plants was reported to have ten times the B.O.D. of ordinary domestic sewage (Rhoads, 1965). In most cases where effluents are discharged into a municipal sewage system, the charge for handling by the municipal system is based upon the volume of waste in terms of pounds of B.O.D. in excess of that in normal sewage, and the proportion of suspended solids in excess of normal sewage. During the past decade many systems have been used by food processors to reduce the B.O.D. of effluents from plants, to reduce sewage costs, to conserve water, and to make available solid wastes for other purposes.

In addition to reducing the total amount of waste discharge from a plant, primarily by in-plant recycling, the processor can pretreat the effluents in an attempt to reduce the B.O.D. and suspended solids content. Biological filters using rock, gravel, cinder, sand, or synthetic materials have been successfully used. Although filters have been used in many plants, the high capital investment has been a deterring factor to some extent. Another method of in-plant treatment is chemical precipitation, usually consisting of chemical coagulants of lime followed by either ferrous sulfate or alum. However, most systems can only reduce the B.O.D. in most vegetable wastes by 50%, and by 25% in fruit wastes. Trickling filter towers, physical screening, and chemical treatments (either batch or continuous flow systems) have also been used in cannery plants to treat the waste products produced. The resulting sludge has also been used in livestock feeding, before or after drying. Drying of the sludge is seldom mechanical but is usually an open air type of drying.

Although lagoons in the United States have been replaced by more sophisticated— and more costly— methods, food processing wastes are held in storage lagoons in many parts of the world. This allows partial or complete decomposition of the wastes, after which the waters may be discharged to a septic system or used for irrigation. However, for small or even large plants, where sufficient well drained land is available, the effluents can be taken directly to the land through irrigation systems, without prior lagooning.

A series of de-watering investigations were conducted with solid bowl basket centrifuges (Esbelt, 1976). Fruit processing waste and activated sludge treatment system products were de-watered and the biological sludge solids used as cattle feeds. Different rations were designed to contain 2.3, 4.6, and 8.9% sludge solids on a dry weight basis. The biological solids contained 39.1% crude protein and 3.2% crude fiber. There were no significant differences in weight gain, feed efficiency, or carcass quality, and no apparent harmful effects from the cattle being fed biological solids.

In a long-term project at Pennsylvania State University, Sopper (1978) concluded that in the application of effluent from municipal sewage systems applied to widely differing plant species (from trees to forage crops), the soil acts as an effective "living" filter. The filtering action of the soil retains nutrients for plant growth and these are immediately available for plant growth. Application of sewage effluents over a ten year period has resulted in essentially no harmful concentrations of such potentially dangerous elements as cadmium, lead, mercury, nitrate-nitrogen, phosphates, and other compounds (Sopper, 1976; Sidle and Sopper, 1976). The primary objection to the deposition of effluents on fields used for pasture, hay, or feed crops is that heavy metal contaminants can be concentrated in animal tissues, representing a possible health hazard from the consumption of the resulting animal products. Naturally, the same concern may be expressed in relation to the deposition of heavy metals in vegetable crops grown on the effluent treated soils and intended for direct human consumption.

The types of land and crops which might benefit from irrigation with processing plant effluents range from forage crops to woodlands and include grain crops such as corn and oats. However, perennial forage crops are most frequently treated with effluents. In practically every part of the world, yields of forages from perennial grasses fluctuate with the season of the year and the amount of rainfall. Using forage crops as an example, yields are generally more predictable in the spring than in the summer because soil moisture and temperature in the spring are usually more favorable for plant growth. In many areas, the size of cattle herds is determined primarily by the amount and availability of summer forage. Overstocking of pasture in summer leads to overgrazing with generally reduced weight gains per animal production per hectare, and can severely damage grass stands. As an example of the dependence of cattle weight gains on the season of the year, Bryant et al (1965) and Wright et al (1968) found that average daily gains were nearly always higher in the spring than in the summer. Autumn weight gains were also appreciably greater in areas of increased autumn rainfall compared to the amount of summer rainfall. An advantage of using effluents as irrigation for forage crops is that the removal of grass extends the life of the disposal system and the grass stands, because nitrogen, phosphorus, and other nutrients may accumulate in the soil if the forage is not removed (Adriano et al, 1975). In Delaware, yields of mixed orchard grass (Dactylas glomerata L.) and tall fescue (*Festuca arundinacea Schreb.*) were increased through the use of irrigated effluents. An average of 58.6 millimeters of effluent (not including 22.4 millimeters of rainfall) was supplied weekly from May to November. Spray irrigation with effluent increased dry matter yield by 113% (7737 kg compared to 3629 kg in the control) in the two year trial, compared to non-irrigated controls. There was essentially no difference in the micro- and macro-mineral element composition, with the exception of a slight increase in sodium and decreased phosphorus contents in the effluent-irrigated forages (Jung *et al*, 1978).

Solid Wastes

Until recently, landfills have been the favored method of disposal through the soil, but more restrictive environmental regulations and decreasing availability of approved sites have decreased the use of landfills as a means of disposal. An example of the changes from landfills to thin-layer spreading and soil incorporation was reported by Reed *et al* (1973). The process of soil incorporation is a combination of aerobic and anaerobic degradation of waste materials following spreading and soil incorporation via disking. A 930 hectare site, consisting of flat, westerly sloping, poorly drained, basin clay soil of marginal agricultural value was utilized for the study. During the processing season, 375 tonnes of wet cannery waste were applied per hectare and spread by large trucks. As pointed out by these researchers, the soil must be of a consistency to allow disking, and some sites would not be suitable because of a hard non-workable surface.

The key elements of successful management of this type of waste disposal system were reported by Reed *et al* (1973) as being:

1. Development of a canning industry organization to manage the waste disposal program on a continuous basis.

2. A study of the recycling site requirements and choosing a site which would not be classified as having public nuisance or health hazards.

3. Conscientious on-site management by a knowledgeable manager to maintain health department sanitation standards.

4. Recognition of the need to study and research the long term potential benefits and hazards of this method of recycling on a particular site.

The amounts of nitrogen, phosphorus, potassium, sodium, calcium, and magnesium contained in the various waste mixtures used by Reed *et al* (1973) are presented in Table 10.3. Naturally, the percentage of nitrogen present is not sufficient to provide a large amount of nitrogen for non-leguminous forage plants, although at the rates of application used in this study, the concentration of nitrogen should stimulate additional plant growth. The addition of 1760 tonnes of wet waste material, or 175 tonnes of dry matter, should add approximately 78 kg of nitrogen per hectare. This is a significant amount of nitrogen for

TABLE 10.3. Analysis of Fruit Waste for Soil Recycling.[1]

Material[2]	Water Content[3] (%)	Total N(%)	Total P(%)	Total S(%)	Soluble Rations (ppm)			
					Na	Ca	Mg	K
Peach-pear	86.2	0.98	0.14	0.08	16.9	33.8	72.6	82.3
Tomato mixed	89.9	1.84	0.31	0.22	13.7	20.4	23.6	125.3
Peach-pear	88.7	0.82	0.13	0.17	8.3	15.3	11.3	94.4
Peach	85.4	0.61	0.10	0.07	10.5	16.5	51.7	89.2
Mixed fruit	89.4	1.00	0.15	0.15	20.0	28.2	15.8	67.9
Peach-pear	82.7	0.60	0.09	0.09	11.3	20.1	79.7	77.3
Peach	84.8	0.89	0.13	0.06	7.2	16.3	18.3	105.3
Pear	87.2	1.37	0.18	0.37	23.6	32.4	14.4	120.6
Peach-pear	89.0	0.91	0.09	0.03	23.2	9.1	13.7	99.2
Mixed fruit	88.4	0.70	0.11	0.07	37.2	34.3	16.4	65.8
Mean	87.1	0.97	0.14	0.13	17.18	22.64	31.74	92.72

[1] Each figure represents a mean of 9 analyses (3 samples × 3 replications). Reed *et al* 1973.
[2] Mixed fruit samples were peach, pear, plum, grape, and cherries.
[3] Calculated on a fresh weight basis; all other data on a dry weight basis.

non-leguminous crops. The economic value of the inorganic matter contained in the waste material should be determined and considered as partially offsetting the costs of disposal. It is more difficult to assign a value to the organic material from a soil improvement point of view except to determine increases in the value of the crop produced and the associated land appreciation.

One of the disadvantages of ground spreading and incorporation is the possibility of an increasing vermin population. However, the combination of deep disking (approximately 20 centimeters) and thin spreading (10 centimeters or less) should prevent this problem from arising.

It was concluded by Reed *et al* (1973) that the total cost of this method of disposal was quite competitive with other methods of disposal, including landfills. The investigators observed that under the specific conditions of the recycling site used, there appeared to be no adverse environmental effects. Health regulatory officials monitoring the disposal site reported no fly problems or other health hazards. The compactness of the soil (clay), consisting of several layers of subsoils, restricted downward water percolation and prevented ground water pollution. The primary benefit is the improvement in soil structure in the surface soil (up to 30 centimeters), which increased the removal of salt and sodium ions from the surface soil and the upper subsoil. Although data over a long-term period on the cumulative effects of soil incorporation of cannery waste on plant growth are not available, studies, including application rates of 0 to 1930 tonnes of wet tomato waste per hectare and from 0 to 3505 tonnes of mixed fruit waste per hectare, have been conducted. The results for the different rates of application of tomato wastes showed a significant curvilinear response with a decline in plant material yields at the higher rates of application. The maximum growth response from tomato waste occurred at an application rate of 1380 tonnes/ha (Reed *et al*, 1973).

Composting

Traditionally, composting has been a major waste handling and disposal process in Europe and several other countries. Basically, composting is the mechanical aeration of waste material with the objective of decreasing the bulk and improving the fertilizer value of the waste material. Little research has been done using composted wastes as livestock feeds or on the composting of cannery wastes. However, the wide variety of materials which have been composted and then recycled through livestock or used as fertilizers suggests that composting may be an effective treatment and storage method for these wastes. The primary use of composted waste materials has been as fertilizer, but the value to society of using these materials as feeds of high energy and protein content may be greater than disposal through the soil.

In many parts of Europe, composted materials have been traditionally used to reclaim sandy, marginal soil, as well as to assist in the control of erosion on steep slopes. Naturally, composting has wider uses than simply as a means of treating cannery wastes. The philosophy of combining wastes of widely differing origins into one large composting program, whether for feeding to livestock or for soil disposal, has generally not been practised in the United States. As an example of the value of compost on steep vineyard slopes in Germany, Hart (1968) reported that 198 tonnes of compost per hectare reduced the volume of runoff per hectare from approximately 49,000 to 28,000 liters. With 400 tonnes of compost per hectare, runoff from the same type of slope was approximately 1870 l/ha. Similar comparisons were obtained when cubic meters of soil lost per hectare per year was measured.

Several different types of composting systems were described by Hart (1968) and practically all of these systems used municipal garbage and other waste materials from municipal sources, with essentially all the compost produced being used by home owners for gardens, flower pots, shrubbery, etc. Using the procedures described by Hart (1968), it should be possible to compost cannery and food processing waste should this be required, the final product providing an organic "mulch" for agricultural land.

REFERENCES

Adriano, D. C., L. T. Novak, A. E. Erikson, A. R. Wolcott, and B. G. Ellis. 1975. "Effect of long term land disposal by spray irrigation of food processing wastes on some chemical properties of the soil and sub-surface water." *J. Environ. Qual.* **4**:242–248.

Ammerman, C. B., R. Hendrikson, G. M. Hall, J. F. Easley, and P. E. Loggins. 1965. "The nutritive value of various fractions of citrus pulp and the effect of drying temperature on the nutritive value of citrus pulp." *Proc. Fla. State Hort. Soc.* **78**:307–310.

A.O.A.C. 1975. *Official Methods of Analysis.* Association of Official Analytical Chemists, Washington, D.C.

Arnott, G. W. and H. H. K. Lim. 1966. "Animal feedingstuffs in Malaysia. 2. Quality of rice bran and polishings." *Malaysian Agric. J.* **45**:387–403.

Baker, F. S. 1953. "Citrus molasses, dried citrus pulp, citrus meal and black strap molasses in steer fattening rations." *N. Fla. Exp. Sta. Mimeo Rept.* **55**.

Baker, F. S. 1955. "Steer fattening trials in north Florida." *Fla. Agr. Exp. Sta. Circ.* **S-89**.

Banse, H. J. and Strauch. 1966. "The importance of prefermentation in composting." *Compost Sci.* **7**:17–23.

Bovard, K. P., B. M. Priode, G. E. Whitmore, and A. J. Ackerman. 1961. "DDT residues in the internal fat of cattle fed contaminated apple pomace." *J. Anim. Sci.* **20**:824–826.

Bovard, K. P., T. S. Rumsey, R. R. Oltjen, J. P. Fontenot, and B. M. Priode. 1977. "Supplementation of apple pomace with non-protein nitrogen for gestating beef cows 11. Skeletal abnormalities of calves." *J. Anim. Sci.* **45**:523–531.

Brown, P. B. 1954. "Methods of supplementing beef cattle rations with urea and ammoniated molasses." *S. Agric. Workers Proc.* **51**:64.

Bryant, H. T., R. D. Hammes, R. E. Blaser, and J. P. Fontenot. 1965. "Effect of stocking pressure on animal and acre output." *Agron. J.* **57**:273–276.

Castillo, L. S., F. B. Aglibut, T. A. Javier, A. L. Gerpacio, G. V. Garcia, R. B. Ruyaoan, and B. B. Ramin. 1964. "Camote (1 promotea batalas) and cassava tuber silage as replacements for corn in swine growing-fattening rations." *Philippine Agric.* **47**:460–474.

Chapman, H. L., R. W. Kidder, and S. W. Plank. 1953. "Comparative feeding value of citrus molasses, cane molasses, ground snap corn and dried citrus pulp for fattening steers on pasture." *Fla. Agr. Exp. Sta. Bull.* **531**.

Chapman, H. L. and C. E. Haines. 1957. "Feeding value of dried tomato pulp and dried celery tops for fattening steers." *J. Anim. Sci.* **18**:1530.

Dronowat, D. S., R. W. Stanley, E. Cobb, and K. Morita. 1966. "Effects of feeding limited roughage and a comparison between loose and pelleted pineapple hay on milk production, milk constituents and fatty acid composition of milk fat." *J. Dairy Sci.* **49**:28–31.

Esbelt, L. A. 1976. *Food Cannery Wastes Activated Sludge as a Cattle Feed.* EPA-600/2-76-253. U.S. Environmental Protection Agency, Cincinnati, Ohio.

Fontenot, J. P. 1977. *Symposium on Alternatives in Animal Waste Utilization.* Annual American Society of Animal Science Meetings, Madison, Wisconsin.

Goering, H. K., C. H. Gordon, R. W. Heinkou, D. R. Waldo, P. J. Van Soest, and L. W. Smith. 1972. "Analytical estimates of nitrogen digestibility in heat damaged forages." *J. Dairy Sci.* **55**:1275–1280.

Goering, H. K. and P. J. Van Soest. 1970. *Forage Fiber Analysis. USDA Handbook No. 379.* Washington, D.C.

Hampden, K. A., L. L. Wilson, T. A. Long, and W. L. Palmer. 1977. "Ensiling of tomato pulper waste for a ruminant feed." *Compost Sci.* **18**:22–24.

Hart, A. 1968. *Solid Waste Management Composting—European Activity and American Potential.* Report **SW-2c.** Solid Wastes Program, U.S. Department of Health, Education and Welfare, Cincinnati, Ohio.

Hinman, D. D. and E. A. Sauter. 1978. "Handling potato waste for beef cattle feeding." *Univ. of Idaho Ag. Exp. Sta. Bull. No.* **425.**

Howes, A. D. and E. A. Sauter. 1975. "Preserving high moisture by-products." Presented at the Northwest Nutrition Conference, Portland, Oregon.

Jung, G. A., G. A. Pearson, R. E. Fowler, D. M. Mitchell, R. E. Kocher, and E. H. Quigley. 1978. "Use of food processing effluent for forage and beef production." Unpublished data. Pennsylvania State University, University Park.

Knoll, K. H. 1961. "Public health and refuse disposal." *Compost Sci.* **2**:35–38.

Lansburg, T. J. 1958. "The composition and digestibility of some conserved fodder crops for dry season feeding in Ghana." *Tropical Agric. (St. Augustina)* **35**:114–118.

Morrison, F. B. 1959. *Feeds and Feeding.* The Morrison Publishing Co., Clinton, Iowa.

N. A. S. 1971. *Atlas of Nutritional Data on U.S. and Canadian Feeds.* National Academy of Science, Washington, D.C.

Nix, J. E. 1975. *Grain-fed versus Grass-fed Beef Production.* ERS Circ. **602.** USDA, Washington, D.C.

N.R.C. 1977. *World Food and Nutrition Study. The Potential and Contribution of Research.* National Academy of Science, Washington, D.C.

Otagaki, K. K., G. P. Lofgreen, E. Cobb, and G. G. Dull. 1961. "Net energy of pineapple bran and pineapple hay when fed to lactating dairy cows." *J. Dairy Sci.* **44**:491–497.

Peacock, F. M. and W. G. Kirk. 1959. "Comparative feeding value of dried citrus pulp, corn feed meal, and ground snapped corn for fattening steers in drylot."

Bull. 616, Agric. Exp. Sta. Inst. Food and Agric. Sci. University of Florida, Gainesville.

Randel, P. F. 1966. "Feeding lactating dairy cows concentrates and sugar cane bagasse as compared with conventional rations." *Univ. Puerto Rico J. Agric.* **50**:255–269.

Reed, A. D., W. E. Wildman, W. S. Seyman, R. S. Ayers, J. D. Prato, and R. S. Rauschkolb. 1973. "Soil recycling of cannery wastes." *Calif. Agric.* **27**:6–9. University of California Division of Agricultural Science.

Rhoads, A. T. 1965. *Disposal Systems for Food Processors.* Maryland Processors Report. National Canners Association, Washington, D.C.

Sidle, R. C. and W. E. Sopper. 1976. "Cadmium distribution in forest ecosystems irrigated with treated municipal waste water and sludge." *J. Environ. Qual.* **5**: 419–422.

Sopper, W. E. 1976. "Renovation of municipal wastewater for groundwater recharge by the living filter method." *Biol. Control of Water Poll.* **31**:269–279. University of Pennsylvania Press.

Sopper, W. E. 1978. "The living filter project." *Water Poll. Control Association of Pennsylvania Magazine* **X1**:6–8.

Tilley, J. M. A. and R. A. Terry. 1963. "A two stage technique for the *in vitro* digestion of forage crops." *J. Brit. Grassland Soc.* **18**:104–111.

Trenkle, A. and R. L. Willham. 1977. "Beef production efficiency." *Science* **198**: 1009–1015.

Wilson, L. L. and M. C. Borger. 1969. "An evaluation of the factors affecting the feasibility of recycling solid wastes through livestock as a horticultural waste disposal method for Pennsylvania." Pennsylvania State University, University Park.

Wilson, L. L., D. A. Kurtz, M. C. Rugh, L. E. Chase, J. H. Ziegler, H. Varela-Alvarez, and M. L. Borger. 1971. "Adipose tissue concentrations of certain pesticides in steers fed apple waste during different parts of the finishing period." *J. Anim. Sci.* **33**:1356–1360.

Wilson, L. L., D. A. Kurtz, J. H. Ziegler, M. C. Rugh, J. L. Watkins, T. A. Long, M. L. Borger, and J. D. Sink. 1970. "Accumulations of certain pesticides in adipose tissues and performance of Angus, Hereford and Holstein steers fed apple processing wastes." *J. Anim. Sci.* **31**:112–117.

Wright, M. J., G. A. Jung, C. S. Brown, A. M. Decker, R. D. Wakefield, and K. E. Varney. 1968. "Management of productivity of perennial grasses in the Northeast **11**. Smooth bromegrass." *W. Va. Agric. Exp. Sta. Bull.* **554T**.

Yeck, R. G., L. W. Smith, and C. C. Calvert. 1975. "Recovery of nutrients from animal wastes. An overview of existing options and potentials for use in feed." *Proc. 3rd Inter. Symp. Animal Wastes.* St. Joseph, Missouri, pp. 192–194.

ll

Coffee and tea wastes

J. H. Topps, Ph.D., F.R.I.C.

Head of Division of Chemistry and Biochemistry, School of Agriculture, Aberdeen, Scotland.

COFFEE WASTES

Coffee pulp and coffee hulls are the two main by-products that arise from the production of coffee beans from coffee cherries. Both have been widely investigated over many years for agricultural use. More recently, another coffee by-product, coffee grounds, has become available in increasing amounts due to the widening popularity of instant or soluble coffee, and it has been examined as an animal feed. Before describing the work that has been undertaken on the three by-products, it may be beneficial to those not familiar with coffee production and processing to briefly describe the systems by which they are obtained.

Coffee cherries or berries are preferably picked when they are red in color and fully ripe in order to give the best quality coffee. Normally, each cherry contains two seeds or beans enveloped by four coats. These coats (in the order of outside to inside) are the pericarp (or fruit wall); a sticky, slimy substance which is often called the pulp; a stiffer outer covering of the beans themselves, known as the parchment; and a thin, almost transparent inner lining covering the seeds which is called the silver skin or seed testa. All these coats have to be removed to prepare the beans for human use. This is achieved in one of two ways—by the older or dry method, or by the modern or wet method (which is often referred to as the West Indian process).

In the dry method, the cherries are either dried on the trees or picked and spread in a layer a few centimeters deep on stone drying grounds known as barbecues. During the first two days, the cherries are frequently turned so that all are exposed to the sun, after which precautions are taken to avoid wetting. The pulpy layer ferments and slowly dries, and the whole process requires about 15 to 25 days. When thoroughly dry, the cherries can be stored. Next, the beans are separated from the dried, shrunken coats. This has been done in a primitive manner by pounding in a mortar, but a hulling machine is now commonly employed. In either process, the dried covering is broken and the beans are set free. Later, the beans are separated from the coats, leaving behind large amounts of mixed residues. Although this method is simple and requires inexpensive machinery, it is becoming obsolete since it yields a low quality product known in the trade as "hard coffee" and it is too dependent on the persistence of fine, settled weather over a considerable period of time.

In the wet method, the cherries are pulped as soon as possible after picking; that is, within 24 hours, to avoid fermentation. They are placed in large tanks full of water immediately after harvesting to separate the immature from the ripe cherries. The former are light and float on the surface, and the latter sink to the bottom from where they are drawn through pipes leading to the pulpers. There are various types of pulpers, all designed on much the same principle. The cherries cannot pass through them without being reduced to a pulp by the rasping action of a revolving cylinder or other moving surface. The mixture of beans and pulp is passed into a vat full of water, where mechanical stirring causes the heavy beans to settle to the bottom and the lighter pulp to be carried away by an overflow of water. The beans are drawn off in a stream of water and then separated by being passed through a form of sieve. The slimy layer which still adheres to the beans is removed by fermentation with yeasts and bacteria in a special vat. This usually takes 18 to 24 hours, but the process may be hastened by adding enzyme preparations or 2% NaOH solution. Washing of the beans with successive amounts of water together with agitation and stirring leaves them completely free of mucilage. They are then dried until yellow-green in color and either sold as such—that is, covered by the last two coats—or de-hulled and polished before being shipped and sold. For the de-hulling or thresh-ing, a variety of machines are employed, all of which are designed to crack the parchment without injuring the beans. This is most readily achieved if the material is thoroughly dry, since the parchment is then brittle and relatively easily broken. Furthermore, the beans that result store well, provided the rela-

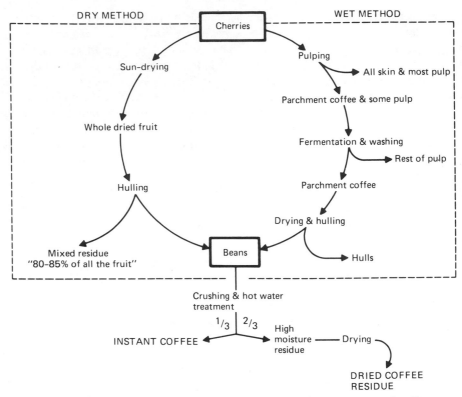

Fig. 11.1. A flow diagram outlining the steps in the processing of coffee.

tive humidity is not high. The light pieces of parchment are separated by winnowing from the heavy beans. Finally, simple rubbing and winnowing removes the silver skin and leaves clean beans as ordinary, unroasted coffee. Both the dry and wet methods are shown in outline in Fig. 11.1.

Instant coffee is made by treating beans, usually a mixture of different varieties, with boiling water. The solution obtained is freeze-dried to give instant or soluble coffee. However, a considerable amount of residue or grounds remains which contains about 80% water. This is dried by heating to give dried coffee residue.

From 1 tonne of cherries, about 430 kg of coffee pulp are obtained, which, in terms of dry matter, represents about 30% of the original cherry. De-hulling at the later stage yields about 60 kg of coffee hulls, equivalent to approximately 40 kg of dry matter, which is close to 12% of that in the cherry. The yield of beans is about 390 kg, which contains about 190 kg of dry matter (i.e., about 55% of that in the cherry). In the manufacture of instant coffee, about 67% of the dry matter in the beans is left in the residue.

Coffee Pulp

The high water content of coffee pulp causes difficulties in its transportation, handling, preservation and direct use as an animal feed. To overcome this problem, two approaches have been considered. At the Institute of Nutrition of Central America and Panama in Guatemala, Bressani et al (1975) have examined ensiling as a method of storing the pulp for feeding to animals. The pulp is allowed to dehydrate to a water content of about 65% by spreading on a clean surface. With sufficient sunshine and air movement, the dehydration is achieved in 8 to 10 hours. It is then ensiled in a trench silo along with 3 to 5% sugarcane molasses. Analysis has shown that there is very little change in nutritive value, but there are significant and advantageous decreases in caffeine and total tannin content. On removing from the silo, the pulp silage is initially readily accepted by animals. Unfortunately, exposure to air causes it to darken in color and become less palatable, and the voluntary intake by animals decreases markedly. The dark coloration is due to oxidation of polyphenols present in coffee pulp, so it is possible that this could be controlled by introducing inhibitors of the activity of polyphenol oxidases. Another possibility to overcome this palatability problem is to produce a mixed silage of pulp with another feed such as grass or green maize. In a laboratory study, Murillo et al (1976) ensiled coffee pulp with equal amounts of either Napier grass (*Pennisetum purpureum*) or maize fodder. An appreciable amount of molasses, about 16%, was added to each mixture and the products were considered to have a better quality than coffee pulp silage. There was too little of the mixed silage to carry out animal feeding trials.

The other approach used to facilitate handling and preservation of coffee pulp is dehydration either in the sun or by heat. To dehydrate in the sun, the pulp is spread on a clean surface to a thickness of about 5 to 8 centimeters. It is turned three to four times daily and is dried to a moisture content of about 12% in about 24 to 32 hours. During the drying, the color changes from red to black, but the product appears to store well and lose some of its caffeine. Sun dehydration is a cheap way to dehydrate pulp, but a large area of surface is needed and the change in color is associated with a decrease in the voluntary intake by the animal. Drying with heated air has been examined and it can be a speedy process—approximately 60 minutes. The results of tests on this product have yet to be published. A modification of this drying method is grinding the wet pulp, followed by continuous centrifugation to remove some of the water, and then dehydrating in a drum drier. The product has a better appearance than the sun-dried material and may be more palatable.

Chemical Composition. The composition of the dry matter in coffee pulp as indicated by a number of chemical analyses is shown in Table 11.1. Fresh coffee pulp invariably contains between 60% and 75% water, and the dehydrated

TABLE 11.1.   Chemical Composition of Coffee Pulp.

	Range (g/kg dry matter)
Crude protein (N × 6.25)	90.1–128.1
Ether extract	21–29
Crude fiber	146–240
Ash	64–96
Nitrogen-free extract	534–678
Calcium	6.34
Phosphorus	1.33
Magnesium	trace
Sodium	1.14
Potassium	20.2
Zinc	4.6 (mg/kg)
Copper	5.7 (mg/kg)
Manganese	7.2 (mg/kg)

material has a moisture content of about 12%. Dehydration causes some changes in the composition of the dry matter. There appears to be some loss in the nitrogen-free extract, which may be due to removal of a little of the more soluble carbohydrates, either by fermentation or in solution through seepage. In general, these chemical data show that coffee pulp has a nutritive value equivalent to that of an average quality roughage such as hay or grass silage. The work of Bressani *et al* (1975) indicates that of the total fibrous constituents in pulp as measured by the neutral or acid detergent fiber determination, approximately half is in the form of lignin. This relatively high lignin content is responsible for the crude fiber being considerably less than the total fiber fraction and it strongly suggests that the digestibility of pulp in ruminants is not high and in pigs is very low. It would appear that at least 25% of the total protein is associated with lignin and may not be completely available. The availability of the protein may also be affected by the tannin in coffee pulp. It is well known that tannins combine with protein and render them less available for microbial fermentation and enzymic digestion in the animal's gut. Work is needed to evaluate the adverse effect, if any, of tannins. Caffeine is also present in coffee pulp, being about 1.5% of the dry matter, and is known to have a diuretic effect that is likely to adversely affect the nitrogen retention of animals given appreciable amounts of the by-product. Fortunately, the storage of pulp— either ensiled or in the dehydrated form—for several months reduces the caffeine content. A reduction of 70% after 15 months of storage has been recorded by Bressani *et al* (1972). Fermentation, autoclaving, and heating also reduce the caffeine content. Nevertheless, the presence of appreciable amounts of caffeine in fresh pulp is considered to be a disadvantage and it may reduce the value of the pulp by limiting the intake of animals. For this reason, two forms of processing for the removal of caffeine have been examined. In one, successive washings with water have been shown to achieve almost complete removal, and in the other, laboratory studies with the fungus *Penicillium crustosum*, which has been reported by Schwimmer and Kurtzman (1972) to utilize caffeine, could be fruitful. Both the nutritional and economical feasibility of either process has yet to be reported in detail.

The results of Bressani *et al* (1972) that give the amino acid composition of the protein in coffee pulp indicate that it has a relatively high quality for non-ruminants. It appears to compare favorably with that in soybean meal, but is even more deficient in the sulfur-containing amino acids. Such a limitation may be enhanced if the tannins in pulp make some of the amino acids partly unavailable to the animal.

Coffee Pulp as a Feed for Ruminants and Pigs. To evaluate the material for ruminants, most of the studies have been carried out with growing and fattening cattle given either sun-dehydrated pulp or coffee pulp silage (Bressani *et al*, 1975). These studies have shown that in production rations, the weight gain and food intake of cattle decreased as the level of pulp in the diet increased. These effects were more pronounced when the level was greater than 20%. In this respect, a similar result was obtained in earlier work with lactating cows (Choussy, 1944; Work *et al*, 1946). The inclusion of coffee pulp at levels below 20% did not affect milk production. There is some evidence, reviewed by Bressani *et al* (1975), that the depressing effect of coffee pulp on the weight gain of cattle is less marked when the level of dietary protein is high. If this is correct, it would seem that the value of pulp is limited by poor utilization of the protein it contains, which is consistent with comments made earlier concerning the characteristic of tannins. Without doubt, pulp for ruminant feeding needs to be classified as an average quality roughage, and to avoid discernible reductions in animal performance, its inclusion in production rations should be 20% or less. At higher levels, Cabezas *et al* (1974) have shown that young steer previously adapted to dried coffee pulp for 102 days gained weight, while those which had not received the pulp lost weight. Such an adaptation with time may occur, but may be less important in cattle given the lower levels of pulp.

Very few studies have been made to evaluate coffee pulp as a pig feed, probably because the composition indicates that it is less suitable for this purpose than for feeding to ruminants. In Guatemala, the work of Rosales (1973) with growing pigs has shown that at three different physiological ages, the inclusion of sun-dehydrated and ground coffee pulp at dietary levels of 8.2, 16.4, and 24.6% caused a decrease in growth rate and efficiency of feed utilization which was greater as the pulp in the diet increased. Similar but more striking results were obtained by Malynicz (1974) in Papua, New Guinea. When dried coffee pulp was given as 10, 20, or 30% of the diet in place of sorghum, the daily gain of growing pigs was 213, 159, and 113g, respectively, compared with 495g for controls. From these results, it may be concluded that coffee pulp is not a suitable feed for growing pigs unless some reduction in growth rate and fat deposition is desirable, as in the production of lean bacon pigs.

Other Uses of Coffee Pulp. Other than its traditional use as an organic fertilizer in coffee plantations, which provides a route for recycling some of the nutrients in coffee cherries, the pulp has been used to make coffee molasses and has been investigated as part of a supplementary feed for fish, *Tilapia aurea*. Coffee molasses is made by drying the sugars from the mucilage and pulp. Buitrago *et al* (1970) have found it has a value a little less that of sugarcane molasses when given in diets based on maize at a level of 30% for growing pigs. The pulp itself gave promising results when given in mixtures containing cotton seed meal with and without urea to *Tilapia* in El Salvador. Further work is needed to specify the amounts of pulp which can be used in relation to other dietary ingredients, including vitamin and mineral supplements.

Coffee Hulls

Coffee hulls are a dry, stable material that is easy to store and handle. Its decomposition is very slow because it is covered by a waxy film and it has a high lignin content.

Chemical Composition. The composition of the dry matter of coffee hulls, as indicated by a number of chemical analyses, is shown in Table 11.2. Coffee hulls contain between 5 and 10% water. The protein content is very low, similar to the amount found in very poor quality roughage such as wheat straw or mature,

TABLE 11.2.   Chemical Composition of Coffee Hulls.

	Range (g/kg dry matter)
Crude protein (N X 6.25)	26.3–101.7
Ether extract	6.5–20.2
Crube fiber	297–754
Ash	5.4–75.3
Nitrogen-free extract	204–515
Calcium	1.61
Phosphorus	0.30

tropical standing hay, and the fibrousness of the by-product is pronounced. The work of Bressani *et al* (1975) has shown that the total amount of cell wall constituents in hulls, measured as neutral detergent fiber, may be as high as 88% of the dry matter. Of this, half is cellulose, and approximately one-quarter and one-fifth are hemicellulose and lignin, respectively. This high lignin content renders the by-product very indigestible and it may be described as a filler in animal feeds. Its inclusion in almost any diet is likely to lower the overall nutritive value. Attempts have been made to improve the value of hulls by chemical treatment causing delignification. The use of either sodium or calcium hydroxide solutions of three different strengths has been shown to cause an appreciable reduction in content of acid detergent fiber, especially at the highest concentration of 10% (Bressani *et al*, 1975).

Coffee Hulls as a Feed for Ruminants.   The very high fiber and very low protein content make coffee hulls an unsuitable feed for non-ruminant animals. Several workers have evaluated the by-product as a constituent of diets for cattle. In Cuba, Rodrigues & Zamora (1970) compared hulls with rice husks as the main roughage source in complete diets for dairy cows. Those receiving the hulls gave lower yields than did the cows given rice husks. Ledger & Tillman (1972), in Kenya, examined the effect of replacing ground maize in a highly concentrated fattening diet with 10, 20, or 30% coffee hulls, given to Boran steer for 89 days. The animals were fed individually to appetite. At the 10 or 20% level of inclusion, feed intake, weight gain, and efficiency of feed conversion were unaffected, but at the 30% level all were reduced. Similar results were obtained by Jarquin *et al* (1974) in Guatemala, using Holstein calves, approximately 3 months old and about 90 kg in weight. They were given diets with no coffee hulls and 15 or 30%. Rate of growth decreased as the hulls in the diet increased. Intake of diet containing 30% hulls was less than that with none. The effect of adding 1.5% urea to the 30% coffee hulls diet and at another time increasing the molasses content from 20 to 28.5% were examined. The performance of the calves eating these modified diets remained below that of those receiving the diet without hulls. From all these trials it may be concluded that hulls may be included in diets for older cattle up to levels of 20% without appreciably affecting their performance. Any decrease that did occur would be difficult to discern in practice, and in the experimental situation would not be statistically significant unless a large number of animals were used. For young cattle, the inclusion of hulls may be more detrimental, and levels less than 20% may give a perceptible decrease in performance.

Other Uses of Coffee Hulls.   Coffee hulls have been used as a fuel. On coffee plantations, large amounts have been spread on fields, where it slowly decomposes and acts as a soil conditioner.

Dried Coffee Residue

There are several alternative names for this by-product from the manufacture of soluble or instant coffee. It has been referred to as extracted coffee bean meal, coffee residue, coffee grounds, spent coffee grounds, and, in certain commercial circles in the United Kingdom, Cherco. All these names apply to the same product, but there are variations in the composition of this material which may be related to differences in the original beans and in the drying of the residue. When dry, the residue is a stable, brown, powdery material which stores well and is relatively easy to handle.

Chemical Composition. The composition of the dry matter of dried coffee residue, as indicated by a number of chemical analyses, is shown in Table 11.3. The residue contains between 5 and 10% moisture. It has an average protein content, approximately 10% dry matter, but a high fiber and a relatively high fat and gross energy content. Its total mineral or ash content is very low, with none of the major minerals being present in more than small amounts. The studies of Ali (1976) have shown that the dry matter contains 73% acid detergent fiber, 48% cellulose, 23% lignin, and traces of tannin (0.12%) and caffeine (0.05%). The fat or oil in dried coffee residue is predominantly triglycerides, which are made up of 50 to 60% unsaturated fatty acids, including a high content of linoleic acids. Measurements of *in vitro* digestibility by Ali (1976) have given very low values, which indicate it may be poorly digested by rumi-

TABLE 11.3. Chemical Composition of Dried
Coffee Residue

	Range (g/kg dry matter)
Crude protein (N X 6.25)	102–130
Ether extract	225–283
Crude fiber	466–510
Ash	7–8
Nitrogen-free extract	143–168
Calcium	0.55–0.65
Phosphorus	0.50–0.70
Magnesium	0.15–0.25
Sodium	0.40–0.60
Potassium	0.35–0.45
Zinc	9–12 (mg/kg)
Copper	25–29 (mg/kg)
Manganese	29–36 (mg/kg)

nants. Since the high lignin content may be partly responsible for the low digestibility, the effect of treating with alkali and other delignifying agents has been studied (Ali, 1976). Considerable improvements in digestibility measured *in vitro* have been obtained, but the treated material is still mainly indigestible.

Dried Coffee Residue as a Feed for Ruminants. To date, most of the work undertaken to examine the residue as a practical feed has been carried out with ruminants. Conflicting results have been obtained when the digestibility of the by-product was determined using sheep and, to a lesser extent, cattle. In Italy, Fornaroli and Perotti (1974) found that adding the residue—which had a relatively high moisture content—to lucerne hay lowered the digestibility in sheep of some chemical constituents of the diet. At a high level of inclusion of the residue, the digestibility of lipids was low, and at both levels of inclusion, the digestibility of fiber in the diet was markedly depressed. Similar results were obtained by Ali *et al* (1977) in the United Kingdom. Dried coffee residue was included in a concentrate diet at 7.5 and 12.5%, and the concentrate only and the mixtures were fed with hay in the ratio of 6 to 4. Increasing the amount of coffee residue in the diet caused a decrease in overall digestibility of the diet and a more pronounced decrease in fat digestibility. From these results, the digestibility of the residue was calculated by difference, assuming there were no associative effects between it and the remainder of the diet. The values obtained were low, especially for fat and nitrogen-free extract. Contrary results have been obtained in Canada by McNiven *et al* (1977). The digestibilities in sheep of two diets, one without coffee residues and the other with 25% coffee residues were compared, and, except for protein, found to be higher in the coffee diet. In this work, the coffee residue was calculated to have 76% total digestible nutrients. It should be noted that in all three studies, the protein in dried coffee residue was found to be poorly digested. However, the digestible energy value and, to a large extent, the nutritive value of dried coffee residue depends on its energy digestibility, which is greatly influenced by the extent of digestion of fat. It is possible that the drying process, if excess heat is used or if it is uncontrolled, may adversely effect the nature of the fat. Such an effect may explain the difference between the results of Ali (1976) and those of Campbell *et al* (1976) with cattle; Ali found that inclusion of dried coffee residue depressed the digestibility of dietary fat, while Campbell *et al* obtained an increase. A few feeding trials have been carried out in which diets with and without dried coffee residue have been compared. Mather and Agpar (1956) used lactating Holstein cows and found that the residue had no significant effect on milk yield, butter fat percentage, or flavor of milk, but the body weights of the cows were significantly reduced, probably due to low intakes. In the same study, feed intake and growth rate of calves were reduced when the residue was included in calf starter rations. Ali (1976) included up to about 7% of dried coffee residue in the diet of fattening steer without adversely affecting the overall performance. However, the rate of growth of steer in this trial may not have been high enough to fully test the diets.

In the majority of these studies with sheep and cattle, the dietary inclusion of dried coffee residue in more than very small levels caused a marked diuretic effect. For example, Campbell *et al* (1976) found that 10% or 20% inclusion of the residue more than trebled or quadrupled, respectively, the urinary output of Holstein steers. The diuresis resulted in increased losses of urinary nitrogen and sodium. It is doubtful that this effect is due to caffeine, since the caffeine content appears to be too low to cause pronounced diuresis. Further work is needed to establish the likely cause. Another problem with dried coffee residue is that the material is not readily acceptable or palatable to cattle or sheep. In all the studies considered, the authors make a special mention of lowered intakes as the level of inclusion increased or the procedures used to induce the animals to eat the residue or both. In summary, it may be concluded that unless dried coffee residue has a high fat digestibility, it should be classified as a very low quality concentrate. When included in conventional diets at levels of more than a few percent, it is likely to depress intake and the digestibility of some of the constituents, in particular that of protein. At dietary levels of 10% or more it has given a pronounced diuretic effect. For these reasons, the inclusion of dried coffee residue in ruminant diets should be no more than about 5% or less. At these levels, it will have the advantage of increasing the total energy content of the diet, and possibly the digestible energy value if the fat fraction is well digested.

Other Uses of Dried Coffee Residue.   Attempts to show that dried coffee residue is a suitable feed for poultry or for other non-ruminant animals by carrying out rat tests have been singularly unsuccessful. Hammond (1944) and Carew *et al* (1967) showed that rations containing the residue had a deleterious effect on growing chicks. Similar adverse effects were obtained by Ali (1976) and Campbell *et al* (1976) when adult and weanling rats, respectively, were given diets containing dried coffee residue. Weight gains or growth rate were retarded, and at the higher levels of inclusion, a high proportion of the rats died. These results suggest that dried coffee residue is not suitable for feeding to non-ruminant farm livestock.

With its high calorific value, it has some use as a fuel. Alternatively, it is akin to coffee hulls as a soil conditioner, and if economically worthwhile, it could be a small industrial source of oil and fat.

TEA WASTE

Very little can be said about the wastes from tea. The dust which arises from the manufacture of tea can act as a useful organic fertilizer. It has a composition similar to that of tea leaves (Eden, 1958) and, as such, would supply a useful amount of nitrogen and potassium to subsequent crops, but only small amounts of the other essential major elements. In recent years, the consumption of instant tea has increased, especially in North America, and the spent tea leaves, which contain approximately 50% moisture, have been marketed as a mulch. Recently Croyle *et al* (1974) have examined the potential of this by-product as an animal feed and for composting. Chemical analysis showed that the dry matter in the spent leaves has a relatively high crude protein content (26.4%) but the material is very fibrous (acid detergent fiber, 64%) and much of the fiber is lignified (lignin, 36%). Furthermore, much of the nitrogen appears to be bound to the cell wall constituents, which would make the crude protein fraction less available for fermentation and enzyme digestion. For these reasons, and since the by-product may contain appreciable residues of tannins and caffeine, it is unlikely to be a suitable material for feeding to farm animals, with the possible exception of small inclusions in some diets for ruminants. To make an acceptable compost, Croyle *et al* (1974) have suggested that it should be moistened to 60% water content, inoculated with a cellulolytic organism, and then treated with certain minerals to make it appropriate for nursery or garden crops. It remains to be seen whether such processing is worthwhile.

REFERENCES

Ali, M. M. 1976. "Studies of dried coffee residue as a component of diets for cattle and sheep." PhD Thesis, University of Aberdeen.

Ali, M. M., J. H. Topps, and T. B. Miller. 1977. "Evaluation of dried coffee residue as a component of diets for ruminants." *Proc. Nutr. Soc.* **36**:67A.

Bressani, R., M. T. Cabezas, R. Jarquin, and B. Murillo. 1975. "The use of coffee processing waste as animal feed." *Proc. Conf. on Animal Feeds of Tropical and Subtropical Origin.* Tropical Products Institute, London, pp. 107–117.

Bressani, R., E. Estrada, and R. Jarquin. 1972. "Coffee pulp and husks. 1. Chemical composition and amino acid content of protein of the pulp." *Turrialba* **22**:299–304.

Buitrago, J., H. Calle, J. Gallo, and M. A. Corzo. 1970. "Coffee molasses in diets for growing and finishing pigs." *Revista Instituto Colombiano Agropenario* **5**:407–410.

Cabezas, M. T., B. Murillo, R. Jarquin, J. M. Gonzalez, E. Estrada, and R. Bressani. 1974. "Coffee pulp and husks. 6. Adaptation by cattle to coffee pulp." *Turrialba* **24**:160–167.

Campbell, T. W., E. E. Bartley, R. M. Bechtle, and A. D. Dayton. 1976. "Coffee grounds. 1. Effects of coffee grounds on ration digestibility and diuresis in cattle, on *in vitro* rumen fermentation, and on rat growth." *J. Dairy Sci.* **59**:1452–1460.

Carew, L. B., G. H. Alvarez, and R. O. M. Martin. 1967. "Studies with coffee oil meal in diets for growing chicks." *Poult. Sci.* **46**:930–935.

Choussy, F. 1944. "Coffee pulp in the diet of dairy cows." *Anales del Instituto Tecnologico de El Salvador* **1**:265–271.

Croyle, R. D., L. L. Wilson and T. A. Long. 1974. "Potential of spent tea leaves for animal feeds and composting." *Compost Sci.* **15**:28–30.

Eden, T. 1958. *Tea.* London, Longmans.

Fornaroli, D. and L. Perotti. 1974. "Expended coffee for livestock feeding." *Rivista di Zootecnia e Veterinaria* **6**:495–503.

Hammond, J. C. 1944. "Dried coffee grounds unsuitable for use in the diet of growing chickens." *Poult. Sci.* **23**:454–455.

Jarquin, R., B. Murillo, J. M. Gonzalez, and R. Bressani. 1974. "Coffee pulp and husks. 7. Coffee husks for feeding ruminants." *Turrialba* **24**:168–172.

Ledger, H. P. and A. D. Tillman. 1972. "Utilization of coffee hulls in cattle fattening rations." *E. Afric. Agric. For. J.* **37**:234–336.

Malynicz, G. L. 1974. "Plantation crop by-products for growing pigs." *Papua New Guinea Agric. J.* **25**:20–22.

Mather, R. E. and W. P. Agpar. 1956. "Dried extracted coffee meal as a feed for dairy cattle." *J. Dairy Sci.* **39**:938.

McNiven, M., P. Flipot, J. D. Summers, and S. Leeson. 1977. "The feeding value of spent coffee grounds for ruminants." *Nutr. Rep. Int.* **15**:99–103.

Murillo, B., L. Daqui, M. T. Cabezas, and R. Bressani. 1976. "Coffee pulp and hulls II. Chemical composition of coffee pulp ensiled with Napier grass and maize." *Archivos Latinoamericanos de Nutricion* **26**:33–45.

Rodrigues, V. and A. Zamora. 1970. *Institute of Animal Science Report, 1967–70.* University of Havana, Cuba, p. 53.

Rosales, F. 1973. "The use of dehydrated coffee pulp in the feeding of pigs." MSc Thesis, Universidad de San Carlos, Guatemala.

Schwimmer, S. and R. H. Kurtzman. 1972. "Fungal decaffeination of roast coffee infusions." *J. Food Sci.* **37**:921–924.

Work, S. H., M. L. Van Severen, and L. Escalon. 1946. "Preliminary information on the value of coffee pulp substituted for maize in the rations of milking cows." *Cafe de El Salvador* **16**:773–780.

12

The use of fish and shellfish wastes as fertilizers and feedstuffs

G. R. Swanson,[1]
E. G. Dudley[2] and
K. J. Williamson[3]

Department of Civil Engineering, Oregon State University, Corvallis, Oregon

INTRODUCTION

The present world marine fish and shellfish harvest is about 70 million tonnes/yr. In the United States, domestic landings amount to only about 4% of this total, although imports increase the amount to about 7% (National Marine Fisheries Service, 1978).

Product yields in fish and shellfish processing range from nearly 100% for whole rendered fish to a low of 15% for some crabs. The portions not incorporated as final products or by-products become waste. Using an average wastage rate of 30%,

[1] Graduate student.
[2] Graduate student.
[3] Associate professor.

the total worldwide waste from seafood processing amounts to about 20 million tonnes and the United States seafood waste to almost 2 million tonnes/yr.

This waste is removed from seafood processing plants as either waste water or solid waste. Waste water (fresh or sea water) solutions contain dissolved materials (proteins and breakdown products) and suspended solids consisting of bone, shell, or flesh, and foreign material. Solid wastes consist of flesh, shell, bone, cartilage, and viscera. Many of these waste materials have sufficient nutrients to be valuable as fertilizers and feedstuffs.

The recovery of seafood wastes as fertilizers and feedstuffs is generally directed at the solid waste fraction, since most waste water solutions (washwaters, etc.) are too dilute for recovery. Increasingly, however, waste water streams are being treated to meet effluent requirements for discharges. Typically, screening, settling or dissolved air flotation methods are utilized, and this results in the production of additional quantities of solid waste.

SEAFOOD PROCESSING AND SOLID WASTE GENERATION

The nature and quantities of solid wastes generated from seafood processing are provided below with the descriptions of processes typically used for major fish and shellfish species. Table 12.1 summarizes the total landings, typical product yields, waste fractions, and peak seasons for the various types of processing plants.

Bottom Fish

Bottom fish are most commonly filleted and frozen for shipment. Most plants processing fillets use mechanized equipment, although some small plants use hand filleting. A typical bottom fish process is depicted in Fig. 12.1.

Solid wastes from filleting and skinning are usually rendered for pet food or animal meal. These wastes constitute 35 to 75% of the total processed weight, with the higher percentages more typical of filleting operations (U.S. Environmental Protection Agency, 1975).

Catfish

Catfish are harvested primarily in the southern and south-central states. Since 1965, the production of farm catfish has increased steadily. Presently, over 60% of the catfish harvest is from farm ponds or raceways.

A typical farmed catfish process is shown in Fig. 12.2. After beheading, eviscerating, and cleaning, the larger fish are cut into steaks or filleted, and smaller fish are packaged whole. Presently, all catfish are marketed either fresh or frozen.

TABLE 12.1. United States Seafood Processing Solid Waste.

Category	Average 1972–76 Landing (10^6 kg)	Product Yield (%)	Solid Waste (%)	Peak Season
Fish:	1850 (1977)			
Bottom fish	238 (1972)	20–40[1]	55–75	Summer
Catfish	72 (1975)	63[2]	37	Fall to Spring
Halibut	10	90[3]	10	May to September
Menhaden and Anchovies	877:115	28–36[4]	15[5]	June to September
Salmon	105	62–68[6]	16–27	July to August
		65–80[2]	21–23	
Tuna	181	40–50[7]	49–59	Summer
Shellfish:	516 (1977)			
Clams and Oysters	46:24[8]	10–15	82–90	Year round
				Fall to Winter
Crabs				
Blue	61	9–16	51	Spring
Dungeness	10	17–27	50–60	December to March
Alaska King	43	25–36	57	Aug. to September
Tanner	27	10–20	50–60	January to May
Shrimp	171	12–20	65–85	Varies with location
		80[9]	15	

[1] Filleting process; fish flesh operations may achieve higher yields.
[2] Farm raised catfish processing.
[3] Fresh/frozen processing.
[4] Rendering process; both fish meal and fish oil are considered final products.
[5] Represents concentrated fish solubles which are commercially marketed.
[6] Canning process.
[7] Does not include red meat (8 to 10% of tuna) which is processed for pet food.
[8] Excluding shell weight, total harvested weight is 5 to 10 times this value.
[9] If beheaded at sea.

Solid wastes are processed into pet food, fish feed, or fish meal where processing facilities are available and into landfills where such facilities are not. Waste water treatment is usually not practiced.

Halibut

The halibut is a large fish, and commercially landed sizes vary from 10 to over 30 kg. They are caught near the sea bottom using baited longlines. The major halibut fishery is centered in the Pacific Northwest and Alaska, with the regulated commercial season extending from May through September.

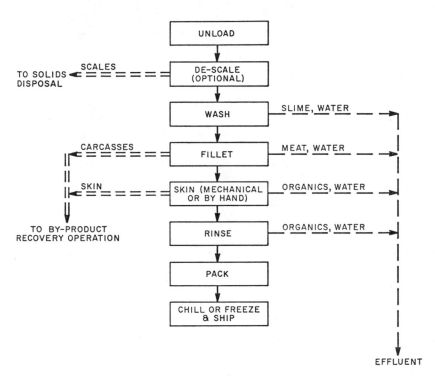

Fig. 12.1. Typical bottom fish process.

A typical halibut process is shown in Fig. 12.3. Smaller fish (under 27 kg) are usually frozen whole, and larger fish are butchered to remove four large sections of flesh called fletches.

Halibut are generally gutted at sea; therefore, solid waste quantities generated in processing are relatively small. Also, the edible cheeks are removed from the heads, bagged, and frozen. Most of the solid waste from Pacific Coast plants is ground and bagged for the pet or animal food market. Solid wastes in Alaska are used for bait or discarded.

Menhaden and Anchovies

Menhaden and anchovies are the major species of small, oily fishes which are harvested primarily for fish meal production. Menhaden is harvested on both the Atlantic and Gulf Coasts and anchovies only on the west coast.

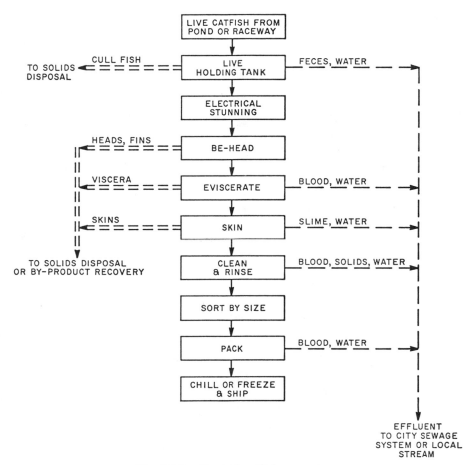

Fig. 12.2. Typical catfish process.

Ninety-nine percent of the menhaden landed in the United States and most of the anchovies are rendered for fish meal, oil, and fish solubles. A typical fish meal production process is shown in Fig. 12.4.

Essentially all of the fish is processed into fish meal or by-products, leaving no solid waste. The fish meal is primarily utilized as a protein supplement in animal feeds. The fish oil has a variety of commercial and industrial uses. Stickwater from the rendering process is evaporated at larger plants from a consistency of 5 to 8% solids to a fish solubles concentrate containing 50% solids. Fish solubles

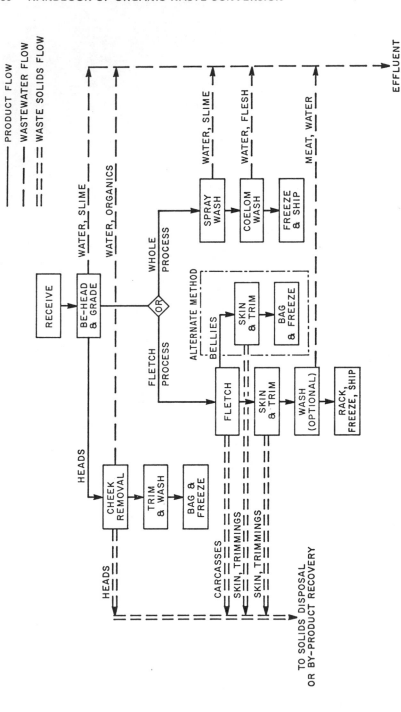

Fig. 12.3.   Typical halibut process.

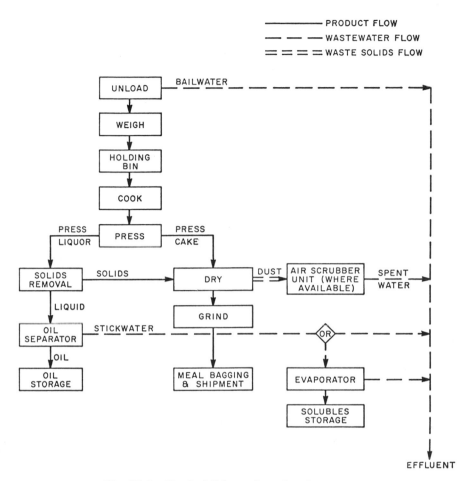

Fig. 12.4.   Typical fish meal production process.

are combined with the fish meal for use as animal feed or marketed as a liquid fertilizer.

Salmon

Five species of Pacific salmon are presently harvested in Alaska, Oregon, and Washington. Most salmon is canned; however, fresh freezing is increasing in popularity.

A typical salmon canning process is depicted in Fig. 12.5. Either hand or mechanical butchering is employed to remove the heads, tails, fins, and viscera. Product yield is slightly greater with hand butchering. Most salmon are mechanically packed.

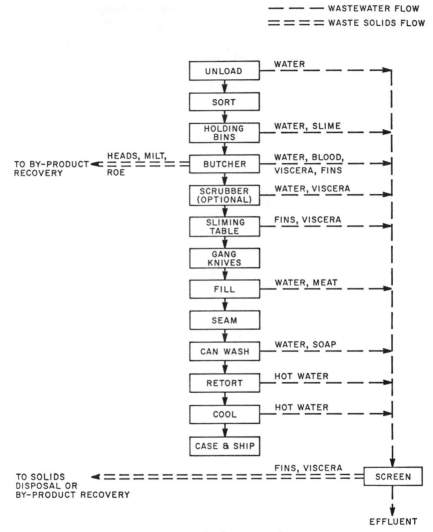

Fig. 12.5. Typical salmon canning process.

Milt and roe, constituting about 8% of the salmon, are further processed into by-products. Roe is salted and packed in boxes or sold for bait. Milt is frozen, packed in boxes, and shipped to other plants for processing. Most plants in the northwest send other recoverable solid wastes (heads, fins, viscera) to rendering plants for fish meal, pet food, or mink food production. Alaska plants have generally discharged those wastes to receiving waters.

Tuna

The annual consumption of tuna in the United States far surpasses that of any other seafood. Essentially all of the tuna harvest is canned. Nearly all of the 300 million kg of tuna consumed in the United States in 1972 was domestically canned. However, only 34% was packed from domestic landings.

A typical tuna canning process is shown in Fig. 12.6. Tuna are harvested by

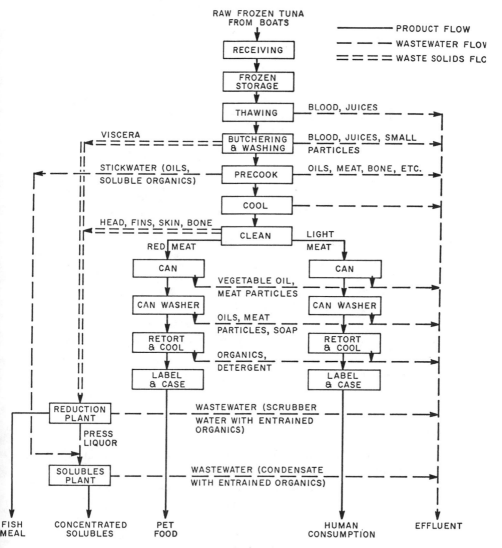

Fig. 12.6. Tuna process.

line or net and frozen on board the fishing vessel. After thawing at the processing plant, the tuna are butchered, precooked, and cleaned before being packed in cans. Dark (red) meat, amounting to about 6 to 10% of the tuna, is processed for pet food, and white meat is used for human consumption.

Due to the large size of most tuna processing plants, extensive by-product recovery is normally practiced. Viscera and other solid wastes are further processed to fish meal. Concentrated solubles are marketed as an animal feed additive or used to produce tuna oil or other by-products. Generally, only about 1% of the tuna is wasted.

Clams and Oysters

The most important types of clams harvested are the surf, hard, and soft clams. About 87% of the clam harvest occurs in the mid-Atlantic region, and most of the harvest is surf clams. Most oyster are harvested in the Chesapeake Bay and Gulf Coast regions. Both clams and oysters are harvested year round.

The processing of surf clams consists of three basic operations: shucking, de-bellying, and packing or freezing (see Fig. 12.7). The clams are either mechanically or hand shucked, whereas de-bellying is often manual. The processing of oysters is similar but less complicated, since viscera are not removed. The two basic operations, shucking and packing, are most commonly manual. Oyster and clam meat is typically fresh packed or frozen.

The primary solid waste generated in clam and oyster processing is the shell. The clam shell is 75 to 80% of the total clam weight (with shell) and the belly, 7 to 10%. Thus, the solid waste from clam processing amounts to five to ten times the final product weight. Oyster processing solid wastes (shell only) are a similarly large fraction of the total weight.

Clam and oyster shells are used as fill for roadbeds, as a media for shellfish larvae attachments, or as landfill. Oyster shells are also used commercially as poultry food, fertilizers, concrete, and pharmaceuticals. Clam bellies are used for bait or animal food, or they are ground and disposed of into municipal sewer systems or receiving waters.

Crabs

The blue crab comprises about 70% of United States crab production and is harvested along the Gulf of Mexico and Atlantic coasts, principally in the Chesapeake Bay area. The remaining crab harvest takes place mainly on the Pacific Coast, where Dungeness crab is the leading species, followed by Alaskan king crab.

A conventional blue crab canning process is shown in Fig. 12.8. Product yields for crab processing are typically very low, ranging from 9 to 15% for blue crabs.

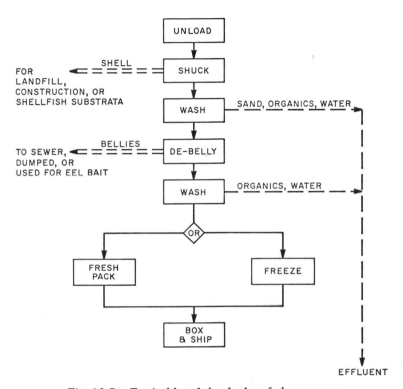

Fig. 12.7.   Typical hand-shucked surf clam process.

From 25 to 35% of the crab weight is lost in the steam cooking process, resulting in concentrated condensates from the crab cookers.

From 50 to 60% of the crab is solid waste. The major portion of this is exoskeleton, along with varying amounts of flesh attached to the shell. The protein concentration of crab waste is considered low compared to visceral fish wastes, but processing into animal feed is practiced. The large Alaska plants generally grind solid wastes for use as a by-product, whereas West Coast plants do not.

Shrimp

Shrimp are harvested in the coastal waters of all the major regions in this country and are a principal part of the seafoods industry in the Gulf of Mexico and South

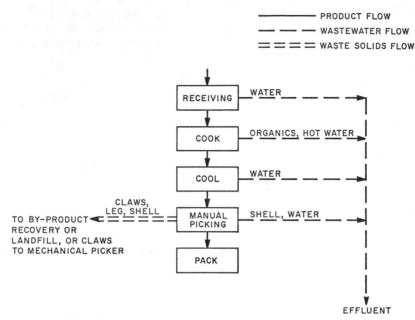

Fig. 12.8.   Conventional blue crab canning process.

Atlantic areas. Breaded and frozen shrimp are the primary products, although about 10% of the total harvest is canned.

Shrimp processing is depicted in Fig. 12.9. The shrimp are frozen on board when long storage times are necessary. Peeling is generally mechanical, but may be done manually along with breading in a fish breading process.

The wastage rate for shrimp processing is very high, especially with mechanical peelers (U.S. Environmental Protection Agency, 1974). The heads, shells, and other solid wastes are typically incorporated into the waste water flow in processing. This waste has traditionally been discharged untreated to ocean waters, but screening is now generally employed to meet water pollution control regulations. Screened solids are landfilled or, in some cases, processed to fish meal or made available to farmers as fertilizer. By-product recovery from shrimp processing solid wastes has generally not been practiced in the past, although many potential uses are being developed, such as extraction of the chitosan (Chitin), a versatile industrial material.

SEAFOOD WASTE CHARACTERISTICS

The practicality and economic attractiveness of recovering seafood wastes as fertilizers and feedstuffs is directly dependent on the waste composition, including

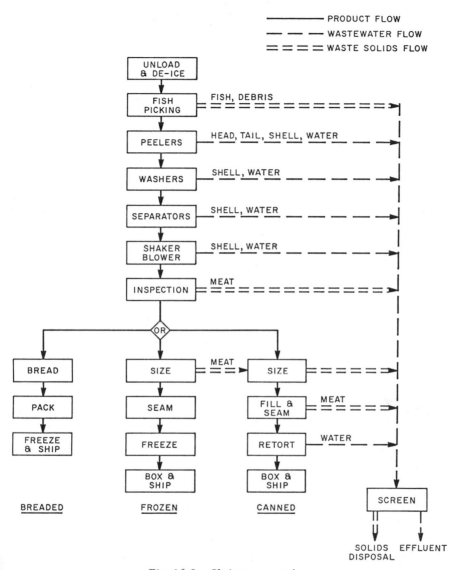

Fig. 12.9. Shrimp processing.

protein, fat, mineral, and moisture content. The characteristics of seafoods and seafood wastes are summarized below.

Fish are commonly placed in the category of "animal proteins" because of their high nutritional value for human and animal consumption. Meats from fish and shellfish, regardless of origin, have similar nutritional properties and contain 15

TABLE 12.2. Typical Composition of Fish and Shellfish (portion normally utilized).

Type	Protein (%)	Fat (%)	CHO (%)	Moisture (%)	Ash (%)
Fish:					
Sole	16.7	0.8	0	81.3	1.2
Rockfish	18.9	1.8	0	78.9	1.2
Cod	17.6	0.3	0	81.2	1.2
Catfish	17.5	3.1	0	78.0	1.3
Halibut	20.9	1.2	0	76.5	1.4
Menhaden	18.7	10.2	0	67.9	3.8
Anchovy	15–20	5–15	0	–	–
Herring	17.4	2–11	0	70.0	2.1
Salmon	19–22	13–15	0	64.0	1.4
Tuna	25.2	4.1	0	70.5	1.3
Shellfish:					
Clams (meat only)	14.0	1.9	1.3	80.8	2.0
Oysters	8–11	2.0	3–6	79–85	1.8
Crab	17.3	1.9	0.5	78.5	1.8
Shrimp	18.1	0.8	1.5	78.2	1.4

Source: U.S. Environmental Protection Agency. 1975.

to 20% protein. Also, the amino acid content of fish is very similar to that in mammalian flesh, so they contain all the essential amino acids. The typical composition of various fish and shellfish (portion normally utilized) is shown in Table 12.2.

On a dry basis, the fish and shellfish listed in Table 12.2 generally contain from 60 to 90% protein. The small, oily fishes usually processed directly for fish meal—such as menhaden, anchovies, and herring—have a high fat (lipid) content and less protein, on a dry basis, than do other fishes. Fish meal produced from these fishes typically contains about 60% protein.

Fish lipids (fat) consist primarily of the desirable polyunsaturated fatty acids. However, these lipids tend to oxidize quite rapidly, resulting in rapid fish degradation.

Fish processing solid wastes are very similar in composition to the edible portion. Stansby and Olcott (1963) reported on the composition of edible and waste fractions of dover sole, as shown in Table 12.3. As shown, the non-edible (waste) fraction has nearly as much protein as does the edible portion. A problem with fish wastes, however, is the higher lipid content. These lipids are subject to rapid oxidation (spoiling), and fish wastes, therefore, are highly putrescible unless dried or stabilized in some other manner.

TABLE 12.3. Approximate Composition of Dover Sole.

Constituent	Whole Fish	Edible Portion	Non-Edible Portion*
Moisture (%)	81.9	83.6	81.2
Lipid (%)	3.5	0.8	4.4
Protein (%)	12.7	15.2	11.7
Ash (%)	2.7	1.1	3.5

*All parts except flesh.
Source: Stansby and Olcott. 1963.

As shown in Table 12.3, the protein content of fish wastes can be expected to be slightly less than that of whole fish and will generally be in the range of 10 to 15% for unprocessed fish wastes and from 30 to 65% on a dry basis. Based on 15% nitrogen in fish protein, the protein-nitrogen content of fish wastes is about 2% for unprocessed wastes and from 6 to 10% on a dry basis. Fryer and Simmons (1977) reported approximately 4% phosphate (as P_2O_5) and about 1% potash (K_2O), dry basis, for fish wastes. A typical fertilizer composition of fish wastes can be approximated as $2:1:0$ (N:P:K) for unprocessed wastes and $8:4:1$ on a dry basis. Fish wastes also contain a large assortment of other minerals and trace elements required by plants.

In contrast to fish wastes, shellfish wastes consist primarily of the exoskeleton (shell). Shells from crustacea, depending on species and time of year, generally contain 25 to 40% protein, 15 to 25% chitin, and 40 to 50% calcium carbonate (U.S. Environmental Protection Agency, 1975). Chitin is an insoluble polysaccharide structural material that serves as a binder in the shell and contains 6.9% nitrogen. Calcium carbonate is an alkaline material similar to lime.

Because of their low protein and high mineral and fiber content, processed shellfish wastes are not as desirable as fish wastes for animal feed. Shellfish wastes, particularly crab and shrimp wastes, are not commonly utilized at present, but are typically disposed in landfills.

Costa (1977) reviewed the literature on crab and shrimp wastes and found compositions ranging from 11 to 42% protein, 9 to 42% chitin, and 36 to 58% calcium carbonate, dry basis, and 57 to 88% moisture for fresh waste. Shrimp wastes contained 5.4 to 7.9% nitrogen, 2.1 to 2.9% phosphorous, and 1.6% potassium. Crab wastes contain 4.9 to 7.3% nitrogen, 0.6 to 1.8% phosphorous, and 0.4 to 1.3% potassium. Both wastes contained significant amounts of nutrient and non-nutrient trace elements.

Typical values for shrimp and crab wastes appear to be 60 to 80% moisture for fresh waste and, on a dry basis, about 30% protein, 30% chitin, and 40% $CaCO_3$; in relation to the fertilizer value, these wastes are about 6% nitrogen, 2% phosphorous, and 1% potassium, or $6:4:1$ (N: P as P_2O_5: K as K_2O). Costa and Gard-

TABLE 12.4.   Typical Composition of Oregon Crab and Shrimp Processing Wastes.

Constituent	Fresh (%)		Dry Basis (%)	
	Shrimp Waste	Crab Waste	Shrimp Waste	Crab Waste
Nitrogen	1.3	1.6	5.6	4.4
Phosphate (P_2O_5)	1.0	1.2	4.3	3.3
Potash (K_2O)	0.05	0.30	0.22	0.83
Sulfur	0.06	0.18	0.26	0.50
Lime ($CaCO_3$ equivalent)	6.0	15	26	42
Magnesium	0.18	0.33	0.78	0.92
Water	77	64	—	—

Source: Costa and Gardner. 1978.

ner (1978) reported the typical Oregon crab and shrimp processing waste characteristics shown in Table 12.4. The $CaCO_3$ content of shellfish wastes acts similarly to lime and is very beneficial for acid soils. Shellfish wastes, particularly shrimp wastes, are relatively deficient in potassium.

USE OF SEAFOOD WASTES AS FERTILIZERS

The use of fish as fertilizer in the United States dates back to the early Pilgrims. The first recorded use was at Plymouth, Massachusetts, where herring were planted along with Indian corn in 1621, a practice supposedly acquired from the Indians. Evidence uncovered by Ceci (1975), however, strongly suggested that fish fertilizer was not a native North American practice, but was conveyed by the Indians from one group of European settlers to another.

   Use of organic materials such as scrap fish and fish wastes as fertilizer was the popular practice until the early twentieth century when inorganic fertilizer technology developed. By 1930, low cost inorganic fertilizers had replaced most organic materials. From 1950 to 1970, the dominance of inorganic fertilizers on the commercial market was nearly complete. Fish waste fertilizers were produced only by isolated small plants for specialty applications. In recent years, energy and environmental concerns and rapid price increases for inorganic fertilizers have renewed interest in the use of waste organic materials of all kinds for fertilizers.

Processing and Application

Seafood wastes ·may be applied as fertilizer either unprocessed (fresh) or processed (dried, composted, etc.) Unprocessed seafood wastes have many undesirable characteristics, such as:

1. High water content (60 to 90%).
2. Extreme liability to becoming putrid, due to the high lipid content.
3. Difficult application, requiring specialized equipment.
4. Seasonal production (see Table 12.1).

The high water content makes economical transportation possible only for very short distances (less than 25 km).

Because of their putrescibility, unprocessed seafood wastes have to be removed from the processing plant and applied immediately. Surface applied wastes can create odors and attract insects and rodents; therefore, these must be immediately tilled into the soil. Other possible application techniques are the plow-furrow-cover method, in which wastes are deposited in a plowed furrow and immediately covered, and subsurface injection, in which the waste is injected 15 to 20 cm beneath the soil surface (Mantell, 1975).

Seafood wastes can generally be applied uniformly with a manure spreader (Costa, 1977). However, grinding of the waste at the processing plant will result in a more homogeneous material that is easier to apply.

The seasonal nature of the seafoods processing industry may conflict with the farmers' requirements for fertilizer. The practicality of using seafood wastes in a particular locale will depend upon the demand for fertilizer during the seasonal seafood processing periods. However, the wastes may be used as a soil conditioner during all seasons because the nutrients in seafood wastes are organic and will be released slowly.

Where sufficient local agricultural land is available, fertilizer demand is sufficient during processing seasons, and cooperative arrangements between seafood processors and farmers can be developed, the use of unprocessed seafood wastes as fertilizers may be a very economically attractive method for processors to dispose of wastes and for farmers to supply nutrients to their crops. This method is especially applicable to small, isolated seafood processing plants, where shipment to a centralized waste processing plant is not practical.

Processing of seafood wastes may be practiced to increase the stability, reduce the water content, increase nutrient concentrations, and produce a fertilizer which is convenient to apply. In a dry, stable state, storage and transportation over relatively long distances to available markets is economically practical. The primary methods by which seafood wastes may be processed include drying, composting, and extraction of valuable components.

The most direct approach to producing a dry, stable fertilizer is drying. Drying and grinding of seafood wastes into a fertilizer-grade meal would be very similar to a fish meal process (see Fig. 12.4). Fryer and Simmons (1977) list this product as a 10:4:1 fertilizer.

Composting is a method of converting organic matter into a drier, non-odiferous form through bacterial action. The organic matter is dumped into piles (windrows), kept moist, and allowed to stabilize over a period of several weeks. Seafood wastes, however, are generally too watery for direct incorporation into compost piles and need to be de-watered or mixed with drier fibrous materials before composting. Also, the market for compost is essentially limited to lawn, garden, and specialty uses. Many other, less valuable organic wastes are better suited for compost.

Use of fish soluble (a waste product of fish meal processing) as a specialty liquid fertilizer has been practical for more than 30 years. By hydrolyzing the proteins in seafood wastes, most of the nutrients can be extracted in liquid form to produce a similar product. A commercial operation in Minnesota presently uses such a process on waste carp to produce an enriched 8:4:4 liquid fertilizer. Hydrolysis is accomplished in a steam cooking process (Fryer and Simmons, 1977).

Fish Wastes

As mentioned previously, fish wastes are only slightly lower in protein and nutritional value than the portion normally utilized. Extensive recovery of fish wastes as animal feed by-products is presently practiced and is becoming profitable even for smaller fish processing plants. Also, research and development is being directed at increasing product yields in fish processing and at further recovery of protein by-products such as fish flour, fish meal, and fish protein concentrate from fish wastes.

The market value of dried, ground, and bagged fish wastes (fish meal) as fertilizer is only a fraction of the value as animal feed and it is not likely that the relative prices will change sufficiently to make use as fertilizer attractive. Consequently, the use of fish wastes as fertilizer will generally be limited to a relatively small amount of fish wastes produced at small, isolated plants where waste processing facilities are not practical and to fish wastes which are not suitable as animal feed, such as concentrated fish solubles from fish meal production.

Shellfish Wastes

Unlike fish wastes, only a small portion of the shellfish wastes presently generated are being recovered as by-products. This is particularly true of crab and shrimp wastes and is the result of two factors. First, practical and profitable by-product markets that can utilize the quantities of waste available have not developed. While some wastes are processed to shellfish waste meal and some for chitin, these by-products have very limited markets. Second, crab and shrimp processing plants are typically small and are scattered along the coastal regions throughout the country. Therefore, extensive by-product recovery facilities are not economically available to individual plants and centralized processing facilities are not practical. As a result of these factors, the marketing of shellfish wastes as fertilizer may be an attractive option for many crab and shrimp processors.

Costa (1977) evaluated the use of crab and shrimp wastes in Oregon coastal soils. A greenhouse experiment was conducted to determine the effects of grinding, surface application versus incorporation into the soil, and waste applications versus equivalent inorganic fertilizer application. The fertilizer materials were applied onto two coastal soils and two pasture crops were subsequently grown. Grinding crab waste significantly increased forage yields when the waste was surface applied, but not when incorporated with the soil. However, unground shrimp waste gave significantly higher forage yields. Shrimp wastes produced at least equal orchard grass yields when compared to an inorganic fertilizer with similar nitrogen, phosphorous, sulfur, and lime content, but they resulted in depleted soil potassium at the end of the growing season.

Encouraged by the fertilizer value of shrimp wastes, a group of growers in Lincoln County, Oregon have established the Coastal Farmers Cooperative to distribute shellfish processing wastes. The cooperative has entered into contractual agreements with processors in Newport to remove unprocessed shellfish wastes from the processing plants on a regular basis. In Oregon, shrimp wastes are being more extensively used for fertilizer as the peak production period, May through August, coincides with growers' requirements for fertilizer. Peak production periods for crab wastes are December to April, when fields are too wet for spreading vehicles. Shellfish wastes are generally made available to farmers at no charge.

TABLE 12.5 Comparison of Heavy Metal Content in Municipal Sludge versus Shrimp and Crab Wastes.

Constituent	Municipal Sludge (mg/kg dry weight) Mean	Municipal Sludge (mg/kg dry weight) Range	Shrimp and Crab Wastes (mg/kg dry weight)
Silver, Ag	225	0–960	1.1
Boron, B	430	200–1430	23–31
Barium, Ba	1460	0–3000	9–41
Cadmium, Cd	87	0–1100	—
Cobalt, Co	350	0–800	0.5
Chromium, Cr	1800	22–30,000	1–27
Copper, Cu	1250	45–16,000	34–117
Mercury, Hg	7	0.1–89	1.3
Manganese, Mn	1190	100–8800	12–400+
Selenium, Se	26	10–180	5
Strontium, Sr	440	0–2230	190–200+
Zinc, Zn	3480	50–28,000	59–264

Comparison with Other Organic Fertilizers

As discussed above, seafood wastes make excellent organic fertilizers. Furthermore, shellfish wastes appear to be the most available wastes for this use. However, the attractiveness of using shellfish wastes as fertilizers depends upon the quality, availability, and price of other organic fertilizers.

The primary potential competitors in the bulk organic fertilizer market are manures and sewage sludges. Domestic animals produce over one billion tonnes of fecal waste per year. However, these wastes are very watery (70 to 90% moisture), are difficult to dry, are highly odorous, and are typically only equivalent in nutrients to a 1:1:0 fertilizer. Furthermore, over 50% of these animal wastes are generated in concentrated growing areas close to urban areas.

Municipal sewage sludges are increasingly being applied to the land. These sludges are typically anaerobically digested to kill pathogens and stabilize the sludge. Digested sludge contains, on a dry basis, about 3% nitrogen, 2.5% phosphorous (as P_2O_5), and 1% potassium (as K_2O) (U.S. Environmental Protection Agency, 1977). Considering that digested sludge typically contains about 5% solids, the nutrient content in the liquid sludge is only about 0.15, 0.12, and 0.05% nitrogen, phosphorous, and potassium, respectively. Additionally, municipal sludges contain relatively high concentrations of heavy metals and may contain toxic organic compounds. A comparison of the typical heavy metal content in municipal sludge (U.S. Environmental Protection Agency, 1977) and that in crab and shrimp wastes (Costa, 1977) is shown in Table 12.5. Clearly, shellfish wastes are vastly superior to liquid digested sludge as an organic fertilizer.

USE OF SEAFOOD WASTES AS FEEDSTUFFS

Worldwide production of various fish materials that are considered unsuitable for direct human use is enormous. Steinberg (1973) noted that the total industrial fisheries volume in the United States exceeds that of the fish food industry by more than five times. Since many of these "waste" materials are exceedingly rich in nutrients, especially protein, it becomes clear that use of this material as feedstuffs could lead to substantial improvements in human nutrition. Further stimulus to the development of appropriate technologies results from the many economic and environmental benefits that may also be realized.

In this section, the possibilities for the use of fisheries by-products as feedstuffs for the production by animals of meat, milk, eggs, and fur are briefly examined. An attempt is made to present pertinent information which points to the current "state of the art" of feeding fisheries wastes to productive animals, and to emphasize those aspects of greatest use to those not directly associated with the field of animal nutrition. Although great volumes of these materials are currently utilized in the feeding of pets in industrial countries, this aspect is not specifically considered. Also excluded from explicit discussion are the many economic concerns which constitute the essence of decision making in this area.

Types of Fisheries Feedstuffs and Their Characteristics

Meals. The most thoroughly understood class of fisheries products used for animal feeding are the meals which are derived from either whole or waste parts (heads, skeletons, and viscera) of a large variety of pelagic teleost species. Meals from whole fish have been used extensively as feedstuffs over many years. However, the meals are given relatively extensive treatment here for a number of reasons. First, the literature concerning fish meal is very extensive, and newly developed products and their nutritional attributes are often discussed in relation to those of fish meal. Second, many problems associated with feeding fish wastes were first identified and subsequently elucidated during work with fish meals, and the same limitations are frequently encountered in feeding other fish by-products. Finally, materials more usually considered wastes are often added back to meals in hopes of supplementing their nutritional value.

Fish meals have gained wide acceptance as feedstuffs, especially as supplements to non-ruminant rations based on grains such as corn and wheat. Brody (1965) listed several of the favorable qualities of fish meals, including 1) exceptional value as sources of high quality proteins, especially the amino acids lysine and methionine, which are often limiting in cereals; 2) richness in B-complex vitamins, as well as vitamins A and D in residual oils; 3) high levels of such macroelements as Ca and P and many of the micro-elements, particularly Fe, Cu, and I; 4) low levels of fiber; and 5) generally high availability of nutrients in all fractions. So-called unidentified growth factors (UGF) are also present and this is frequently emphasized. That very large quantities of these meals are available was indicated by Tarr and Biely (1973), who estimated that as of 1973, 31% of the fish landed annually are processed into fish meal directly, and James (1975), who stated that about 40% of the total annual tonnage of fish taken ultimately are marketed as fish meal.

Numerous authors have published data describing the composition of fish meals (for example, Stansby, 1963, and Braude, 1962) and representative examples are given in Tables 12.6 and 12.7. Since fish processed into meals vary as to species (among the most common are herring, menhaden, anchovy, sardine, mackerel, and several species of "whitefish" such as hake and codfish), seasons of capture, and method of reduction employed, published analytical values may be expected to vary widely. Fishing regulations generally ensure that fish are not taken during the breeding season, a time when fat content is relatively low and nucleoproteins high (Tarr and Biely, 1973). Thus, nutrient content for a given species will generally not vary so much as between species values, and meal quality, especially as it concerns the availability of nutrients, will depend primar-

TABLE 12.6. Composition in Proximate Analysis of Fish Meals. (%)

Meal	Portion of Fish Used	Protein	Oil	Ash	Moisture
Herring	Whole fish and scrap	60–77	8–12	8–11	6–10
Menhaden	Whole fish	50–65	5–15	16–25	5–12
Ocean perch	Fillet scrap	50–60	8–12	20–24	5–10
Salmon	Cannery scrap	55–65	10–14	12–30	6–10
Tuna	Cannery scrap	60–65	5–10	20–23	7–10

Source: Sanford and Lee. 1960.

TABLE 12.7. B-Vitamin Content of Fish Meal.

Vitamin	mg/lb	mg/kg
Riboflavin	3	6.5
Niacin	30	65
Pantothenic acid	3	6.5
Choline	1,500	3,261
B-12	0.1–0.33	0.2–0.72

Source: Brody. 1965. (Reprinted with permission of the Avi Publishing Company, P.O. Box 831, Westport, Connecticut 06880.)

ily upon the method of processing. In order to further encourage uniformity of composition and quality among fish meals on the market, the American Feed Manufacturers Association has published a set of recommended standards (Stansby, 1963). Brody (1965) has summarized the content in fish meals of major proximate components as follows:

Protein   55 to 70% by weight, commonly between 60 and 65%.
Lipids   5 to 10%. If levels are too low, the meal may tend to be dusty, and if too high, oxidation, overheating, and spoilage can occur. About 8% is ideal.
Ash   Exceedingly variable; depending on raw materials used, can be between 12 and 33%. Levels ranging from about 15 to 20% are most suitable in typical feeding situations.
Fiber   Less than 1%.
Water   6 to 10%, with approximately 8% being ideal. Too much moisture encourages mold and spoilage; too little can lead to excessive heating and oxidation.

Whale and eviscerated dogfish (*Squalus spp.*) have been prepared and tested experimentally against those made from teleost species. The very high oil content and the sandpaper-like texture of the skin of dogfish make processing into meal difficult. The harsh conditions required for preparation have resulted in a consistently inferior quality compared to other meals in poultry rations (March *et al*, 1971) and salmon feeds (Satia and Bramon, 1975). These meals have low lysine availability for poultry feeds. Soderquist *et al* (1970) described meals made from fish viscera and catfish wastes that suggested suitability for use as animal feeds. However, neither these nor dogfish meals are commonly encountered in feed formulations at present.

A number of meals prepared from the processing wastes of several crustacean species, including shrimp, crabs, and crayfish, are also showing promise as feedstuffs. In addition to the more usual formulations suited for poultry, swine, and mink, several new markets for these products have become apparent. Watkins (1976) noted that a burgeoning shrimp and prawn cultivation industry in the southeastern United States has successfully utilized diets containing up to 30% shrimp meal. Also, chitosan, a de-acetylated derivative of chitin, has been found to have value as a coagulating agent, whereby organic solids which otherwise would be lost may be recovered from a variety of food processing effluents.

Crustacean wastes are made up largely of protein, minerals, and chitin, the latter being a polymer of N-acetyl-D-glucosamine units, containing about 6.8% nitrogen. The composition of meals derived from these wastes are discussed in detail by Watkins (1976), who summarized their content as being 25 to 50% corrected crude protein, 10 to 30% chitin, 20 to 40% ash, and 2 to 10% fat. Lovell *et al* (1968) indicated that values obtained for fiber can provide an adequate index of chitin content. Crude shrimp wastes usually have a considerably higher protein/ash ratio than crab or crayfish wastes, and thus would seem to be more favorable as feed ingredients for most purposes. Shrimp and certain species of crab also contain high levels of a red carotenoid pigment, astaxanthin, which, when included in poultry or fish rations, can alter the pigmentation of skin and/ or flesh (Chawan and Gerry, 1974). An example of proximate analysis data for freshwater crayfish waste meal is given as Table 12.8. This product is of particular importance commercially, as along the south Gulf Coast, where waste materials comprise some 85% of the catch liveweight (Lovell *et al*, 1968). It should be noted that the chitin is as unavailable as a source of energy or nitrogen to nonruminants as cellulose is of energy, in that neither is susceptible to enzymatic hydrolysis in these animals.

Oils. Fish oils are probably the most valuable of the products obtained from fish rendering, having many industrial uses beyond those as animal feedstuffs. As feed ingredients, however, they have been a valuable source both of energy and of vitamins A and D, although with the wide acceptance of synthetic vitamin preparations, this aspect of their value has been diminished (Scott *et al*, 1976). Many species of fish of particularly high oil content, such as herring, menhaden, anchovies, and the cartilagenous fishes, store large quantities of oil in muscle

TABLE 12.8.   Average Composition of Fresh-water Crayfish Waste Meal.

Component	Quantity
Minerals	
Magnesium (ppm)	2656.0
Manganese (ppm)	157.0
Potassium (ppm)	1400.0
Iodine (ppm)	1313.0
Iron (ppm)	8.8
Calcium (%)	18.1
Phosphorous (%)	1.2
Corrected crude protein (%)	32.2
Ether extract (%)	4.9
Chitin (%)	14.1
Crude fiber (%)	14.2
Ash (%)	29.0

tissues as well as in the liver, where it tends to be concentrated in "whitefish" species such as codfish, haddock, and hake. Brody (1965) has summarized the chemical nature of fish oils from many species, but it may be mentioned that compared to most vegetable oils they are highly unsaturated (of the order of 75% of the fatty acids), with fatty acids tending to be longer in chain length, unsaturated ones predominantly ranging from C_{16} to C_{22}. Unsaturated fatty acids are more readily digestible than saturated ones in the diets of non-ruminants, but this advantage may often be offset by their greater chemical reactivity and susceptibility to oxidation during storage. For this and other reasons, their levels in poultry and swine rations must be adjusted closely.

Fish Protein Concentrate (FPC).   This powdered product (sometimes called fish flour), which may be made only from whole fish or fleshy parts and not from offal, is strictly defined as to composition by federal regulations (Soderquist *et al*, 1970), and its high quality and cost have effectively restricted it to use as a protein supplement for humans (Steinberg, 1973). Stipulations as to composition are that it consists of at least 75% protein, not more than 0.5% lipids, and less than 10% moisture. Ash content varies from 2 to 25%, depending upon whether or not skeletons were removed before processing; thus, protein and ash percentages tend to vary inversely (Northrup Services, Inc., 1973). Lipid content depends upon the efficiency of extraction below the 0.5% maximum, which, in turn, depends upon the method used. A very low lipid content (i.e., 0.1% or less) seems to enhance the stability of that which does remain. The vitamin content is very low. Although the isopropyl alcohol extraction method appears to be in use most commonly, there are indications that ethylene dichloride extraction or enzyme hydrolysis (the latter leading to a product commonly called "enzyme hydrolyzates") may yield products of inferior nutritional quality (Northrup Services, Inc., 1973). Soderquist *et al* (1970) provided a detailed discussion of the technologies and products which have been developed.

Many studies have shown improvements in growth and efficiency of protein utilization when FPC is used at low levels in diets. Toma and James (1975) cited other workers (Dubrow and Stillings, 1971; Stillings, 1967; Stilling *et al*, 1969) who have determined that cystine and methionine are often limiting, but high levels of lysine and other essential amino acids make FPC a good complement to cereals, and even diets with soybeans as a major protein source have shown improved performance with some FPC added (Sidwell *et al*, 1970, in Toma and James, 1975). Optimum levels of use are most often reported to be under 3% of the diet or under 8% of the total protein in the ration. Further information concerning the composition and nutritional adequacy of these materials is given in the F.A.O. publication compiled by Northrup Services, Inc. (1973).

There is evidence that other kinds of fisheries wastes may also be valuable in the form of protein concentrations. Toma and James (1975) reported that in rats, shrimp waste protein "promoted rat growth 80% as efficiently as casein" and that the protein quality of a soybean based diet was improved by 74%. He suggested that this formulation should make an excellent supplement in animal feeds.

Condensed Fish Solubles (CFS). Materials classified as condensed fish solubles (CFS) are produced when the "stickwater" resulting from fish processing is recovered and concentrated. Procedures for its production were first developed for large-scale use as a means of water pollution abatement, but since the mid-1940's, their great value as feed constituents has been recognized, especially for poultry (Stansby, 1963). In some instances, these solubles have been added back to fish meals, resulting in so-called "whole meals," a procedure which substantially increases the nutritional value of the meals so supplemented.

Before treatment, stickwater usually contains 6 to 10% dry matter. Following condensation, dry matter values increase to between 40 and 50%, of which as much as 35% is crude protein. However, it is often the case that over half of the fraction determined as protein is actually non-protein nitrogen, a fact which has suggested an abundance of uses in the feeding of ruminants (Braude, 1962). Thus, amino acid content may often be relatively low, and is in any case extremely variable. Fish solubles are most often added to non-ruminant rations as sources of B-vitamins, micro-elements, and UGF. In mashes fed to poultry, they are often included for the additional reason that they help reduce the dustiness

TABLE 12.9. Typical Analysis of Condensed Fish
Solubles.

Parameter	Value
Total solids	50.43%
Ash	8.86%
Fat	4.8%
Crude protein (N × 6.25)	33.85%
Specific gravity at 20°C	1.20
pH	4.5

TABLE 12.10. Comparison of Herring Solubles to Herring Meals.

	Dry Matter (%)	Riboflavin (mg/kg)	Nicotinic Acid (mg/kg)	Pantothenic Acid (mg/kg)	Vitamin B-12 (mg/kg)
Vacuum condensed herring solubles	50	14	175	100	0.8
Ordinary herring meal	90	5	68	17	0.23

Source: Braude. 1962.

of the ration. A typical proximate analysis and a comparison of condensed fish solubles and herring meal as to B-vitamin content are included as Tables 12.9 and 12.10, respectively.

Fish Silage. Fish silage is a product of liquifaction at an acid pH of fish or fish wastes by naturally occurring enzymes. Because this method of handling is very simple relative to those previously mentioned, requiring only a mincer, acid, and drums for storage, it would seem particularly adaptable for use in underdeveloped countries. However, it is not easy to transport and may contain as little as 15% protein (James, 1975). Soderquist *et al* (1970) indicated that several acids have been utilized experimentally in silage preparations, but Whittemore and Taylor (1976) noted that use of formic acid has a number of advantages, including 1) stabilization of pH at about 4, which seems optimal for the activity of the enzymes involved; 2) effective restriction of bacterial growth and spoilage; and 3) removal of neutralization requirements prior to feeding. Another method for ensiling has been employed in Sweden, and involves fermentation using an acid resistant *Lactobacillus*, which is added to a fish slurry and warmed slightly. The lactic acid produced effects preservation (James, 1975). It has been found that, by whatever means it is produced, extraction of oil before ensiling leads to a product with more favorable storage and nutritional properties. Unfortunately, due to the great variability of composition among silages, a typical analysis is not available. As might be expected, fish silages have been found most useful in the feeding of swine.

Raw Fish. Feedstuffs containing fish from which the oil has not been extracted are notoriously unstable under conditions of storage. However, whole fish or fish viscera have been used with good success as the basis for mink rations for some years. The only processing generally has involved washing, freezing, and—following transport to the mink ranch—grinding prior to feeding in combination with other ingredients (Stout, 1973). Sometimes grinding precedes packaging and freezing, with the frozen packages being moved to the ranch for use as needed (Sanford, 1957).

Quality Evaluation and Limitations in Use of Fisheries Derived Feedstuffs

The discussion above has focused on fisheries by-product feedstuffs in terms of their proximate chemical characteristics, with some further reference to their amino acid, vitamin, and possible UGF contents. However, the nutritional value of feed depends not only on nutrient composition, but also upon digestibility and availability to the animal. Interactions of nutrients are often complex, as is particularly the case among mineral elements (Underwood, 1977). For example, Grau and Carroll (1958) and Ousterhout and Snyder (1962) have aptly considered this question for the case of proteins and the statement of Grau and Carroll (1958) apply as well in principle to other proximate fractions.

"... no single figure can be devised to represent the value of a protein because proteins are not entities, but a collection of various amino acids combined in various ways and present in various combinations."

Particularly where materials as complex and variable chemically as those described are concerned, it is of the greatest importance that accurate assessments of the nutritional value of *what is actually being fed* be available. Both chemical and biological methods for such determinations have been developed and are discussed in detail in the literature (see Tarr and Biely, 1973, or Stansby, 1963). Biological assay, such as that reported by Ousterhout and Snyder (1962), using chicks for protein adequacy assessment, are especially valuable. Since most fisheries feedstuffs are used in protein supplementation, this fraction is usually of greatest interest, and several indices are in use which can facilitate comparisons among protein sources. Two of the simplest and most commonly seen (as discussed by Scott *et al*, 1976) are given below.

1. PER—"protein efficiency ratio." This is the earliest index developed for use with proteins, and is derived from the weight gain of the animal divided by the protein intake.

$$PER = \frac{\text{weight gain}}{\text{protein intake}}$$

2. BV—"biological value." Usually determined through nitrogen balance studies, it is the percentage of digested and absorbed N which is not excreted in the urine. This is an excellent reflection of amino acid balance, with "complete" proteins such as whole egg protein set at about 100; meats, 72 to 79; and cereals, 50 to 68.

The above values were obtained using diets containing the evaluated protein at sub-optimal levels for the assay species, so that determinations do not reflect protein excess rather than amino acid imbalance.

Factors Affecting Quality. Proteins which are present in fisheries sources are inherently of very high biological value. Therefore, the challenge has been to preserve their high value as they pass from the processing plant to storage and finally into animal rations. The most intractable source of difficulties has proven to be the oil fraction, which, as has been noted, is highly unsaturated and very liable to undergo oxidative degeneration. Products such as CFS and FPC, being comparatively low in oil content, are much more stable than meals, which must be handled with great care.

Deterioration of fish oils can occur at any time during the use cycle and may be brought about by chemical, enzymatic, or microbial activity. Upon chemical analysis, oils which have oxidized or meals containing them have elevated levels of free fatty acids (FFA) and lower iodine numbers than does the fresh material (Brody, 1965). Biological analyses using vertebrates generally show poor growth and, in some instances, evidence of toxicity, and very often the organoleptic quality of meat, eggs, or other products are diminished. Voluntary intake is reduced, doubtless due in part to poor palatability, but also implicated have been direct toxicities of peroxides or polymerized fatty acids, reduced protein availability, and subsequent deficiency as well as induced vitamin deficiencies, especially of vitamin E (Carpenter et al, 1963). Details of certain of these pathologies are given below.

Oxidative deterioration of these oils is encouraged by a number of factors, which may further interact with conditions of processing and storage. Lipases and lipoxidases both apparently occur naturally in fish tissues, and heating—to a point which often occurs as moist fish wastes are allowed to stand—may accelerate their activity. Also, simple atmospheric oxidation is accelerated when various metals (Fe, Cu, etc.) or organic compounds, like peroxides, are present. It is possible, but costly, to remove FFA from meals by alkali processing, but the end-products of lipoxidase activity are more difficult to eliminate. Both enzymes also reduce the vitamin A (retinol) potency of the preparation. Fortunately, both are readily inactivated by heating to 176 to 212°F for 15 to 20 minutes (Brody, 1965).

However, excessive heating during processing may also lead to impaired feed value of meals. Bissett and Tarr (1954), using enzyme digestion as an index of availability, found that lysine, arginine, and methionine were all more available in raw herring flesh and presscake than in heated meals, and that heating for 30 to 60 minutes at 180°C caused much more damage than at 154°C for all essential amino acids. Lea *et al* (1960) reported that herring meal stored in air for 3 months at 10°C lost 4% of the available lysine, as determined chemically. When it was subsequently heated for 30 hours at 100°C in nitrogen, a further loss of 12% occurred. In additional work, a sample of herring meal which was overheated and darkened badly during bulk storage lost 60% of the available lysine and resulted in poor growth when fed to chickens. Fresh and pre-oxidized meals heated in the laboratory for 30 hours in the absence of oxygen also darkened, and at 115°C lost 19 and 27% of lysine availability, respectively. At 130°C, 67 and 69% of the lysine were lost. Biely *et al* (1955) obtained particularly informative results with herring meal in varying oil extraction (oil contents of meals averaging about 7.4 and about 0.21% for unextracted and extracted meals, respectively), heat treatment (149°C or not at all) and length of heating (60, 120, or 180 minutes). They found that oil content did not affect the nutritive value of the unheated meals, while heating for 120 minutes damaged only the original and not the extracted meal. Heating for three hours, however, reduced the availability of all essential amino acids in both meals. Addition of fresh herring oil to rations containing extracted meal produced results insignificantly different from those reported as length of heating was varied. A relatively small reduction in nutritive value was also observed in meals heated for one hour. These results, combined with those from many other similar studies (see Clandinin, 1949, and Carpenter *et al*, 1963) have led to the conclusions that 1) heating for relatively short periods of time at moderately high temperatures damages heat-labile vitamins, and 2) the most serious losses of nutritive value occur as the result of Maillard-type reactions between carbonyl and free amino groups, especially the ϵ - NH_2 in lysine. Particularly severe heat treatment may also lead to reactions among amino acids themselves.

Alternatives (or additions) to the use of heat for controlling oxidation of lipids also exist. A large number of compounds are available which, when used in concentrations often of 0.1% or less, are quite effective as antioxidants. These fall into two broad classes: 1) those which occur naturally, such as tocopherol (forms of vitamin E), phosphatides, ascorbic acid and its esters, and selenium, and 2) synthetic compounds such as ethoxyquin, hydroquinone, and butylated hydroxytoluene (BHT) (Brody, 1965). Tarr and Biely (1973) reviewed results from a number of studies with antioxidants. For example, March *et al* (1965) found that the digestibility of lipids in fresh herring meal was 80% that of good herring oil and 91% if heated with antioxidants. Following 11 months of storage, these values became 70 and 82%, respectively. March *et al* (1965) also showed that meals treated with ethoxyquin or BHT and stored for 9 to 11 months were better protein supplements than untreated ones, although ethoxyquin at 0.05% seemed to afford little protection from loss of lysine availability. Lea, *et al* (1960) reported that BHT, monobutylhydroquinone (MBHQ), and especially diphenyl-*p*-phenylenediamine (DPPD) and 6-ethoxy-2,2,4-trimethyl-1,2-dihydroquinoline (Santoquin R) reduced the rate of oxidation of the oil contained in herring meal. In using red or pink salmon waste as mink feeds in Alaska, Leekley *et al* (1962) found that while both DPPD and BHT were effective in controlling oxidation, DPPD had a significant harmful effect on reproduction.

It should be stressed that deterioration of fisheries derived feedstuffs is not attributable only to oxidative processes as they have been discussed thus far. Briefly, psychrotrophic micro-organisms such as *Achromobacter*, *Pseudomonas*, and others may often be involved. Botulism, a response to a toxin elaborated by certain species of *Clostridia*, is not unknown (Stansby, 1963). Chilling to the freezing point of muscle ($-1°C$) or treating with bacteriostatic agents such as sodium nitrite, formaldehyde, or antibiotics have been more or less effective against these organisms, although in some cases difficulties have been associated with the use of certain of these compounds. Also, freezing and thawing cause a certain amount of damage to proteins and tend to increase the toughness and diminish the water holding capacity of the material. Dripping results in appreciable losses both of protein and water soluble vitamins. Tarr and Biely (1973) and Morris *et al* (1970) discussed the problem of *Salmonella* contamination, which has often arisen as a result of inadequate methods of handling during the processing of fish meals.

Common Toxicoses.   Several of the more common problems associated with the handling of fisheries by-products have been discussed in addition to some means of solution. Next, several specific types of toxicity which have been observed in feeding these materials are described. It will be noted that these syndromes were first observed and subsequently studied in mink (*Mustela vision*), particularly along the Pacific Northwest Coast. This is not surprising since it is an area where mink ranching has been intensively undertaken for many years (though the industry is presently in decline due to economic pressures), partly because of the ready availability of marine fisheries by-products.

<u>**Steatitis ("Yellow Fat" or "Wet Belly" disease).**</u>   The common names for this syndrome very adequately describe the typical signs of advanced cases. In addition to causing extensive mortality, abdominal and subcutaneous edema often

render the pelt worthless. This disease has been confirmed to be the result of feeding high levels of trienoic fatty acids, especially linolenic. Lalor *et al* (1951), for example, showed that linseed oil added to mink diets to contain 3% linolenic acid produced the typical pathologies. Interestingly, this disease was also relatively common before fish wastes were extensively utilized as mink rations. However, the horse meat used as the primary ingredient at that time contained linolenic acid in the fat at levels of up to 16% (Lalor *et al*, 1951). Vitamin E (α-tocopherol) is known to exert a significant protective effect, similar to that of exudative diathesis in chicks. Steatitis is also well known in swine. However, the most effective means of control is careful monitoring of the levels and quality of unsaturated fats in diets.

Hepatotoxicosis. It was observed by Stout *et al* (1968) that in mink which had been fed a diet including herring meal, even if for only several months, a degeneration and hemorrhaging of the liver occurred which resulted in sudden death. A search for the cause ultimately pointed to the use of sodium nitrite by fishermen, most commonly in Scandanavia, as a preservative for their catches. Trimethylamine, a product of bacterial decomposition, reacts with nitrite, forming the toxin dimethylnitrosamine, which in turn produced the lesions. This particular application of sodium nitrite has—it is hoped—been discontinued, especially as nitrosamines are considered potential human carcinogens (Stout, 1973).

"Cotton fur." Another abnormality in mink associated with feeding fish products is characterized by poor growth and emaciation, anemia, and an absence of pigmentation of the underfur. While mortality does not often result, the value of the pelt is greatly diminished. This syndrome has been shown to be induced by another product of the breakdown of trimethylamine in fish, formaldehyde, and is a physiological (not a nutritional) iron deficiency (Stout, 1973). That is, iron administered parenterally restores normal body levels, whereas dietary supplementation has no effect, indicating either that adsorption of iron from the gut is blocked or that the iron is "bound" by the formaldehyde and is unavailable. Stout *et al* (1960) reported that two species of fish, whiting (*Merluccius bilinearis*) and hake (*Merluccius productus*), included in otherwise adequate diets, increase the incidence of this abnormality, but it could be substantially reduced by heating at 88°C for 5 minutes. They also suggested that certain strains of mink were more predisposed to the problem than others.

Thiamine (vitamin B$_1$) deficiency. This syndrome has long been known in a wide variety of animal species—in humans as beri-beri, in chicks as polyneuritis, in small carnivores as Chastek's paralysis (after the fox-rancher among whose animals it was first recognized). Although it may be caused by an absolute deficiency, it is generally the case among domestic or captive animals that it is induced by the presence of a thiaminase enzyme in feeds. Typical symptoms are severe anorexia (inappetence), emaciation and weakness, incoordination, paralysis (characterized in chicks by a "star-gazing" posture), and death. Parenteral thiamine administration results in a dramatic response, as ingestion presumably would if the animal could be induced to eat. Again, heat processing for a couple of minutes at 88 to 93°C denatures the thiaminase enzyme and eliminates the problem. Stout (1973) noted that the affected feedstuffs may also be fed to mink on an alternate day basis as a means of avoiding this deficiency.

Feeding Fisheries Products for Animal Production

Having covered briefly the nature of feedstuffs made from fisheries wastes, important aspects of their preparation from the standpoint of nutritional quality, and several problems which have been associated with their use, the discussion turns to the ways in which these feedstuffs may be fed to animals for production purposes. It cannot be stressed too greatly that actual feeding decisions should be made only after the available nutritional and economic information has been applied to the specific situation at hand.

Mink. Mink rations are high in protein (30 to 45% dry matter basis) and fat (18 to 28%) and low in carbohydrate compared to other animal rations. In the Pacific Northwest, about 70% of these diets consist of animal products, with the remaining 30% coming from plant sources, primarily cereals. Traditionally, mink rations have been fed at 60 to 70% moisture as a semi-solid mass, following thawing of the meat ingredients, grinding, and mixing, although Stout (1973) indicated that new formulations involving fish products as a dehydrated "fish flour" (FPC), as silage, or as a component in complete ration pellets dispensed with self-feeders are all showing promise. At this time, however, the usual ration involves wet processed and frozen whole fish or fish wastes.

A wide variety of fish species are suitable for incorporation into mink diets. Hake has been used successfully at a 25 to 30% level, although it has been less successful when used as a meal (*Pacific Hake*, 1970). Whole hake averaged 14.7% protein, 3.2% lipid, and 3.0% ash. However, a large proportion of hake taken are parasitized by the myrosporidian *Kudoa sp.*, the cysts of which develop in muscle tissue and reduce its quality. Leekley *et al* (1952) reported on the use of Alaska salmon cannery waste in mink diets, and found that a high percentage of salmon heads was detrimental to young mink if fed from the time kits could eat until they were about ten weeks old. For older animals, however, raw frozen pink salmon waste was an adequate basis for good gains. Roberts and Kirk (1964) studied the use of several species of freshwater fish in mink diets, fed raw and either with or without an added dry meal mix based on ground wheat, fish meal, and soybean meal. Animals fed suckers developed thiamine deficiencies only two days after the trial began. Further, animals on fish plus meal diets consumed an average of 25.4% more dry matter than the others, though this was at least partly due to the higher dry matter of the diets including meals. Apparent digestion coefficients (non-corrected for endogenous sources) for dry matter, energy, and nitrogen were significantly higher (P < .01) for suckers only diets than for suckers plus meal; however, nitrogen retention was higher (n.s.) for animals fed the meals. This suggests that perhaps an amino acid imbalance in the fish could be partially compensated for by inclusion of the meal. Very similar results pertain to sunfish based diets. Diets of ling only, however, were superior to ling plus meal for all these parameters.

Crustacean wastes have also received some attention as feedstuffs for mink. Watkins (1976) reported on the uses of king crab and shrimp waste, the former being included as a meal at up to 10% of the diet replacing equal amounts of fur seal meal. The crab wastes reduced mean final weight and tended to increase feed consumption, indicating reduced efficiency of conversion. At 3% crab meal, pelts were superior, but at other levels they were of lower value than those from control animals. Shrimp waste was used to replace, on a wet basis, 45, 45, and 30% halibut waste, salmon heads, and halibut waste plus salmon heads, respectively. Mink fed the shrimp meal rations showed 11.9, 1.0, and 3.5% reduction in mean final weights and 26, 20.6, and 19.0% greater feed consumption than respective controls. In his own work, Watkins (1976) substituted the fish carcass component with the following materials at the protein replacement levels indicated: shrimp meal, 10 and 20%; sieved shrimp, 10 and 20%; and crab protein concentrate, 2.75%. Results showed that the control diet, based upon 55% fish carcass, produced the greatest final weights and rates of gain across both sexes and strains of mink. Groups fed lower levels of shrimp performed better than those fed higher. However, it was noted that, unlike controls, waste fed animals increased their intake as the trial progressed. This was attributed to the fact that lipid levels were not balanced across treatments, with the result that the animals continually were adapting to the rations and attempting to balance energy intake. It was suggested that a further problem with the higher levels of shrimp was an excessive amount of dietary calcium.

Poultry. As has been mentioned, fish wastes as meals have long played an important role in poultry feeding as high quality sources of proteins, vitamins, minerals, and UGF. They have been used successfully at levels ranging up to about 10% in the diets of grower, layer, and breeder chickens (Tarr and Biely, 1973), but their increased cost and the availability of synthetic amino acids have reduced the extent to which they have been used in recent years. Numerous papers have been published which indicate that when properly supplemented, diets based on soy protein can support optimum performance in poultry (see, for example, Tarr and Biely, 1973; Biely, 1970; and Aitken *et al*, 1969). However, as feed products based upon fisheries materials which are truly wastes continue to be developed, they will no doubt find a place in poultry rations.

Amino acid balance is a highly important factor in realizing efficient gains in poultry. This has been shown by Schumaier and McGinnis (1969), who fed chicks diets containing whole herring meal as the sole protein source and where most of the energy was provided by glucose, at 64.35% of the basal diet. When 4, 8, or 12% additional protein from fish meal was substituted into the ration, growth was improved as the protein level reached 30% and did not decline even when 34% protein from the fish meal was being fed. However, it was found that chicks grew as well with 22% CP in the diet when 20% of the diet was corn as they did when the much higher CP levels from fish protein only were fed. Further, no non-protein supplements were able to elevate growth rates again when added to the basal fish meal/glucose diet. Therefore, it was concluded that both the very high fish protein levels and the supplementary corn were able to compensate for amino acid inadequacies, and that herring meal was not suitable for use as a sole protein source for chickens.

While this conclusion probably does apply to fish meals from virtually all sources, this is not to say that all are necessarily of equal value. Two comparative studies can serve to illustrate this point. Berg (1969) tested the relative feeding value of meals made from British Columbia herring, Pacific hake, Puget Sound hake, and Peruvian anchovy when each was fed to fryer chicks at 5% of a corn/ soybean diet with or without 5% meat meal. For all diets, eight week weights and feed efficiencies were reported to be excellent, although herring showed an insignificant superiority on all comparisons. However, when compared at levels of 20 to 24.3% of the diet, there were important differences, even though all diets contained 22% CP. The herring meal supported growth comparable to that seen at lower levels of use, whereas both the anchovy and hake meals produced marked depressions in growth rate. Further studies showed that the chicks had a dislike for anchovy meal, although a lot containing 10% lipids was preferred to one with 6%. Herring and hake were apparently equally acceptable. In a study comparing egg production among diets including meals from different fish species fed at 5 or 10%, Bearse (1971) also obtained lower production with anchovy diets, and feed efficiency was lowest in the anchovy and Norwegian herring rations. The British Columbia herring supported hatchability superior to hake and anchovy, as well as to the control soybean based diet when meals were fed at 5% of the diet.

An interesting problem among breeder chickens in which fish meals have been implicated is a large increase in mortality seen at about day 19 of incubation. Among other hypotheses that had been offered was one suggesting that improperly handled meals might have caused a reduced vitamin A potency in the diet, although experimentally induced vitamin A deficiency in layers is known to cause mortality on or about day 3 (Scott et al, 1976). Coles (1956) was able to demonstrate a substantial reduction in hatchability in eggs from birds fed either 5 or 10% whitefish meal, due to greatly increased mortality around day 19. Of the birds that did hatch, much higher losses were observed among birds whose dams had been fed whitefish, and the predominant cause of death was yolk sac infection traceable to spores originating in the fish meal. This problem does not seem to be of great import in the more current literature, however.

A large amount of work has been done on the feeding of crustacean wastes to poultry. Much of this was prior to World War II and is reviewed by Watkins (1976). The general trends of this early research were that meals from shrimp or crab could produce results comparable to those obtained with fish meal when used at levels of approximately 10% in the diets. In some cases, shrimp meal has been distinctly superior (Fronda, et al., 1934; Francisco, et al., 1934; Watkins. 1976), while crab meal has occasionally been inferior to fish meals, possibly due to excessive calcium content or poor protein availability. In a case where crab waste was fed on an equal protein basis with red fish meal, and calcium levels were equalized, there were no differences between groups as to egg production, feed efficiency, hatchability or egg quality (Watkins, 1976). Supplemental lysine and/or methionine have sometimes led to increases in production as compared to control shrimp meal rations. In a study using turkey poults, Potter and Shelton (1972) found that crab meal added to a corn/soybean meal based ration at 3 and 6% increased both weight gain and feed consumption by over 5% as compared to the controls, while the levels were not significantly different from each other. They were both inferior to the control ration from the standpoint of efficiency of utilization.

Crustacean wastes showed promise also as a source of pigments in poultry rations and, as such, are useful in accomodating the preferences of consumers. Chawan and Gerry (1974) found that skin pigmentation scores increased exponentially with increases in total ration carotenoids. When corn was the basis of the ration, 4% shrimp waste meal produced skin pigmentation which was unacceptably intense, whereas a wheat based diet could be supplemented with more than 6% shrimp meal without detriment. Watkins (1976) noted that these pigments are also transferred to egg yolks through feeding to layers. The proximate composition for shrimp meal is given in Table 12.11.

As has been mentioned, CFS have often been used in supplementing poultry rations with B-vitamins, micro-elements, and UGF. That the protein they contain is inadequate for supporting growth has been clearly demonstrated by March et al (1961), who determined a gross protein value for condensed herring solubles to be 67, versus 93 and 98 for herring meal samples. A similar conclusion was reached by Laksesvela (Baelum, 1963), also in comparisons of herring meal with herring solubles using chicks. Lassen et al (1949) found that improperly prepared CFS that had undergone some spoilage due to a lipid content of over 8% retarded growth in chicks. Lassen et al (1949) also observed

TABLE 12.11. Composition of Shrimp Waste.

Crude protein	46%
Calcium	9.5%
Phosphorus	2.4%
Amino acids	
Arginine	2.21%
Lysine	2.35%
Methionine	0.84%
Histidine	0.75%
Leucine	2.67%
Isoleucine	1.84%
Phenylalanine	1.87%
Threonine	1.58%
Valine	2.24%
Glycine	2.20%
Total carotenoids	81.7 mg/kg

Source: Chawan and Gerry. 1974.

that the non-protein nitrogen content increased from 1.32 to 4.67% as deterioration progressed and that certain amino acids (especially valine, methionine, and cystine) were destroyed to a much greater extent than others.

Finally, the matter of organoleptic taints being imparted to poultry products from fish derived feeds has been widely discussed. By holding dietary levels of fish meals to below 10% or, more to the point, oils to 1% or less during critical production stages, the problem can usually be avoided. Also, supplementing with vitamin E or other antioxidants can be helpful (James, 1975; Edwards and May, 1956).

Swine.   Fish and offal meals have been included in swine rations for many years, and results with them have often been superior to those obtained with diets lacking them. In many cases, it has been difficult to determine the causes of this superiority because of the differences among formulations and their preparation in the experiments which might otherwise offer valuable comparisons.

One particularly revealing experiment which is often cited is that by Laksesvela (1961), in which herring meal containing 10% oil was fed at 3, 4, 5, 6, 8, 10, and 12% of the ration, replacing extracted soybean meal on the basis of digestible CP. Barley meal and oat hulls were further varied so as to equalize energy values among rations. The amount of herring meal in each of the rations was reduced gradually from the initial levels given above as slaughter approached, and the meal was withdrawn completely when animals attained weights of 80 kg. This practice is a common measure taken to reduce the likelihood of off-flavors in the meat. The most pertinent results may be summarized as follows:

1. Pigs grew faster when fed diets with herring meal as compared to all-vegetable control diets.

2. Although the best response was at the 6 to 8% initial level, or 13 to 16% of total CP, within the same treatment the growth rate was highest at the higher level of herring meal (earlier in the experiment). The correlation between level of herring meal and growth response was $r = +.983$.

3. As to efficiency of utilization, substantial improvements are suggested by the fact that at the 6% initial level, total feed intake was 5% less than controls in reaching the same slaughter weight of 90 kg. At 12%, 7% of the feed was spared.

4. Quantitative indications of carcass quality (backfat thickness, organoleptic tests, iodine number of fat extracts) did not show significant differences among treatments. However, there was a tendency toward the detection of "oily" or "slightly fishy taste," as well as softened fat, at initial levels of 10 to 12%.

Several interesting suggestions might explain the above results. A superior amino acid balance might be expected to have been at least partly responsible, although, unfortunately, synthetic amino acids were not available at the time these trials were run. Higher levels of zinc in the fish meals were clearly involved as well, in that gilts on the basal diet prior to the experiment developed skin lesions characteristic of parakeratosis, which cleared up most rapidly in animals later fed the herring meal rations. Differences in vitamin B_{12} content were also thought to have been important.

Madsen *et al* (1965) showed the importance of proper processing in the preparation of fish meals for use in feeding swine. In one trial, four groups of pigs were fed a control diet and three treatments, the latter all consisting of 5% fish meal (FM), with FM_2 and FM_3 having been heated to 120°C for different lengths of time. It was determined that FM diets 1, 2, and 3 had 6.5, 4.6, and 2.1 g available lysine per 16 g N—or 87.4, 76.7, and 65.5% of the total lysine content, respectively. Again in summary, results are given below.

1. Dry matter and N intakes for the three treatments were very similar, but FM_3-fed animals had about 10% less pepsin-HCl digestible N and 20% less apparent digestible N than FM_1-fed animals. The average apparent dietary N digestibility figures for animals on FM_1, FM_2, and FM_3 were 76, 65, and 61%, respectively.

2. Groups fed FM_2 and FM_3 grew at rates far inferior to FM_1-fed animals. The experiment was ended when all animals had consumed an equal quantity of dry matter, and mean group weights for 1, 2, and 3, respectively, were 50.5 kg, 40.0 kg and 31.0 kg. These differences were highly significant. Thus, excess heat treatment again results in a nutritionally inferior product.

There has been some interest in the effects on swine reproduction of rations containing fish meal, and generally, indications are that they are favorable.

Palmer *et al* (1970) conducted a study over two reproductive cycles through two generations of animals in which a whole menhaden meal was added at 6.67% of the ration to a basal diet. Analyses showed "no large differences" among rations as to protein, amino acid, or mineral content, although lipid percentage was somewhat higher in the diet with FM. Animals fed the FM containing ration had significantly higher rates of gain, an effect which was even more pronounced in the second generation of animals; significantly more total and live pigs were farrowed (average increase of 0.8 and 0.9 pigs, respectively), with a slightly higher average birth weight of live pigs (1.28 ± 0.03 versus 1.23 ± 0.03 kg), and a significantly greater average birth weight for live pig litters. Effects as to litter size at weaning, pig weaning weights, or pig weight gains from birth to weaning were not significant.

Watkins (1976) noted that little information is available on feeding crustacean waste to swine, although the fact that swine formulations do not need to be dried and ground suggests that research in this area is definitely warranted. He does cite several studies in which shrimp meal has been tested, tending to give results comparing favorably with fish meal or tankage. So far as flavor defects are concerned, the lower fat content of crustacean meals than most fish meals would be an important advantage, although swine adapt especially well to diets with high lipid levels.

As has been mentioned, swine show great promise in being able to utilize fish silages, a technology which would likely be applied to crustacean derived materials as well. Whittemore and Taylor (1976) found that ensiled herring offal fed to swine as 25% of the dry matter in a ration based on barley meal reduced total dry matter intake somewhat as compared to controls, but improved digestibility coefficients for dry matter, gross energy and nitrogen significantly. Nitrogen balance and retention were also more favorable.

Fish. In many parts of the United States, the aquaculture industry has experienced a resurgence in recent years, and it is increasingly being forced to turn to more economical sources of high quality protein, such as fish wastes. The fact that diets for fish almost invariably exceed 40% protein serves to emphasize the great demand this industry will increasingly have for these materials. The best known and most successful formulations for fish, such as the Oregon Moist Pellet (OMP) specify large proportions of fisheries products (Church, 1977), and dry, pelleted fish feeds typically contain about 50% FM in starting and 30% in growing rations (Tarr and Biely, 1973).

As an example of the value that fish wastes may have in rations for fish, a study by Satia and Bramon (1975) may be mentioned. Using OMP as a control, by-products from six species of fish were fed to Coho salmon alevins at about 63% of the diet by weight. Rations were fed at 4% of the body weight of the fish from the experimental group with the lowest average weight. Table 12.12

TABLE 12.12.   Proximate Composition and Amino Acid Content of Diets Fed to Young Coho Salmon.

	Flatfish	True Cod	Evis-cerated Dogfish	Salmon Carcass	Whole Dogfish	Herring	Oregon Moist Pellet	Require-ments[1]
Protein (N × 6.25) (%)	42.7	45.2	42.9	43.4	40.6	43.9	44.6	–
Lipid (%)	6.9	7.1	8.6	7.4	10.5	9.2	8.9	–
Ash (%)	10.2	9.0	10.0	8.6	10.3	7.8	7.4	–
Moisture (%)	32.0	30.4	30.8	32.6	31.6	31.8	31.5	–
Carbohydrate (%)	8.2	8.3	7.7	8.0	7.0	7.3	7.4	–
Amino Acid (%)								
Alanine	2.2	2.4	1.9	2.0	2.1	2.1	2.1	–
Arginine	2.0	3.3	1.9	2.3	2.0	2.0	2.0	2.5
Aspartic acid	2.6	3.1	2.4	2.8	2.9	3.1	3.2	–
Glutamic acid	4.1	5.0	4.2	4.4	4.7	5.0	5.0	–
Glycine	3.1	3.2	3.0	2.7	2.9	2.1	2.1	–
Histidine	0.6	0.7	0.6	0.7	0.7	0.6	0.8	0.7
Isoleucine	1.3	1.4	1.1	1.3	1.4	1.5	1.6	1.0
Leucine	2.0	2.3	1.8	2.0	2.1	2.5	2.5	1.5
Lysine	1.9	2.4	1.7	2.1	1.7	2.4	2.5	2.1
Methionine	0.8	0.9	0.6	0.8	1.0	0.9	0.9	0.5
Phenylalanine	1.0	1.2	0.9	1.1	1.1	1.3	1.4	2.0
Proline	2.0	2.2	2.0	1.8	2.0	1.7	1.8	0.8
Serine	1.5	1.7	1.2	1.3	1.4	1.4	1.4	–
Threonine	1.2	1.5	1.1	1.3	1.4	1.4	1.5	–
Tryptophane[2]	–	–	–	–	–	–	–	0.2
Tyrosine	0.9	1.1	0.7	0.9	1.0	1.1	1.1	–
Valine	1.4	1.7	1.2	1.4	1.5	1.7	1.8	1.5

[1] Requirements for chinook salmon.
[2] Not determined.
Source: Satia and Bramon. 1975.

gives the proximate composition of the diets used. Their results show that there were excellent gains and efficiencies of conversion for all diets save those based on dogfish, and mortality was low in all cases. Generally, essential amino acid complements of the diets were in excess of published requirements with the exception of phenylalanine, arginine, and lysine, but phenylalanine should not be limiting, given that tyrosine can exert a sparing effect on its levels in the diet. All diets also contained suitably high amounts of polyunsaturated fatty acids, which are essential for salmonids. Economically, all the experimental diets, again excluding dogfish, showed the possibility of being much less expensive than OMP.

Crustacean wastes are often included in many fish diets (see OMP formula), primarily as sources of pigments, as has been seen for poultry. Saito and Regier (1971) noted that such materials as raw fresh shrimp waste, vacuum dried wastes, and high temperature dried wastes have been used. In their study, a vacuum dried material from carapace and abdominal shells of pink shrimp (*Pandalus norealis*) was fed at either 20 or 30% of a pelleted mixture. This experimental diet was determined to have the following pigment content: astaxanthin, 10 mg/g dry matter, astacene, 0.91 mg/g dry matter, and non-carotenoidal, 1.08 mg/g dry matter. Another diet was made by a similar process from snow crab (*Chinoecetes opilio*) carapace and leg shells, and had a pigment analysis of lutein-like carotenoid, 0.01 mg/100 g dry matter, astaxanthin, 0.47 mg/100 g dry matter, and astacene, 0.17 mg/100 g dry matter. This was included at 20% of dry matter intake, also in pellets, and both were fed to brook trout. No conclusions could be drawn concerning growth rates or feed efficiency, except to note that there was no apparent growth inhibition. The 20% crab diet could not be distinguished from controls as to the lipid content of the flesh or coloration of either the flesh or skin. However, the shrimp based diets caused the lipid content of the flesh to increase with time (at 4, 8, and 12 weeks). After 8 weeks of the experiment, shrimp fed fish had pinkish-orange flesh and the color intensity could be correlated with the crude carotenoid content. The group fed the 30% shrimp diet had the same coloration at 8 weeks as the 20% shrimp fed fish at 12 weeks. Also, the detectable carotenoids in the flesh became more varied with time, enabling the details of the deposition process to be discussed.

Steele (1971) fed a shrimp meal made from viscera and meat particles at 15% dry matter intake to trout, and noted increased pigmentation of both skin and muscle as well as superior flavor and desirability of the flesh. He reported that for pigmentation of muscular tissue, the 15% shrimp diet and a pigment-lipid extract of the meal were equally effective, while for external pigmentation, the extracted supplement increased coloration more rapidly.

Cattle. To this point, the discussion has been limited to the ways in which fisheries by-products could be used as feeds for non-ruminant, or "simple-stomached" animals, In consideration of ruminants, however, one is dealing with a very different sort of digestive physiology and anatomy, one which is more adaptable in many ways and much less fastidious. Ruminants have nutritional requirements, as do non-ruminants, but often their needs can be met through the feeding of materials which are of virtually no use to non-ruminant animals.

Given that fish meals are generally fed for their value in protein and B-vitamin supplementation, it will be no surprise that they are of little but academic interest in ruminant nutrition. This is because of the ability of the rumen microorganisms to synthesize both high quality proteins and B-vitamins from very simple organic precursors. There is some evidence, though, that fish meal proteins may undergo ruminal degradation at a slower rate than plant proteins. This would be advantageous from the standpoint of efficient utilization of nutrients if it were the case that the insoluble protein passed through the rumen to be hydrolyzed in the abomasum, provided that the rumen micro-organisms were not then left with a reserve of nitrogen inadequate to meet their requirements.

Seoane and Moore (1969) studied this relationship in work with four fistulated steer fed anchovy meal substituting for soybean meal at 0, 33, 67, and 100% in a 22% CP supplement. There were no differences among rations as to dry matter intake, palatability, ruminal pH, or apparent digestibilities for dry matter, organic matter, cellulose, or ether extract. However, the fish meal caused a highly significant linear reduction in CP digestibility: at 0, 33, 67, and 100% substitution, digestible CP values were 53.7, 53.1, 51.6, and 49.4%, respectively. Rumen ammonia levels at 0, 2, and 4 hours after feeding were significantly reduced by the FM, and the rate of increase in total volatile fatty acid (VFA) concentrations following feeding were much reduced at 67 and 100% substitution. All these results indicate that, at least for this fish meal, much of the protein indeed did pass beyond the rumen before being utilized. Additionally, VFA proportions among rations were similar with the exception of valerate, which was decreased by increasing FM percentage to a highly significant extent at all sampling times. This could be an important result, as valerate is known to stimulate cellulose digestion by rumen bacteria. There was no measurable effect on cellulose breakdown reported in this study, however, nor is it obvious what the implications for production might be.

Fish solubles have most often been used in the feeding of cattle as additions to liquid supplements, which otherwise consist mainly of urea (a nitrogen source for bacterial protein synthesis), mineral and vitamin mixes, mendicants, and carriers such as molasses, which also contribute calories. Huber and Slade (1967) noted that use of solubles in this way "has improved both daily gains and feed efficiencies in beef cattle." Velloso et al (1971) indicated that while this may be the case, the evidence overall is rather equivocal.

Fish flour or protein concentrate would perhaps be expected to be uneconomical in any ruminant feeding situation. Interestingly, this is not the case for ruminant neonates, which for several weeks are functionally non-ruminants. A large amount a research has been done in hopes of finding less costly sources of protein for use in milk replacers than dried milk itself, and FPC shows some promise here. Huber and Slade (1967) found that for calves, fish flour included at 20 and 40% of dietary protein in a replacer fed from 3 to 45 days of age did not depress daily gains or feed efficiency, but at 60 and 67%, significant decreases were seen and at 100% death resulted. Although there seemed to be no interaction between fat and protein substitution levels, a linear increase in gains was noted as the fat content of the replacers was increased from 10 to 20% on a dry matter basis. Digestibilities of dry matter, CP, ether extract, and ash decreased as the amount of fish flour increased. Makdani *et al* (1971) also studied FPC in milk replacers for calves, concluding that poor performances with certain of these diets was attributable to poor protein quality relative to dried skim milk, in some instances aggravated by marginal vitamin E deficiencies and the presence of certain contaminants.

Finally, it may be mentioned that materials such as oyster shells may be of some value to large ruminants, both as supplemental calcium sources and for their contribution, as roughage, in maintaining proper rumen function.

CONCLUSIONS

Extensive research has been completed on the applicability of seafood wastes for fertilizers and feedstuffs. This research has shown that excellent opportunities exist in both areas, although numerous problems have been noted. It is hoped that these opportunities will continue to be exploited through the sound and novel application of both engineering and nutritional expertise.

REFERENCES

Aitken, J. R., J. Biely, A. Nikolaiczuk, A. R. Robblee, and S. J. Slinger. 1969. "Comparison of laying rations with and without animal protein at the Canadian random sample egg production test." *Brit. Poult. Sci.* **10**:247–253.

Baelum, J. 1962. "Fish and fishery products in poultry rations." *Fish in Nutrition.* E. Heen and R. Kreuzer (Eds.). Fishing News (Books), London, pp. 356–363.

Bearse, E. 1971. "Fish meals in rations of white leghorn laying and breeding chickens." *Feedstuffs* **43**:30–33.

Berg, L. R. 1969. "Comparison of various fish meals as ingredients in chick rations. *Proc. Pacific Northwest Animal Nutrition Conf.*, p. 60.

Biely, J. 1970. "Substitution of animal protein supplements with soybean meal in laying rations." *Feedstuffs* **42**:18–21.

Biely, J., B. E. March, and H. L. A. Tarr. 1955. "The nutritive value of herring meal 3. The effect of heat treatment and storage temperature as related to oil content." *Poult. Sci.* **34**:1274–1279.

Bissett, H. M. and H. L. A. Tarr. 1954. "The nutritive value of herring meals 2. The effect of heat on availability of essential amino acids." *Poult. Sci.* **33**:250, 254.

Braude, R. 1962. "Fish and fishery products in pig nutrition." *Fish in Nutrition.* E. Heen and R. Kreuzer (Eds.). Fishing News (Books,), London, pp. 332–352.

Brody, J. 1965. *Fishery By-products Technology*. The Avi Publishing Company, Westport, Connecticut.

Carpenter, K. J., C. H. Lea, and L. J. Parr. 1963. "Chemical and nutritional changes in stored herring meal." *Brit. J. Nutr.* **17**:151–169.

Ceci, L. 1975. "Fish fertilizers. A native North American practice," *Nature* **188**:26–30.

Chawan, C. B. and R. W. Gerry. 1974. "Shrimp waste as a pigment source in broiler diets." *Poult. Sci.* **53**:671–676.

Church, D. C. 1977. *Livestock Feeds and Feeding*. D. C. Church, Corvallis, Oregon.

Clandinin, D. B. 1949. "The effects of methods of processing the nutritive value of herring meals." *Poult. Sci.* **28**:128–133.

Coles, R. 1956. "Observations on the occasional depressing influence of fish meal on the hatchability of hens' eggs." *J. Agric. Sci.* **47**:354–362.

Costa, E. C. 1977. *The Fertilizer Value of Shrimp and Crab Processing Wastes*. M.S. Thesis, Oregon State University.

Costa, R. E. and E. H. Gardner. 1978. "A new look at old fertilizers: shrimp and crab processing wastes." *Oregon State University Sea Grant College Program*, *Pub. No.* **ORESUTL-78-001**.

Dubrow, D. L. and B. R. Stillings. 1971. "Chemical and nutritional characteristics of fish protein concentrate processes from heated whole red hake, *Urophycis chuss*" *Fish. Bull.* **69**:141–144.

Edwards, H. M., and K. N. May. 1965. "Studies with Menhaden oil in practical-type broiler rations." *Poult. Sci.* **44**:685–689.

Francisco, A. A., G. S. Chan and F. M. Fronda. 1934. "Protein supplements in poultry rations. II. Comparative effects of shrimp meal, meat scraps, tankage and fish meal as supplements in rations for laying hens," *Philippine Agric.* **22**:685–697.

Fronda, F. M., A. C. Badelles and J. S. Padilla, 1934. "Protein supplements in poultry rations. I. Comparative studies of the effects of shrimp meal, meal scraps, tankage and fish meal as supplements in rations for growing chickens," *Philippine Agric.* **22**:582–598.

Fryer, L. and D. Simmons. 1977. *Food Power From the Sea*. Mason/Charter, New York.

Grau, C. R. and R. W. Carroll. 1958. "Evaluation of protein quality," *Processed Plant Protein Foodstuffs*, A. M. Altschul (Ed.). Academic Press, New York, pp. 153–189.

Huber, J. T. and L. H. Slade. 1967. "Fish flour as a protein source in calf milk replacers." *J. Dairy Sci.* **50**:1296–1300.

James, D. G. 1975. "Total utilization of fish protein." *J. Aust. Inst. Agric. Sci.* **41**:27–30.

Laksesvela, B. 1961. "Graded levels of herring meal to bacon pigs, effect on growth rate feed efficiency and bacon quality." *J. Agric. Sci.* **56**:307–315.

Lalor, R. J., W. L. Leoschke, and C. A. Alvehjem. 1951. "Yellow fat in the mink." *J. Nutr.* **45**:183–188.

Lassen, J., E. C. Bacon, and H. J. Dunn. 1949. "The relationship of nutritive value of condensed fish solubles to quality of raw material." *Poult. Sci.* **28**:134–140.

Lea, C. H., L. J. Parr, and K. J. Carpenter. 1960. "Chemical and nutritional changes in stored herring meal." *Brit. J. Nutr.* **14**:91–113.

Leekley, J. R., C. A. Cabell, and R. A. Damon. 1962. "Antioxidants and other additives for improving Alaska fish waste for mink feed." *J. Anim. Sci.* **21**:762–765.

Leekley, J. R., R. S. Landgraf, J. F. Bjork, and R. A. Hagevig. 1952. *Salmon Cannery Waste for Mink Feed.* U.S. Fish and Wildlife Service, Fisher Leaflet No. **405**.

Lovell, R. T., J. R. Lafleur, and F. H. Hoskins. 1968. "Nutritional value of freshwater crayfish waste meal." *J. Agric. Food Chem.* **16**:204–207.

Madsen, A., V. C. Mason, and K. Weidner. 1965. "The effect of heated and unheated fish meals on protein synthesis, daily gain and carcass composition of pigs." *Acta Agric. Scand.* **15**:213–234.

Makdani, D. D., J. T. Huber, and R. L. Michael. 1971. "Nutritional value of 1,2-dichloroethane extracted fish protein concentrate for young calves fed milk replacer diets." *J. Dairy Sci.* **54**:886–892.

Mantell, C. L. 1975. *Solid Wastes: Origin, Collection, Processing and Disposal*, John Wiley & Sons, New York.

March, B. E., J. Biely, I. G. Claygett, and H. L. Tarr. 1971. "Dogfish in meals as supplements for poultry rations." *Poult. Sci.* **50**:1072–1076.

March, B. E., J. Biely, J. R. McBride, D. R. Idler, and R. A. MacLeod. 1961. "The protein nutritive value of liquid herring preparations." *J. Fisheries Res. Board of Canada* **19**:113–116.

March, B. E., J. Biely, H. L. A. Tarr, and F. Claygett. 1965. "The effect of antioxidant treatment on the metabolizable energy and protein value of herring meal. *Poult. Sci.* **44**:679–685.

Morris, K., W. T. Martin, W. H. Shelton, J. K. Wells, and P. S. Brachman. 1970. "Salmonellae in fish meal plants. Relative amounts of contamination at various stages of processing and a method of control." *Appl. Microbiol.* **19**:401–408.

National Marine Fisheries Service. 1978. *Fisheries of the United States, 1977.* National Oceanic and Atmospheric Administration, U.S. Department of Commerce, Washington, D.C.

Northrup Services, Inc. 1973. *Fish Protein Concentrate Information Package Parts I, II & III.* Prepared for the National Marine Fisheries Service.

Ousterhout, L. E. and D. G. Snyder. 1962. "Nutritional evaluation of fish meals using four short-term chick tests." *Poult. Sci.* **41**:1753–1757.

Pacific Hake. 1970. U.S. Fish and Wildlife Service, Circular **332**.

Palmer, W. M. H. S. Teague, and A. P. Grifo. 1970. "Effect of whole fish meal on the reproductive performance of swine." *J. Anim. Sci.* **31**:535–539.

Potter, L. M. and J. R. Shelton. 1972. "Evaluation of crab meal as a dietary ingredient for young turkeys." *Brit. Poult. Sci.* **14**:257–261.

Roberts, W. K. and R. J. Kirk. 1964. "Digestibility and nitrogen utilization of raw fish and dry meals by mink." *Amer. J. Vet. Res.* **25**:1746–1750.

Saito, A. and L. W. Regier. 1971. "Pigmentation of brook trout (*Salvelinus fontinalis*) by feeding dried crustacean waste." *J. Fisheries Res. Board of Canada* **28**:509–512.

Sanford, F. B. 1957. "Utilization of fish waste in Northern Oregon for mink feed." *Commercial Fisheries Rev.* **19**:40–42.

Sanford, F. P. and C. Lee, 1960. *U.S. Fish Reduction Industry*, Tech. Leaflet No. 14. Bureau of Commercial Fisheries, Washington, D.C.

Satia, B. P. and E. L. Bramon. 1975. "The value of certain fish processing wastes and dogfish (*Squalus suckleyi*) as food for Coho salmon (*Oncorhynchus kisutch*) fry." *The Progressive Fish Culturist* **37**:76–80.

Schumaier, G. and J. McGinnis. 1969. "Studies with fish meal as the sole source of protein for the growing chick. 1. Effect of different supplements on growth and feed efficiency." *Poult. Sci.* **48**:1462–1467.

Scott, M. L., M. C. Nesheim, and R. J. Young. 1976. *Nutrition of the Chicken.* M. L. Scott and Associates, Ithaca, New York.

Seoane, J. R. and L. E. Moore. 1969. "Effects of fish meal on nutrient digestibility and rumen fermentation of high-roughage rations for cattle." *J. Anim. Sci.* **29**:972–976.

Soderquist, M. R., K. J. Williamson, G. I. Blanton, D. C. Phillips, D. R. Law, and D. L. Crawford. 1970. *Current Practices in Seafood Processing Waste Treatment.* Department of Food Science Technology, Oregon State University, Corvallis.

Sidwell, V. D., B. R. Stillings and G. M. Knobl, 1970. "Fish protein concentrate story. 10. U.S. Bureau of Commercial Fisheries FPCS nutritional quality and use in foods," *Food Technol.* 24:876–882.

Stansby, M. S. 1963. *Industrial Fisheries Technology.* Reinhold, New York.

Stansby, M. S. and H. S. Olcott. 1963. "Composition of fish." *Industrial Fisheries Technology.* M. S. Stansby (Ed.). Reinhold, New York, pp. 339–349.

Steele, R. E. 1971. *Shrimp Processing Waste as a Pigment Source for Rainbow Trout (Salino gairdneri).* M.S. Thesis, Oregon State University.

Steinberg, M. A. 1973. "Technological developments in fish processing and implications for animal feeding." *Alternative Sources of Protein for Animal Production.* National Academy of Sciences, Washington, D.C., pp. 119–129.

Stillings, B. R. 1967. "Nutritional evaluation of fish protein concentrate," *Activities Report Vol.* **19**, Research and Development Associates, Inc., Natick, Massachusetts, pp. 109–117.

Stillings, B. R., O. A. Hummerle and D. G. Snyder, 1969. "Sequence of limiting amino acids in fish protein concentrate produced by isopropyl alcohol extraction of red hake "(*Urophycis chuss*)." *J. Nutr.* **97**:70–78.

Stout, F. M. 1973. "Effect of processing on the nutritional value of mink feeds." *Effect of Processing on the Nutritional Value of Feeds.* National Academy of Sciences, Washington, D.C., pp. 383–392.

Stout, F. M., J. Adair, and J. E. Oldfield. 1968. "Hepatotoxicosis in mink associated with feeding toxic herring meal." *Amer. Fur Breeder* **41**:12–16.

Stout, F. M., J. E. Oldfield, and J. Adair. 1960. "Nature and cause of the cotton-fur abnormality in mink." *J. Nutr.* **70**:421–426.

Tarr, H. L. A. and J. Biely, 1973. "Effect of processing on the nutritional value of fish meal and related products." *Effect of Processing on the Nutritional Value of Feeds.* National Academy of Science, Washington, D.C., pp. 252–281.

Toma, R. B. and W. H. James. 1975. "Nutritional evaluation of protein from shrimp cannery effluent (shrimp waste protein)." *J. Agric. Food Chem.* 23:1168–1171.

Underwood, E. J. 1977. *Trace Elements in Human and Animal Nutrition,* Academic Press, New York.

U.S. Environmental Protection Agency. 1974. "Development document for effluent limitations guidelines and new source performance standards for the catfish, crab, shrimp and tuna segment of the canned and preserved seafood processing point source category." *U.S. Environmental Protection Agency, Report No.* **EPA-440/1-74-020-9**. Washington, D.C.

U.S. Environmental Protection Agency. 1975. "Development document for effluent limitations guidelines and new source performance standards for the fish meal, salmon, bottom fish, clam, oyster, sardine, scallop, herring and abalone segment of the canned and preserved fish and seafood processing industry point source category." *U.S. Environmental Protection Agency, Report No.* EPA 440/1-75-041-9. Washington, D.C.

U.S. Environmental Protection Agency. 1977. "Municipal sludge management: environmental factors." *U.S. Environmental Protection Agency, Report No.* EPA 430/9-77-004 (MCD-28). Washington, D.C.

Velloso, L., T. W. Perry, R. K. Peterson, and W. M. Beeson. 1971. "Effect of dehydrated alfalfa meal and of fish solubles on growth and nitrogen and energy balance of lambs and beef cattle fed a high urea liquid supplement." *J. Anim. Sci.* 32:764–768.

Watkins, B. E. 1976. *Evaluation of Shrimp and Crab Processing Waste as a Feed Supplement for Mink (Mustela vision),* M.S. Thesis, Oregon State University.

Whittemore, C. T. and A. G. Taylor. 1976. "Nutritive value as the growing pig of deoiled liquified herring offal preserved with formic acid (fish silage)." *J. Sci. Food Agric.* 27:239–243.

13

The fertilizing value of tannery sludges

Professor dr. hab. Teofil Mazur and Dr. Jozef Koc

Institute of Agricultural Chemistry,
Academy of Agriculture and Technology,
Olsztyn, Poland

INTRODUCTION

During treatment of tannery waste waters, sludges are obtained which are characterized by a large variety of organic and mineral compounds. The amount of these sludges from one tannery can be up to 12,000 tonnes/yr. Fresh sludges undergo preliminary fermentation on liquid separating plots and then are stored in heaps, where further decomposition processes occur. As a result of this, semi-decomposed material that resembles compost is obtained, which in consistency is similar to damp black earth. Because of these physical properties, it is used reluctantly by farmers for fertilizing purposes. However, its chemical composition indicates the possibility of using it in agriculture. The problems that might arise could be sanitary in nature; for example, the possibility of anthrax when the hides come from infected animals, or too high a concentration of toxic elements (e.g., chromium). But since sanitary regulations do not permit the using of hides from infected animals, these sludges should not pose a health hazard when added to the soils. The content of heavy metals in the sludge depends on the technology of hide tanning and the methods of sewage treatment. Sludges from tanneries using chromium in the treatment process are not suitable for fertilizers as they contain too high a level of chromium.

Trials have been carried out on the use of tannery wastes as fertilizers in Czechoslovakia (Havelka, 1968; Liska and Kozik, 1963; Masoryk, 1968), France (Moulinier and Mazover, 1968), and Poland (Mazur and Szafranek, 1969; Mazur, Ciécko, and Koc, 1972; Szafranek et al, 1973). The positive results of these trials were the basis for undertaking intensive studies on the agricultural uses of these waters. Investigations have included determinations of the physicochemical characteristics of sludges and their effects on crop yields and soil properties (Koc et al, 1976; Mazur and Koc, 1976a,b,c).

PHYSICOCHEMICAL PROPERTIES OF TANNERY SLUDGES

In the initial treatment of tannery sewages, the material is made alkaline with milk of lime and then coagulated with ferrous sulfate or aluminium sulfate. The precipitated sludge contains 95 to 98% water. Water removal is brought about on liquid separating plots or vacuum filters. As a result, a sludge is obtained that contains some 17% dry matter. When problems arise with the draining installations, the sludge obtained may contain much less dry matter; then its use in agriculture is limited and transport costs increase.

The color of drained sludge ranges from gray through brown-gray to black. As a result of decomposition processes, sludges have an unpleasant smell which disappears after spreading them in the field.

Specific gravity of tannery sludge ranges from 1.005 to 1.200 g/cm^3, with an average of 1.069 g/cm^3, and is dependent on the mineral content. The pH of sludge varies from 6.9 to 9.7.

Tannery sludges have high ion absorption capacity (Table 13.1). This accounts for their positive effect on soil sorption properties, which is of special importance in increasing fertility of light soils with poor absorption capacity.

TABLE 13.1. Sorption capacity of tannery sludges (meq/100 g sludge dry matter).

Sludge	Sorption with NH_4Cl + M KCl			Sorption of NH_4^+ With M NH_4Cl	Sorption of PO_4^{-3} With $Ca(H_2PO_4)_2$
	NH_4^+	K^+	$NH_4^+ + K^+$		
Lodygowice	41.8	319.1	360.0	52.3	833
Zywiec	5.2	327.3	332.5	19.7	742
Kepice	47.1	351.0	398.1	44.5	567

TABLE 13.2.   Dry Matter, Organic Matter, and Fat Content of Tannery Sludges.

	Dry Matter (%)	Organic Matter (% dry matter)	Ash (% dry matter)	Fats (% dry matter)
Mean	17.2	57.3	42.7	6.9
Range	2.5–31.2	37.7–88.2	11.8–66.0	1.1–26.2

Dry matter, organic matter, and fat contents are given in Table 13.2. Dry matter is correlated with the consistency of sludges; however, the sludges from vegetable tanneries were found to contain a higher percentage of dry matter than sludges from vegetable-chromium tanneries. Tannery sludges are characterized by large variations in ash content. Great variations in the content of ash and organic compounds are due to variability of tannery sludges that are treated with various amounts of coagulants. There is also a large variation in fat content. Some samples were found to contain over 26% fatty compounds. With additions of 7.5 dry tonnes/ha of these sludges, 2 tonnes/ha fatty compounds are introduced into the soil. In the light of experiments (Zabek, 1958), these amounts have no deleterious effects on soil properties. On the average, the amount of fat compounds incorporated into the soil is nearly four times less than these figures and has no practical importance, as fat decomposition in the soil is quite fast. However, proper aerobic conditions must exist and sufficient calcium is required to neutralize the fatty acids.

The nitrogen and mineral content of tannery sludges is variable (Table 13.3). Based on the mean content of nitrogen and minerals, it may be concluded that

TABLE 13.3.   Chemical Composition of Tannery Sludges (% dry matter).

Element	Mean	Range
Total nitrogen	3.78	1.97–5.67
Ammonium nitrogen	0.25	–
Phosphorus	0.25	0.09–0.49
Potassium	0.09	0.04–0.21
Calcium	3.83	1.00–7.51
Magnesium	0.27	0.01–0.94
Sulfur	2.86	1.22–5.45
Sodium	1.39	0.13–6.37
Iron	0.95	0.001–12.52
Chromium	1.29	0.34–2.80

tannery sludges contain high levels of nitrogen, calcium, sulfur, sodium, iron, and chromium, but have small amounts of phosphorus and very little potassium, as compared with the nitrogen content. The occurrence of ammonium nitrogen as more than 7% of the total nitrogen accounts for the decomposition processes of nitrogen compounds during storage of sludges. Tannery sludges are therefore fertilizers which, when introduced into the soil, undergo decomposition. The released nitrogen is used by crops. The ratio of nitrogen to phosphorus and potassium indicates the need for supplementation of the tannery sludges with phosphorus and potassium, especially on soils that are poor in those elements.

This high content of sodium is due to the use of that element in the preservation and tanning of the hides. Sodium, used in small amounts, has a positive effect on crop yields. However, too high a concentration of this element may lead to the destruction of soil structure or some other unwelcome effect. In such cases, liming of the soil is necessary. Fortunately, tannery sludges contain more calcium than sodium and the negative effect of the latter should be counteracted by the high level of calcium. A desirable constituent of tannery sludges is sulfur, which can have a positive influence on the yield and quality of crops in rural areas.

The element which poses the greatest problem when evaluating tannery sludges is chromium. Tannery sludges from Polish tanneries contain, on an average, 1.29% chromium, but a few samples contain as much as 2.8% of the element. It has been found that if the chromium content does not exceed 3%, it has no detrimental effect on crops (Andrzejewski, 1970; Schreiber, 1963). The toxic effect of chromium on crops was found at high rates of application of the element, amounting to 1000 kg chromium per hectare or 180 mg chromium per pot (Liska and Kozik, 1963; Andrzejewski, 1971). Low rates of chromium were found to have a beneficial influence on the crops (Moulinier and Mazover, 1968; Andrzejewski, 1970, 1971). Bertrand and Wolf (1968) have found that chromium application to soils that were poor in the element increased potato yield by 42%. In tannery sludges, chromium occurs as a trivalent element, a form in which it is less toxic than it is in the hexavalent form. The alkali reaction of tannery sludges ensures low solubility of the chromium (Blood, 1963; Moulinier and Mazover, 1968; Roszyk, 1968, 1969; Andrzejewski, 1970, 1971).

The effects of organic materials as fertilizers are associated with the rate at which they are mineralized in the soil. Data in Fig. 13.1 shows the decomposition of tannery sludges as compared with that of farm manure. The lower rate of CO_2 release from sludges compared with that from farm manure is due to the different content of organic matter undergoing decomposition in those two materials. During six months, under laboratory conditions, 21 to 36% of the total organic matter of tannery sludges was decomposed. The rate of decomposition under field conditions is slower and depends mainly on the soil moisture and temperature. A measure of sludge decomposition is its effect on crop yields.

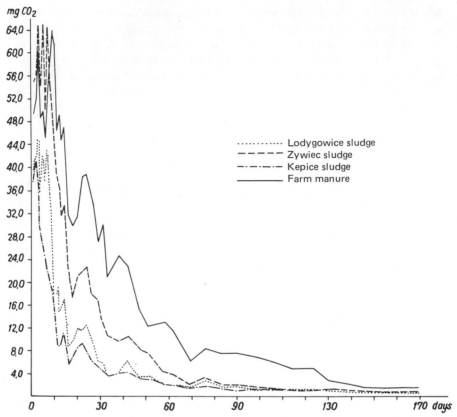

Fig. 13-1. Rate of CO_2 release from soil during decomposition of tannery sludges.

EFFECT OF TANNERY SLUDGES ON CROP YIELD

The first experiments on the fertilizer value of tannery sludges were conducted in pots. It was found that tannery sludges increased yields of barley, oats, buck-wheat, and sunflower. Raw sludge and sludge composted with lime were also studied. The sludge fermented with calcium carbonate had a somewhat better effect on crops than did sludge alone. It was also found that tannery sludges from chromium tanneries, with a high chromium content, had a deleterious effect on plant growth. (Szafranek *et al*, 1973).

In many years of field experiments, the effect of tannery sludges applied every year to a good wheatland soil on the yield of crops in a rotation was studied. Tannery sludges from three tanneries were compared with farm manure. For each crop, 7.5 tonnes of dry matter per hectare of each material was applied and

TABLE 13.4.   Effect of Fertilization with Tannery Sludges on Yield of Crops.

Treatment	Potatoes (tonnes/ha) Tubers	Summer Barley (tonnes/ha)		Horse Beans (tonnes/ha)		Winter Wheat (tonnes/ha)		Feed Units for Four Years (tonnes/ha)
		Grain	Straw	Seed	Straw	Grain	Straw	
No fertilization	27.40	3.64	3.60	4.12	5.06	2.72	2.88	23.628
Lodygowice, full sludge rate	32.90	4.62	4.53	4.66	8.05	3.55	3.98	29.562
Lodygowice, half sludge rate + NPK	40.20	4.62	4.67	4.48	6.85	3.81	4.07	30.936
Zywiec, full sludge rate	33.90	4.57	4.39	4.44	7.16	3.00	3.20	27.602
Zywiec, half sludge rate + NPK	38.70	4.60	4.57	4.63	6.09	3.29	3.35	29.156
Kepice, full sludge rate	28.50	4.02	3.90	4.36	6.77	3.12	3.43	25.986
Kepice, half sludge rate + NPK	37.80	4.26	4.04	4.39	6.25	3.42	3.56	27.323
Farm manure, full rate	33.60	4.42	4.35	4.62	7.05	3.94	4.20	28.897
Farm manure, half rate + NPK	38.72	4.54	4.35	4.70	7.24	3.74	3.83	30.190
LSD P = 0.05	2.60	0.27	0.46	0.16	0.84	0.12	0.13	

supplemented with mineral fertilizers. The yields obtained are presented in Table 13.4. Application of tannery sludges resulted in significant yield increases, and only the sludge from Kepice had no influence on the yield of potatoes and barley straw. Farm manure gave somewhat better results than tannery sludges in the third and fourth application. In the rotation, the sludge from Zywiec gave somewhat smaller amounts and that from Lodygowice higher amounts of oat feed units than did farm manure. The supplement of mineral fertilizers enhanced the action of tannery sludges. A half rate of sludges supplemented with farm manure produced higher amounts of oat feed units than the full rate of the sludges applied alone.

Other farming experiments conducted in different regions of Poland indicate an increased yield when tannery sludges were applied for the fertilization of potatoes (Table 13.5). Tuber yield increased by 17 to 22% under the sludge treatment. The difference between the two materials in their effect on yield was due to the fact that in the same amount of both materials applied (30 tonnes/ha)

TABLE 13.5.   Effect of Fertilization with Tannery Sludges on Potato Yield.

Locality	Treatment	1970 (tonnes/ha)	1971 (tonnes/ha)	1972 (tonnes/ha)	1973 (tonnes/ha)	Mean (tonnes/ha)
Kepice	No organic fertilization	19.8	20.5	14.5	16.2	17.8
	Tannery sludge	18.1	25.2	20.8	22.5	21.7
	Farm manure	17.4	34.3	22.6	25.1	24.9
Lodygowice	No organic fertilization	16.1	25.4	10.0	18.0	17.4
	Tannery sludge	16.0	32.2	11.6	24.0	21.0
	Farm manure	17.5	31.3	15.6	29.0	23.4
Zywiec	No organic fertilization	4.2	23.5	10.0	18.0	13.9
	Tannery sludge	8.9	22.5	11.7	22.0	16.3
	Farm manure	10.6	19.3	15.6	29.0	18.6

TABLE 13.6.   Effect of Fertilization with Tannery Sludges on Dry Matter, Starch, and Vitamin C Content of Potato Tubers.

Treatment	Dry Matter		Starch		Vitamin C (mg 100/g)
	(%)	(tonnes/ha)	(%)	(tonnes/ha)	
No fertilization	22.4	6.12	15.7	4.30	33.9
Lodygowice, full sludge rate	21.5	7.08	15.7	5.17	25.9
Lodygowice, half sludge rate + NPK	20.7	8.32	12.6	5.06	24.7
Zywiec, full sludge rate	22.1	7.48	14.4	4.88	23.9
Zywiec, half sludge rate + NPK	21.4	8.28	13.1	5.07	23.7
Kepice, full sludge rate	22.1	6.30	15.2	4.34	27.2
Kepice, half sludge rate + NPK	21.6	8.17	14.5	5.49	29.0
Farm manure full rate	22.8	7.67	14.6	4.91	28.5
Farm manure, half rate + NPK	21.0	8.14	14.3	5.54	25.7

more nutrients were contained in farm manure, and supplemental commercial fertilizers were not applied to the tannery sludges. Liska and Kozik (1963), in field experiments using tannery sludges, obtained a 10 to 17% higher yield of potatoes than under a moderate mineral fertilization. The combined application of sludges and commercial fertilizers increased tuber yield by 42%.

Application of tannery sludges has no marked effect on the dry matter and starch content of potato tubers (Table 13.6). Commercial fertilizers applied in combination with organic fertilizers reduced the dry matter and starch content, but this decrease had no significant influence on the total yield of dry matter and starch content because tuber yield was increased. Application of both tannery sludges and farm manure reduced the vitamin C content of potato tubers by 25%.

Tannery sludges have a positive effect on the protein content of crops (Table 13.7). The average crude protein yield was increased under the effect of sludges by 95 to 197 kg and that of true protein by 54 to 98 kg/ha. This was due to both increased protein content and increased yields of crops. A still better effect on the yield of protein was obtained with the combined use of tannery sludges and commercial fertilizers. Of the three sludges used, the best results were obtained with the sludge from a vegetable tannery at Lodygowice, and the worst results with the sludge from a vegetable-chromium tannery at Kepice.

Chemical analyses have shown that the application of tannery sludges increases the iron, sodium, and sulfur contents and decreases the potassium content of crops (Mazur and Koc, 1976a,b). However, it is essential to monitor the amount of chromium absorbed by plants (Table 13.8).

Fertilization with tannery sludges resulted in repeated increases of that element in crops. This indicated that more chromium can be absorbed by plants if the available forms of the element are present in large quantities in the soil.

TABLE 13.7. Effect of Fertilization with Tannery Sludges on Protein Content and Protein Yield.

Treatment	Potato Tubers		Barley Grain		Horse Bean		Wheat Grain		Crude Protein Mean Yield (kg/ha)	True Protein Mean Yield (kg/ha)
	CP	TP	CP	TP	CP	TP	CP	TP		
No fertilization	6.88	5.06	10.63	9.81	31.56	26.75	10.69	9.19	673.1	493.4
Lodygowice, full sludge rate	7.56	4.38	11.50	10.38	31.75	26.93	13.00	10.94	870.6	591.2
Lodygowice, hall sludge rate + NPK	9.44	5.75	10.94	10.44	31.38	27.50	12.06	10.19	897.5	621.1
Zywiec, full sludge rate	8.19	5.13	12.00	11.00	32.13	27.90	12.63	10.81	840.1	583.4
Zywiec, half sludge rate + NPK	10.25	6.50	10.94	10.81	32.13	27.81	12.38	9.88	904.8	633.7
Kepice, full sludge rate	7.25	4.81	10.81	10.19	30.63	27.12	13.75	11.63	765.7	547.5
Kepice, full sludge rate + NPK	8.38	5.75	10.88	10.13	33.43	28.06	11.81	10.38	850.7	594.1
Farm manure, full rate	7.19	4.31	10.94	10.00	31.19	27.06	12.81	10.63	851.3	592.8
Farm manure, half rate + NPK	7.25	5.06	10.81	10.19	31.75	26.06	12.44	10.31	863.9	623.6

NOTE: CP—crude protein, in % dry/matter. TP—true protein, in % dry matter.

TABLE 13.8. Effect of Fertilization with Tannery Sludges on Chromium Content of Crops.

Treatment	Potato Tubers (ppm dry matter)	Barley (ppm dry matter)		Horse Bean (ppm dry matter)		Wheat (ppm dry matter)	
		Grain	Straw	Seed	Straw	Grain	Straw
No fertilization	1.05	1.17	0.80	0.30	0.40	0.80	0.80
Lodygowice, full sludge rate	1.00	3.30	1.95	1.00	0.60	1.35	4.00
Lodygowice, half sludge rate + NPK	4.27	0.70	1.45	1.00	0.50	0.90	2.20
Zywiec, full sludge rate	2.20	0.95	2.00	1.45	1.25	2.80	2.00
Zywiec, half sludge rate + NPK	1.20	0.55	2.00	1.15	0.80	1.20	0.80
Kepice, full sludge rate	1.00	1.20	2.20	0.80	0.70	1.70	2.00
Kepice, half sludge rate + NPK	0.80	0.80	1.20	1.30	0.50	1.10	5.00
Farm manure, full rate	0.70	0.70	0.75	0.30	0.40	0.80	1.10
Farm manure, half rate + NPK	0.70	0.70	0.60	0.40	0.40	0.80	1.10

Other workers (Moulinier and Mazover, 1968; Andrzejewski, 1970) have arrived at the same conclusion. An increase in the chromium content of crops was not accompanied by symptoms of physiological disorders and there was no reduction in yield. However, there is still a lack of data on plant tolerances to chromium. It may be assumed that at normal rates of application of organic fertilizers given for crops in a rotation, an excessive concentration of the chromium in the crop will not occur. Moreover, in the case of light soils that are naturally poor in chromium, the element will increase crop yields.

EFFECT OF TANNERY SLUDGES ON SOIL PROPERTIES

Tannery sludges were found to increase the amount of humus compounds in the soil (Table 13.9). Under sludge treatment, there were more humus compounds in

TABLE 13.9.   Effect of Fertilization with Tannery Sludges on Humus Content of Soil.

Treatment	Total	Bitumen	Soluble in 0.1 N Na$_4$P$_2$O$_7$	Humic Acids	Fulvic Acids	Non-hydrolyzing	Nitrogen (mg/100 g soil) Total
No fertilization	934	28	125	87	94	600	101.7
Lodygowice, full sludge rate	1156	88	118	101	99	750	126.6
Lodygowice, full sludge rate + NPK	998	75	132	101	72	618	109.0
Zywiec, full sludge rate	1131	146	125	97	119	644	122.0
Zywiec, half sludge rate + NPK	1026	108	122	101	103	592	110.4
Kepice, full sludge rate	1108	115	115	95	98	685	120.4
Kepice, half sludge rate + NPK	1051	106	120	96	107	622	117.7
Farm manure, full rate	1065	73	122	99	113	658	116.1
Farm manure, half rate + NPK	1025	63	122	96	117	627	107.6

(Carbon (mg/100 g soil) spans the Total, Bitumen, Soluble, Humic Acids, Fulvic Acids, and Non-hydrolyzing columns.)

the soil than under the farm manure treatment. The effect of tannery sludges on the content of fulvic acids was about the same as that of farm manure. However, there was a marked increase in bitumin and the non-hydrolizing fraction under sludge treatment. Total soil nitrogen was also increased by the application of sludges; during four experimental years, its increase amounted to between 14 and 25%.

Application of tannery sludges had no positive effect on the phosphorus, potassium, and magnesium contents of the soil (Table 13.10). Large amounts of nitrogen applied with tannery sludges resulted in an increased uptake of potassium, phosphorus, and magnesium. As a consequence, the low content of these minerals in sludges leads to their loss from the soil. Therefore, it is necessary to apply supplemental commercial fertilizers.

An important problem is the accumulation of chromium in soil to which sludges have been added. Table 13.10 shows that chromium content in the plough layer was markedly increased, and it accumulated in the sparingly soluble forms. Only

TABLE 13.0.   Effect of Fertilization with Tannery Sludges on Available Nutrient and Chromium Content of Soil.

Treatment	pH H$_2$O	pH KCl	P$_2$O$_5$ (mg/100 g soil)	K$_2$O (mg/100 g soil)	MgO (mg/100 g soil)	Cr (mg/100 g soil) Total	Soluble in 2.5% CH$_3$COOH
No fertilization	6.2	5.8	4.9	6.0	2.8	11.2	0.7
Lodygowice, full sludge rate	6.8	6.5	4.5	7.2	2.4	175.0	7.5
Lodygowice, half sludge rate + NPK	6.2	6.0	6.6	10.8	3.0	46.0	1.2
Zywiec, full sludge rate	6.3	5.9	4.5	6.2	2.6	241.0	7.0
Zywiec, half sludge rate + NPK	6.1	5.8	5.8	11.3	3.4	110.0	0.8
Kepice, full sludge rate	6.7	6.4	5.0	8.0	3.2	90.0	1.0
Kepice, half sludge rate + NPK	6.3	6.0	6.1	10.9	3.4	80.0	1.7
Farm manure, full rate	6.5	6.1	6.3	14.0	4.7	11.5	0.6
Farm manure, half rate + NPK	6.3	6.0	7.1	14.8	4.7	11.4	0.6

1 to 4% of the total chromium was found to be soluble in 2.5% acetic acid. However, the processes occurring in the soil tend to change insoluble forms into compounds available to crops.

GENERAL PRINCIPLES OF USING TANNERY SLUDGES

Tannery sludge as an organic fertilizer may be used in agriculture. However, only low chromium content sludges are suitable for this purpose.

The principles of using tannery sludges should be the same as in the use of farm manure; i.e., they should be used for crops responding well to organic fertilizers. These include potatoes, beets and forage, and processing crops.

Application of tannery sludges should be supplemented with commercial fertilizers, especially with potassium and phosphorus. Rates of those fertilizers should be fitted to the nutrient requirements of crops and soil fertility.

Tannery sludges should be used on limed soils. Lime speeds up the decomposition of sludges, particularly of fats contained in them, and reduces chromium solubility.

In regions where tannery sludges are used, it is necessary to continuously monitor the chromium content in soils and crops.

REFERENCES

Andrzejewski, M. 1970. "Wykorzystanie odpadów skóry chromowej jako źródła azotu w nawożenia roślin." *Polish Agric. Ann. A* **96**, *pt. 2:*315–328.

Andrzejewski, M. 1971. "Wpływ nawożenia chromem na plony kilku gatunków roślin i na zawartość chromu w glebie." *Polish Agric. Annu. A* **97**, *pt. 2:*75–97.

Bertrand, D. and A. Wolf. 1968. "Necessite de L'oligo-element chrome pour la culture de la pomme de terre." *Compte Rendu hebd. Seanc. Akad. Sci. Paris D* **226**, *pt. 14:*1494–1495.

Blood, J. W. 1963. "Problems of toxic elements in crops husbandry." *N.A.A.S. Quart. Rev.* **14**, *pt. 59:*97–100.

Havelka, B. 1958. "Vysledku vegetacnych pokusu s hnojenim prumyslovymi odpady." *Sb. Vys. S. Zamed. Les. Brno. A* **2:**155–166.

Koc, J., L. Krefft, and T. Mazur. 1976. "Badania nad wartościa, nawozowa, osadów garbarskich. Cz. I. Chemiczno-fizyczna charakterystyka osadów." *Soil Sci. Ann.* **27**, *pt 1:*103–112.

Liska, J. and J. Kozik. 1963. "Vyuziti odpadnich kozeluskych kalu ke hnojeni." *Za Vysoku Urodu* **9:**347–349.

Masoryk, S. 1968. "Vyuzitie garbarienskych odapow v polnohospodarstve." *Agrochemia* **8**, *pt. 7:* 201–202.

Mazur, T., Z. Ciećko, and J. Koc. 1972. "Chemiczno-rolnicza charakterystyka osadów garbarskich." *Gaz. Woda i Technika Sanitarna* **47**, *pt. 2:* 47–49.

Mazur, T. and J. Koc. 1976a. "Badania nad wartością nawozowa osadów garbarskich. **Cz. II.** Wpływ nawożenia osadami garbarskimi na plon roślin." *Soil Sci. Ann.* **27**, *pt. 1:* 113–122.

Mazur, T. and J. Koc. 1976 b. "Badania nad wartością nawozową osadów garbarskich. **Cz. III.** Wpływ nawożenia osadami na skład chemiczny roślin." *Soil Sci. Ann.* **27**, *pt. 1:* 123–135.

Mazur, T. and J. Koc. 1976c. "Badania nad wartością nawozową osadów garbarskich. **Cz. IV.** Wpływ nawożenia osadami garbarskimi na zmiany chemicznych właściwości gleby." *Soil Sci. Ann.* **27**, *pt. 1:* 137–147.

Mazur, T. and R. Cz. Szafranek. 1969. "Zawartość składników nawozowych i związków próchnicznych w osadach garbarskich." *Zeszyty Naukowe Wyższej Szkoły Rolniczej, Olsztyn* **25**, *pt. 718:* 837–845.

Moulinier, H. and R. Mazover. 1968. "Contribution a l'etude de l'action du chrome sur la croissance des vegetaux." *Ann. Agron.* **19**, *pt. 5:* 553–567.

Roszyk, E. 1968. "Zawartość wanadu, chromu, manganu, kobaltu niklu i miedzi w niektórych glebach Dolnego Śląska wytworzonych z glin pylastych i utworów pylowych. **Cz. I.** Ogólna zawartość mikroskładników." *Soil Sci. Ann.* **19**, *pt. 2:* 223–246.

Roszyk, E. 1969. "Zawartość wanadu, chromu, manganu, kobaltu, niklu i miedzi w niektórych glebach Dolnego Śląska wytworzonych z glin pylastych i utworów pyłowych. **Cz. II.** Składniki rozpuszczalne." *Soil Sci. Ann.* **20**, *pt. 1:* 135–154.

Schreiber, H. 1963. "Zagadnienie osadu w garbarniach." *Przegląd Skórzany* **2:** 27–32.

Szafranek, Cz., T. Mazur, M. Koter, J. Koc, and J. Czapla. 1973. "Działanie osadów garbarskich na plon i skład chemiczny jęczmienia, gryki i łubinu w doświadczeniach wazonowych." *Zesz. nauk. ART Olszt., Rolnictwo* **3:** 179–198.

Ząbek, S. 1958. "Zatłuszczenie gleb łąkowych wskutek nawadniania ściekami miejskimi." *Polish Agric. Ann. F* **72**, *pt. 4:* 559–563.

14

The production of feedstuff biomass from liquid organic wastes by fermentation

P. Gray, B.Sc. and
D. R. Berry, M.Sc., Ph.D.

Department of Applied Microbiology, University of Strathclyde, Glasgow, Scotland.

INTRODUCTION

The concept of using liquid wastes from the food processing industry as a substrate for fermentations is not new (Peppler, 1970). For many years, molasses has been widely used as a carbon source in bakers' yeast, citric acid, and penicillin fermentations (Prescott and Dunn, 1959). However, a renewed awareness of the increasing world population, the increasing cost of energy, and the increased legislation to control pollution of the environment has given a new impetus to the use of fermentation processes to convert waste materials into profitable by-products (Meadows, 1974; Rolfe, 1976). In particular, the direct conversion of waste materials into microbial biomass suitable for animal feeds and for human consumption has received much attention and has recently been the theme of several books (Mateles and Tannenbaum, 1968; Tannenbaum and Wang, 1975; Birch *et al*, 1976).

Since most of the work in this field has been carried out in developed countries, the material in this review is also biased in this direction. However, we wish to emphasize that many developing countries which have problems of malnutrition also have available to them low-grade effluents which could be upgraded by fermentation. Some—though not many—of these are being studied; e.g., coffee waste water in Guatemala, (Rolz, 1975), and palm oil waste water (Imrie and Righelato, 1976). Even such valuable substrates as molasses are discarded as waste in many North African and Middle Eastern countries which are at the same time importing grain to provide feedstuffs for animals. It is unlikely that the technology employed in developed countries is ideal for developing countries (Stanton, 1977), so it is essential that these should be adapted to the requirements of individual situations.

In this review we have discussed some of the alternatives available at each stage of the waste conversion process and some of the criteria which affect the choice of a particular process, rather than presenting a detailed description of any one process. Material relevant to liquid wastes from the brewing, distilling, and wood pulp industries has been excluded, since these are covered elsewhere in this volume.

Nature of Liquid Wastes

Large quantities of liquid waste of an appropriate chemical composition to act as a substrate for microbial fermentations are produced as by-products of the food industry. In the domestic situation, potato water is used to make gravy and is not a waste; however, on the industrial scale, spent liquor from potato processing presents a major disposal problem. This appears to be general for the food industry, where it has been estimated that for every tonne of food produced, an equal amount of carbohydrate waste and effluent is produced (Imrie and Righelato, 1976). In the past, this "waste" has often been disposed of by pumping into rivers or the sea, where it causes pollution, or by treatment in an effluent plant, which increases the cost of the overall process. The scale of the problem is indicated in Table 14.1, which also gives some information on the nature of typical wastes from the food industry. It can be seen that the effluents vary from a B.O.D. of 900 mg/l for some canning wastes to 4500 mg/l for beet processing waste water. They are mostly rich in carbohydrate but do contain nitrogen and phosphate (Tomlinson, 1976a). The $C:N:P$ ratio of a waste is important, since a shortage of one will—if not supplemented—restrict the growth of the organism and hence the uptake of other substrates from the waste water. An ideal substrate contains $C, 100:N, 5:P, 1$ (Espinosa, *et al*, 1977).

TABLE 14.1. Analysis of Industrial Effluents.

Industry	m^3/day	B.O.D. (g/l)	C.O.D. (g/l)	Organic (g/l)	Nitrogen (mg/l)	Phosphate (mg/l)	pH
Caramellising	30	3.2	5.8	2.06	692	0.7	8.3
Beet processing	2054	4.5	5.9	2.265	137	834	7.2
Canning 1	155	0.9	1.76	0.594	35	2.05	5.7
Canning 2	400	3.5	9.45	2.28	66	253	6.9
Potato processing 1	1580	3.25	5.8	2.28	62	68	4.8
Potato processing 2	2300	2.3	3.8	1.51	147	6.07	4.5

Source: Tomlinson. 1976.

Effluents which contain simple sugars (e.g., beet processing waste) are easier to handle than those containing high molecular weight polysaccharides, since they can be assimilated rapidly by a wide variety of micro-organisms (Anderson *et al*, 1975). However, wastes which contain high molecular weight polysac-charides such as starch require a more sophisticated approach, since it is neces-sary to break them down to assimilable monosaccharides and disaccharides before they can be metabolized. This has been achieved using a two stage process in the Symba Process (described later), although a simple route in the future may involve the direct addition of enzymes to the substrate before the fermenta-tion is inoculated. Since starch containing wastes are produced by a wide variety of industries—e.g., those concerned with potato processing, rice products, grain products, and baking (Skogman, 1976)—processes using this two stage approach should find a wide application.

THE FERMENTATION PROCESS

Selection of Micro-organisms

A wide variety of micro-organisms have been studied as possible organisms for biomass production. Although this includes several bacteria and algae, these groups of organisms have not attracted the same interest as filmentous fungi and yeasts. Bacteria are capable of faster growth rates than fungi, and are capable of assimilating a variety of substrates (Sucuru *et al*, 1975), but the increased difficulty and cost of harvesting bacteria, as well as concern at the difficulties of detecting contaminating pathogenic bacteria, appear to have restricted their use in the production of biomass from liquid wastes. Many species of bacteria, (e.g., *Methylococcus* and *Pseudomonas*) are, however, used in the production of biomass from hydrocarbons (Litchfield, 1977), so these problems need not be insuperable. The use of algae for the processing of liquid wastes has been in-vestigated, using both conventional fermentation techniques and lagoon systems of cultivation (Gromov, 1967; Oswald and Golueke, 1968). However, it seems probable that their use will be restricted to the removal of nitrogenous and phosphorus compounds from waste to prevent uncontrolled algal growth in natural waterways (Kosaric *et al*, 1974).

Yeasts are the most widely used organisms for the production of biomass at the present time. A recent estimate has placed the annual production rate of yeast biomass at a million tonnes (Anon, 1977). Although large quantities of *Saccharomyces cerevisiae* are produced in the form of bakers' yeast (Burrows, 1970) and are available as a by-product of the brewing industry, *Candida utilis* (previously named *Torula utilis* and *Torulopsis utilis*), the fodder yeast, which is less sensitive to high concentrations of glucose (Feichter, 1975) has been preferred in processes designed for biomass production (Forage, 1978). *Saccharomyces fragilis*, which has an active lactase, has been used for the fermentation of whey (Peppler, 1970) and *Endomycopsis fibuliger*, which has a good amylase activity, has been used for starch products (Skogman, 1976). Meyrath (1975) recently reported the use of a thermotolerant, highly flocculent strain of *S. kloekerianus* in a biomass process. Other yeasts which have been used for biomass production include *Rhodotorula glutinis* (Tomlinson, 1975), *Candida steatolytica* (Shannon and Stevenson, 1975a,b), *Torula cremoris* (Muller, 1969), and *Trichosporon cutanum* (Peppler, 1970). The acceptability of yeast biomass to man and animals has been demonstrated by prolonged use in traditional processes and products (Peppler, 1970).

Experience of conditions suitable for growing filamentous fungi on a large scale has been gained in the pharmaceutical industry over recent decades. This, combined with the observation that filamentous fungi produce a biomass product which is easy to harvest and has a good texture, has stimulated extensive investigations into the use of these organisms for the conversion of effluent to biomass. The group at Rank Hovis McDougall in the United Kingdom has screened a large number of micro-fungi and determined their amino acid profiles and growth characteristics. Of these, they have chosen a strain of *Fusarium graminearum* as the most suitable organism for biomass production (Anderson *et al*, 1975). Since this has now been subject to extensive nutritional and toxicological testing, this must be a favored organism for future developments in this and other laboratories. The strain of *Aspergillus niger* (M1) used by Tate and Lyle (United Kingdom) (Imrie and Vlitos, 1975) also offers the advantage of having been extensively tested. However, *Aspergillus oryzae*, which has been extensively used in traditional fermentations (Wood and Fook Min, 1975) may also be considered as an acceptable organism. The available evidence suggests that few filamentous fungi or yeasts have an ideal amino acid profile. The recent observation that *Sporotrichum pulverulentum*, which is the imperfect stage of the basidiomycete, *Phanerochaete chrysosporium*, has a good amino acid profile (Hofsten and Ryden, 1975) may indicate that further studies should be carried out on other basidiomycetes. Previous studies on the basidiomycetes have emphasized flavor development rather than nutritional quality (Worgan, 1968). The acceptability of fungi such as yeasts and mushrooms, which have been a traditional component of man's diet, is an asset which should not be ignored in choosing a suitable organism for a biomass process.

Fermentor Design

A fermentor may simply be described as a vessel that provides for the physiological and nutritional requirements which are essential for the establishment of a microbial culture (Hatch, 1975). For an aerobic culture to grow optimally, its requirement for an adequate supply of oxygen must be met (Mukhopadhyay and Ghose, 1976). Of the water soluble nutrients necessary for balanced growth, most may be supplied at levels in excess of the requirements of the organisms (Sinclair and Ryder, 1975). However, oxygen has only a limited solubility in aqueous media. Consequently, the rates of oxygen solution and subsequent transfer to a culture may represent the main growth-limiting factors for the production of single-cell protein (Tannenbaum and Wang, 1975). The problems of supplying sufficient oxygen for growth are increased with increase in fermentor scale (Ovaskainen et al, 1976), such that the costs of aeration may contribute substantially to total fermentation costs (Imrie and Righelato, 1976). Thus, the prime objective of fermentor design is the attainment of maximal rates of oxygen transfer at the minimal power required (Hatch, 1975).

The Stirred Tank Reactor. The preferred and consequently the most widely used fermentor design is the mechanically agitated fully baffled stirred tank reactor (Fig. 14.1; Prokop and Sobotka, 1975). Outlines of possible design configurations of this fermentor have been provided by Miller (1964), Sideman et al (1966), Schaftlein and Russell (1968), and Reith (1970). The absence of reactor baffles allows for vortex formation and results in improved efficiencies of gas transfer on fermentor scale-up (Blakebrough et al, 1967). Culture agitation is generally achieved by the use of an open-blade turbine impeller (Prokop and Sobotka, 1975) although alternative forms of agitator have been described by Steel (1969), Topiwala and Hamer (1973), and Ali et al (1975), in attempts to improve gas transfer rates. Culture agitation helps to maintain a relatively constant environment for growth, by assisting heat transfer and culture dispersion (Pollard and Topiwala, 1976). Agitation additionally enhances the mass transfer of nutrients between the various culture phases, while the mass transfer of oxygen from the gaseous phase to the liquid phase increases with increase of gas liquid contact times (Robinson and Wilke, 1973, Linek et al, 1973). Basically, agitation increases contact time by increasing the available surface area between the gas and liquid phases by reducing bubble coalescence

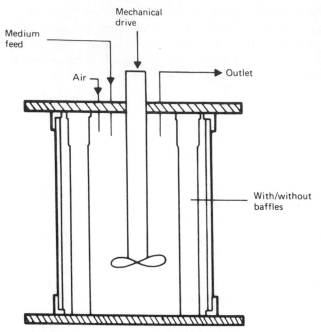

Fig. 14.1.   Typical stirred tank fermentor.

(Radovcich and Moissis, 1962). This results in small bubbles of fairly uniform size. The concomitant increase in culture turbulence also acts to reduce the size of the stationary liquid film, further aiding gas transfer. Fukuda *et al* (1968) demonstrated that the power required to sustain comparable efficiences of oxygen transfer increases on larger-scale fermentations. At the same time, they also reported that the performance of the fermentor (the power required to transfer unit mass of oxygen) decreases.

Stirred tank reactors have been used in the production of single-cell protein (SCP) from waste substrates in processes utilizing yeasts (Meyrath, 1975; Bernstein *et al*, 1977; Hang, 1977a; Forage, 1978) and micro-fungi (Romantschuk, 1975; Imrie and Righelato, 1976).

Tubular Fermentors.   The simple vertical tubular fermentor is a non-mechanically agitated fermentor through which there is an upward co-current flow of gases and medium (Morris *et al*, 1973; (Fig. 14.2). The aspect ratio of the fermentor column (ratio of height to diameter) is minimally 6 to 1 (Morris *et al*, 1973), but may be 10 to 1 (Smith and Greenshields, 1974) or greater (Pace and Righelato, 1976). The design of this fermentor differs from that of the stirred tank in that the aeration and agitation of a culture are achieved simultaneously by the

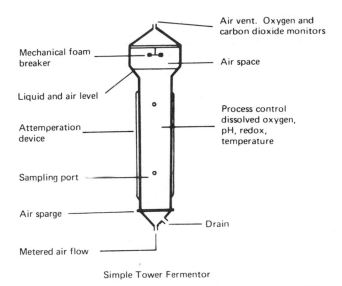

Air vent. Oxygen and
carbon dioxide monitors

Mechanical foam
breaker

Air space

Liquid and air level

Process control
dissolved oxygen,
pH, redox,
temperature

Attemperation
device

Sampling port

Air sparge

Drain

Metered air flow

Simple Tower Fermentor

Fig. 14.2. Tower fermentor. *Source:* Morris *et al.* 1973.

sparging of gas across a perforated or sintered plate situated at the fermentor base. The characteristics of oxygen mass transfer for various designs of tubular fermentor have been described by Murphy *et al* (1959), Yoshida and Akita (1965), Reith *et al* (1968), Freedman and Davidson (1969), Morris *et al* (1973), and Blanch and Bhavaraju (1976). Morris *et al* (1973) discussed gas flow behavior in terms of three basic patterns. At low superficial gas velocities (1 cm/sec), gas flow behavior is characterized by "bubbly flow," during which the rate of oxygen transfer increases almost linearly with increase of flow rates. At a higher superficial gas velocity (3 cm/sec), conditions of "turbulent flow" prevail, and the rates of gas transfer are maximized. For superficial gas velocities greater than 5 cm/sec, "slug flow" is evident and rates of gas transfer decline.

The advantages attributed to simple tubular designs are their ease of construction and operation, made possible by the lack of moving parts (Snell and Schweiger, 1949), so that fermentor running costs may be lower than for designs dependent upon mechanical agitation (Greenshields and Smith, 1974). The use of this fermentor for single-celled biomass production has been outlined by Greenshields and Smith (1971, 1974). The production of fungal biomass from synthetic substrates has been described by Daunter (1972), Cocker (1975), Pannell (1976), and Pace and Righelato (1976), and from waste substrates by Imrie and Righelato (1976) and Espinosa *et al* (1977).

A design of fermentor which is basically similar to the simple vertical tubular fermentor is the multi-stage column fermentor (Fig. 14.3). Here the fermentor column is partitioned by the use of a number of perforated plates into a series

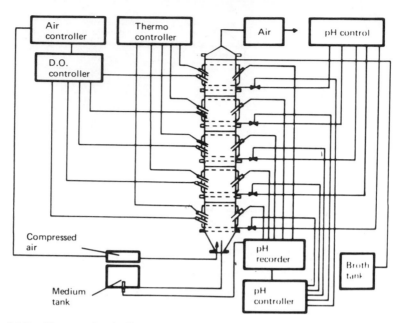

Fig. 14.3. Non-mechanically agitated multi-stage tubular fermentor. *Source:*
Solomons. 1972.

of compartments, each resembling a mixed tank (Prokop *et al*, 1969; Kitai *et al*,
1969; Kitai and Yamagata, 1970; Hsu *et al*, 1975). An integral feature of this
design is the presence of an air space beneath each of the perforated plates,
which serves to break the continuity of the liquid phase, and without which
the multi-stage effect would be minimal (Solomons, 1972). A modification of
this basic design in which the facility for mechanical agitation is introduced
into each of the column stages is described by Falch and Gaden (1969) and
Paca and Gregr (1976). The latter have described co-current liquid/gas flow,
while the former used counter-current flow. For the production of biomass,
however, the former method of operation is favored, as this reduced the tendency toward oxygen limitation experienced during counter-current operation.
Despite the fact that mechanical agitation is provided for all stages, oxygen
limitation may still be apparent for the uppermost reaches (Paca and Gregr,
1976) so that additional air inputs may be required.

Le Francois *et al* (1955) have published a patent describing an "air lift"
fermentor design in which the absence of the need for mechanical agitation
is further established (Fig. 14.4). By sparging air into the base of an internal
draught tube, a difference of pressure is created between this latter section and
an outer annulus which results in a recirculating gas/liquid flow regime. However, Prokop and Sabotka (1975) subsequently indicated that the majority of

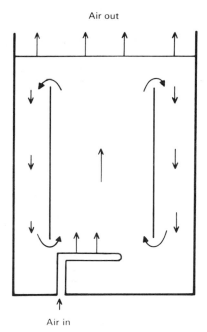

Fig. 14.4. Gas/liquid flow patterns in a draft tube type air-lift fermentor. *Source:* **Kristiansen. (Personal communication)**

processes utilizing this design employ a reversal of this flow pattern. This is simply achieved by sparging air into the base of the outer annulus, which results in improved characteristics of heat transfer. The gas transfer characteristics of this type of air lift design are discussed by Chakravarty *et al* (1973) and Hatch (1973, 1975).

An alternative form of air lift design in which the presence of an external loop replaces the internal draught tube is illustrated in Fig. 14.5 (Anon, 1976). This fermentor design has been termed the "Pressure Cycle" fermentor and is currently used by ICI in the production of bacterial SCP from hydrocarbon substrates. The gas transfer and mixing characteristics of this design are outlined by Lin *et al* (1976), who indicated that this form of fermentor is capable of providing rates of oxygen transfer comparable to those obtained for stirred tank reactors.

Microbial Growth Kinetics in Fermentors

Simple Batch Fermentors. The succession of events described in terms of organism specific growth rate, growth limiting substrate and biomass concentrations, and the yield of biomass that follows the introduction of a microbial

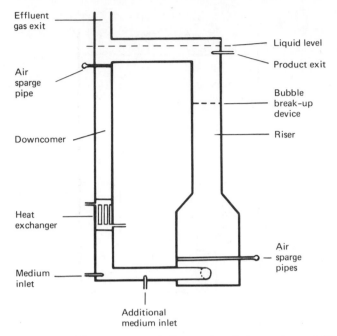

Fig. 14.5. I.C.I. pressure cycle fermentor. *Source:* Anon, 1976.

suspension into an initial limited amount of medium have been collectively termed by Tempest (1970) as the "Microbial Growth Cycle". The formation of biomass during the batch cultivation of organisms has been widely covered, and kinetic models are available which describe the synthesis of biomass by uni-cellular (Monod, 1942; Moser, 1958; Van Uden, 1969) and multi-cellular organisms during growth either as filaments (Pirt and Callow, 1960; Trinci, 1969) or, alternatively, as pellets (Marshall and Alexander, 1960; Pirt, 1966). At the commencement of the cycle, an initial variable "lag" or phase of adaptation is evident (Righelato, 1975; Pamment *et al*, 1978). Subsequently, biomass concentrations increase with time. This "exponential" phase of growth, the duration of which is partly dependent on the initial concentration of growth limiting nutrient, is characterized by rapid substrate utilization and product formation (Hang *et al*, 1974; Hang *et al*, 1975; Hang, 1977a). However, as a consequence of these processes, environmental changes are brought about. Eventually, these changes limit the growth rate of the organism, finally causing the cessation of growth in the stationary phase. This is the direct consequence of the lack of a nutrient or the accumulation of an inhibitory product or some environmental change (Pirt, 1975).

The application of such kinetics to waste utilization has been widely considered and the reader is referred to the work of Thanh and Simard (1973), Hang et al (1973), Shannon and Stevenson (1975a,b), Tomlinson (1976a,b), Weiner and Rhodes (1974), and Green et al (1976).

Continuous Flow Chemostat Cultures. The elements of simple continuous flow chemostat cultures were initially conceived by Monod (1950) and Novick and Szilard (1950) and were fundamentally derived from batch culture dynamics. A simple chemostat culture consists of a culture into which fresh medium is continuously added at a constant rate and where the culture volume is maintained essentially constant by the continuous removal of culture. Except for one nutrient, which is initially present in quantities capable of supporting only a limited amount of growth, the remaining nutrients are present in excess of immediate growth requirements. Typically, growth is initially allowed to proceed as in a batch fermentor until all the growth limiting substrate is consumed; then, at that point, fresh medium is continuously added at a constant rate, so that growth continues. However, as the growth limiting substrate is utilized, this limits the specific rate of growth, until the latter equals the culture removal rate. Eventually, there will be no further changes in the concentrations of the growth limiting substrate and biomass, and the rates of conversion of the growth limiting substrate to biomass will balance the culture removal rate. A "steady state" is now established, where the growth rate is less than the maximum and equal to the culture removal rate; i.e., the number of medium volumes passing through the chemostat per hour. Such a "steady state" is self-adjusting, and consequently, it is possible to maintain culture specific rates of growth at any value from zero to maximum. The reader is referred to Herbert et al (1956), Tempest (1970), Righelato and Elsworth (1970), Solomons (1972), Righelato (1975), and Pirt (1975) for further details.

The operation of a continuous biomass process offers several advantages over a simple batch system. The quantities of biomass produced in continuous cultures are several-fold greater than for batch cultures of comparable volume (Mateles and Tannenbaum, 1968). Solomons (1975) reported that in comparison to a batch fermentation of 36 hours duration, a simple chemostat which is operated at a dilution rate of 0.1 hr^{-1} will produce 3.6 times the product over a similar period of time. Chemostat cultures which are operated for extended periods of time under specified conditions also act to select in favor of mutants or contaminants which may possess improved abilities for substrate utilization (Righelato, 1975). For simple batch cultures, where more than one source of carbon is available, the sequential utilization of these (Monod, 1942) may prolong batch process times. However, it has been shown that when a mixture of two carbon sources is available in a carbon limited simple chemostat culture, both may be simultaneously utilized over a wide range of dilution rates (Baidya et al, 1967; Standing et al, 1972). The ability to control and maintain rates of growth during continuous cultivation also provides for some measure of control over organism nucleic acid levels (Herbert, 1959). Furthermore, because biomass is produced continuously, cell recovery may also be achieved in a continuous fashion. This minimizes harvesting time and thus reduces the times available for biomass autolysis and subsequent protein loss (Solomons, 1975). One disadvantage of simple chemostat cultures is that it is not possible to produce the maximum harvest from a particular substrate while at the same time producing a high quality effluent (Herbert *et al*, 1956; Leudeking, 1967; Forage, 1978). At culture dilution rates which allow maximum biomass production, a reduction in the efficiency of substrate utilization occurs, thus decreasing the quality of the effluent. Alternative chemostat culture systems which partly circumvent such a disadvantage are summarized below.

Simple Chemostats with Biomass Recycle. The recycling of biomass in a chemostat culture gives rise to steady state biomass concentrations greater than those obtained at comparable dilution rates in a simple chemostat (Herbert, 1961). Pirt and Kurowski (1970) described the internal recycle of biomass in a simple single stage chemostat, by the use of an internal yeast biomass filtration system. Such internal feedback may also be realized inadvertently through the use of a reactor of poor design. Solomons (1972) has illustrated this for the stirred tank reactor where recycle is dependent on the design of the culture effluent take-off and the velocities with which the culture passes through this point. Feedback, however, is only described for micro-fungal growth, whereas the growth kinetics of smaller single-celled organisms are similar to those of the simple chemostat. Internal recycling has also been described for the vertical tubular reactor (Pannell, 1976; Pace and Righelato, 1976). In this case, recycling is seen as a balance between the sedimentation of biomass, which acts to increase the amounts of cell mass retained, and the actions of medium and gas flow which act to wash out the biomass. However, these kinetics are only observed during the pelletal form of micro-fungal growth; with filamentous morphologies, conventional continuous stirred tank kinetics are obeyed. Alternatively, feedback may also be facilitated through the use of an external biomass separator. The simplest form of separator is a cell sedimentor, which, when culture effluent passes through, acts to produce both dilute and concentrated streams of biomass. A fraction of the latter is then fed back to the reactor. Meyrath (1975) described such a system for the recycling of yeast biomass, while Lin and Yang (1976), Pohland and Hudson (1976), and Bonotan-Dura and Yang (1976) described the recycling of heterogeneous microbial populations.

The application of single stage recycling kinetics will be of use only when a growth limiting substrate is available in dilute concentrations (Pirt, 1975). This is illustrated by Meyrath (1975), who reported that yields of *Saccharomyces kloeckerianus* were similar to those obtained during commercial yeast production but at only one-tenth of the initial substrate concentration. The use of the continuous vertical tubular reactor for the production of biomass from dilute carbohydrate containing effluents was discussed by Pannell (1976). However, Pace and Righelato (1976) reported that for a similar system operated at culture dilution rates close to or in excess of those required to establish maximum growth rates, the substrate utilization efficiencies compared unfavorably with those obtained in a simple stirred tank chemostat.

Where recycling kinetics are used, the maximum output rate of biomass will exceed that obtained for a non-recycling chemostat. Consequently, improved substrate utilization efficiencies will be possible for culture dilution rates which support greater biomass production. Furthermore, such a system will also be more resistant to culture disturbances, as may be experienced during sudden rises in substrate concentration (Edwards *et al*, 1972).

Multi-stage Chemostats with Single Stream Feeds. A multi-stage system is established when the culture effluent from one chemostat becomes the feed for a second. Where complex substrates containing more than one source of carbon and nitrogen are available, such a system will be of use in order to effect more complete utilization of substrate (Pirt, 1975). A two stage chemostat was described by Church *et al* (1972) in an attempt to improve the efficiency waste utilization following primary digestion in a "Lagooning chemostat". Fencl (1969) described a two stage chemostat system for the production of bakers' yeast from molasses. In this case, the production of biomass was similar for both single and two stage systems, while cell yields were improved for the latter owing to more complete usage of growth substrates. A multi-stage process for the production of bakers' yeast has also been described by Burrows (1970). For two stage chemostats, Pirt (1975) indicated that the concentrations of biomass for both stages will be similar, and unless culture dilution rates for the first stage approach the critical value, the specific growth rate of the organism in the second stage will approach zero. This may be of use in providing some measure of control over organism nucleic acid levels (Fencl, 1969).

An alternative form of the multi-stage chemostat, of interest for the conversion of effluents to biomass, is provided by the multi-stage vertical tubular reactor (Prokop *et al*, 1969; Kitai *et al*, 1969; Kitai and Yamagata, 1970; Hsu *et al*, 1975). The culture kinetics of this system are such that they allow for the equivalent effect of a plug-flow culture to be established in a fully aerobic system (Solomons, 1972; Pirt, 1975). Furthermore, by varying the operating parameters of the multi-stage columns—i.e., backflow, sedimentation of biomass, and feed geometry, it is possible to increase the operating range of such a chemostat (Erickson *et al* 1973). The recycling of biomass in a multi-stage reactor has been described by Kitai *et al* (1969). Recycling kinetics will be of use when they are applied to waste treatment, and Pirt (1975), using the data of Powell and Lowe (1964), indicated that by the use of a high recycling ratio of biomass to effluent, the efficiency of substrate conversion will be significantly improved over that obtained for a simple chemostat with biomass feedback. This will be more significant with increased dilutions of the growth limiting substrate. Furthermore, the maximum output rate or biomass will exceed that for a single chemostat with biomass feedback.

Variable Volume Cultures. A continuous fermentor in which there is a feed stream but no outlet stream has been termed a fed-batch reactor (Yoshida *et al*, 1973). When a portion of the culture is withdrawn at intervals, the culture system becomes a repeated fed-batch culture (Pirt, 1975). Like the chemostat, such a system may be seen to be a continuous culture, where substrate is supplied throughout the course of the fermentation. However, the variation in volume encountered during such a fermentation due to the unequal inlet and outlet medium flow rates distinguishes such a culture from a chemostat culture (Dunn and Mor, 1975).

Fed-batch cultures were initially used in the traditional fermentation industry, where attempts were made to improve the productivities of bakers' yeast (Burrows, 1970) and penicillin (Hockenhull and Mackenzie, 1968). In an attempt to maximize the yields of bakers' yeast during the fermentation of glucose or molasses, Aiba *et al* (1976) discussed the use of a fed-batch culture as a means of overcoming the "glucose effect." The ability to restrict the rates of medium feed to a reactor will also be of advantage when process rates are limited by the rates of oxygen transfer (Pirt, 1975). Furthermore, Blanch and Bhavaraju (1976) suggested that greater reductions in the apparent viscosities of a culture would be possible for processes operated semi-continuously than for those operated fully continuously. In such a case, the rates of oxygen transfer will be considerably improved, allowing for higher biomass levels to be maintained. Fu and Thayer (1975) compared the use of batch and semi-continuous cultures for the production of SCP and indicated that considerably more protein may be produced for a semi-continuous system. Worgan (1976) described a multiple substrate feed, multi-stage culture system suitable for the production of biomass from dilute waste and effluent substances.

Microbial Growth Rates

The requirement of an organism to be capable of a relatively fast rate of growth is frequently cited amongst the criteria which outline an organism selection program. This requirement will be of special importance for the operation of batch processes and it enables fermentor residence times to be minimized. However, for continuous processes where, at high culture dilution rates, limitations in the efficiencies of substrates removed will be apparent (Forage, 1978), this will be of lesser importance (Solomons, 1975).

Many fungi are typically regarded as being slower growing organisms, whereas yeasts and bacteria are represented clearly as fast growers (Worgan, 1976). Maximum cell mass doubling times quoted are 30 minutes for bacteria, 4 hours for yeasts, and 18 hours for fungi (Imrie and Vlitos, 1975). However, as the latter authors indicate, such figures are generalizations, and while there is little information describing the maximal rates of fungal growth (Solomons, 1975), the widely held belief that these organisms are capable of only relatively slow rates of growth is now not wholly justified (Carter and Bull, 1969; Espinosa *et al*, 1977). Certain micro-fungi possess cell mass doubling times of 3.5 hours or less, making them of potential interest for SCP production (Hang *et al*, 1974).

Effects of Temperature

For a process concerned with biomass production, temperatures are often specified which are optimal for the growth of a defined micro-organism (Hang, 1977a,b). An increase of temperature may have the advantage of raising the maximal growth rates of the organism (Trinci, 1969, 1972), and this may be a disadvantage when considering other growth linked process parameters. Hang (1976) demonstrated a requirement for an optimal growth temperature for *C. utilis* during waste brine utilization at 25°C, and at temperatures above and below this, reductions in the yield of cell mass and efficiency of removal of B.O.D. were apparent. Reade and Gregory (1975) similarly specified an optimal growth temperature of 45°C for *Aspergillus fumigatus*. At temperatures above this value, reductions in the yield of protein were reported. However, for simpler biomass processes, where the requirement for rigid fermentation controls appears to be unjustified, a need is often expressed for an organism to be capable of growth over a range of temperatures. Church *et al* (1972) indicated that both *Trichoderma viride* and *Gliocladium deliquescens* are capable of growth over the range 18 to 35°C without severe impairment of their efficiencies of B.O.D. removal. Similarly, Imrie and Vlitos (1975) showed that while the optimum growth temperature for *A. niger* (M1) is 30°C, biomass production and protein yields are largely unchanged at 36°C. Furthermore, the carbohydrate conversion efficiencies of *Fusarium sp.* (M4) are similar from 25°C to almost 40°C, while this organism grows optimally at 35°C (Imrie and Righelato, 1976). However, at lower growth temperatures, the time required to produce an acceptable yield of biomass, will, of necessity, be longer. Fermentation at higher temperatures might be favored not only from the point of view of reducing batch process times, but also for reducing the problems of excess process heat (Cooney and Makiguchi, 1976). Occasionally, a fermentation initially operating at mesophilic growth temperatures may later operate at higher temperatures due to the lack of adequate cooling facilities (Espinosa *et al*, 1977). Wasserman (1960) reported that the yields of *S. fragilis* were largely unchanged at growth temperatures varying from 31 to 43°C during such unchecked temperature rises. However, this mode of operation is unlikely to be suited to the growth of mesophilic yeasts in a simple continuous process at high temperatures, where significant reductions in the yields of biomass are likely to be encountered (Van Uden and Maderia-Lopes, 1976). To offset this, a process utilizing thermotolerant (Meyrath, 1975; Reade and Gregory, 1975) or thermophilic (Sucuru *et al*, 1975) organisms might be preferred.

The saturation level of oxygen in aqueous media at 60°C is considerably lower than that found at 25°C, but the mass transfer coefficient of oxygen is considerably increased at higher temperatures; thus, the overall rates of oxygen transfer at thermophilic growth temperatures may be similar to those reported for mesophilic temperatures (Sucuru *et al*, 1975). Risks of contamination are also reported to be reduced for processes operated at higher temperatures. Reade and Gregory (1975) and Meyrath (1975) indicated that the substrate conversion efficiencies of thermotolerant organisms are similar to those commonly found for mesophilic species. However, for thermophilic species, considerably lower yield values have been reported (Sucuru *et al*, 1975).

During continuous biomass production, Sucuru *et al* (1975) showed that although levels of biomass protein increase with increasing culture residence times, the improved yields of biomass experienced during faster rates of growth allow overall yields of protein to increase simultaneously. Furthermore, the amounts of oxygen consumed per unit quantity of protein produced decrease at higher culture growth rates. However, under these conditions, reductions in the efficiency of substrate removal are apparent. Thus, during any assessment of the process variations of such a system, a balance between the desired operating parameters must be made.

Oxygen Requirements

As the requirements for oxygen during rapid microbial growth are quantitatively similar to those for carbon, an adequate supply must be available in order to prevent reductions in yields of biomass and protein. Mateles and Tannenbaum (1968) reported that most continuous biomass processes operate under conditions of only a slight oxygen excess due to the high costs encountered in supplying this substrate. For batch processes, this is also true, when the peak requirement for oxygen invariably represents only a fraction of total process time (Wasserman, 1960). In an attempt to reduce the oxygen requirements of a batch process, Imrie and Righelato (1976) discussed the growth of *Fusarium sp.* (M4) under conditions of oxygen limitation. In this case, while reduced rates of growth were apparent, the overall yields of biomass were largely unchanged. Ryu and Humphrey (1972) discussed the use of oxygen as a means of achieving process control, and Ryder and Sinclair (1972) indicated that, at least for continuous processes, improved substrate conversion efficiencies might be obtained through limitation of both oxygen and carbon.

The efficiency with which a fermentation is oxygenated will be of considerable economic importance and will be related to the behavior of the culture fluid, which, in turn, will vary with the morphology of the chosen micro-organism (Blanch and Bhavaraju, 1976). Because of the high viscosities of filamentous fungal culture fluids, they are difficult to aerate efficiently (Solomons and Weston, 1961; Espinosa *et al*, 1977). Reductions in the rates of oxygen transfer (Deindoerfer and Gaden, 1955) and increases in the critical dissolved oxygen concentration (Steel and Maxon, 1966) have been reported for these thixotropic cultures. Consequently, a pellet morphology may be chosen for biomass production. In cultures containing pellets which vary in diameter from 0.1 to 1.0 millimeters, the lower viscosities result in easier agitation and improved rates of oxygen transfer (Taguchi, 1971). However, the pellet structure frequently represents a barrier to oxygen diffusion (Bhavaraju and Blanch, 1975). Furthermore, in large pellets, autolysis of the pellet center due to a lack of oxygen may often occur (Phillips, 1966; Hofsten and Ryden, 1975). Consequently, a knowledge of the size of pellet produced during biomass production is essential (Blanch and Bhavaraju, 1976). The possible causes of pellet formation were outlined by Whitaker and Long (1973) and by Metz and Kassen (1977) and will not be touched upon here. For single-celled organisms, studies indicate that such cultures possess the same properties as cultures of fungal pellets (Solomons, 1975; Sanchez-Marroquin, 1977), so that the ease with which they may be oxygenated will be partly related to the concentrations of biomass present in a fermentor (Blanch and Bhavaraju, 1976).

Substrate Conversion Efficiencies

For any biomass process to be successful, it is important that the yields of cell mass and protein produced per unit of growth limiting substrate consumed (usually carbon) are maximized. This will be of special significance where the growth limiting substrate represents an organic pollutant. The necessity for growth to be carbon limited is apparent for both batch and continuous processes and prevents accumulation of carbon, which would result in conditions of nitrogen limitation (Herbert, 1959). When carbon is not growth limiting, the differential assimilation of one carbon compound may be apparent in a simple chemostat culture where more than one such compound is available (Ng and Dawes, 1973). As a waste may contain a mixture of carbon compounds, the operation of a simple continuous system under conditions of nitrogen limitation may act to limit the efficiency of substrate conversion. Furthermore, during the production of fungal biomass, the initiation of secondary metabolism, which may lead to toxin production, is also largely avoided under conditions of carbon limited growth (Bu'lock, 1965).

The similarities of the growth yield coefficients obtained for aerobic micro-organisms are indicative of a maximum carbon conversion efficiency common to these organisms (Worgan, 1972; Righelato, 1975). For micro-fungi, a value of 45 to 52% is considered achievable (Righelato *et al*, 1968; Carter *et al*, 1971). Assuming that the average yeast cell contains 47% carbon, White (1954) estimated that the maximum yield of biomass from glucose would be 57.7%. Values for the substrate conversion efficiencies obtained for these groups of organisms during their growth on waste substrates are given in Table 14.2.

In any discussion of the yields of biomass obtained for any organism during its growth on a particular substrate, a full knowledge of that organism's cellular carbon content is desirable. Lilley (1965) has indicated that values for the cellular carbon content of *Aspergillus niger* may vary from 40 to 63%, and considerable variation in the apparent yields of biomass for this organism may be experienced with differing strains. Many authors, when describing the cell mass levels obtained from a particular substrate, do so in terms of B.O.D. or C.O.D. This may lead to considerable inaccuracy, for wastes frequently contain a variety of potentially assimilable substrates which may produce exaggerated yield values (Shannon and Stevenson, 1975b). In other cases, the presence of non-fermentable substrates may create reductions in the levels of cell mass obtained. Espinosa *et al* (1977) indicated that when a strain of *Verticillium* was grown on molasses, the substrate conversion efficiencies approached the maximum values quoted for micro-fungi. Yet, when the same organism was grown on coffee waste waters, the production of cell mass was considerably lower. For those continuous processes which are operated at culture residence times designed to achieve a more complete substrate removal, reductions in apparent yields may be ascribed to an organism's requirement for maintenance energy. Such requirements are partly attributed to the reduced yields obtained for *S. cerevisiae* during chemostat growth at high temperatures (Van Uden and Maderia-Lopes, 1976). The morphological form of an organism may also create differences in yields of biomass. Litchfield (1968) indicated that biomass yields obtained from fungal pellets are greater than those from the mycelial growth form, while Imrie and Vlitos (1975) reported the contrary to be true for *A. niger* (M1). The choice of organism is equally important for maximizing substrate conversion efficiencies. While traditional whey processes utilize *S. fragilis*, improved lactose conversion efficiencies for *Trichosporon cutaneum* have been reported (Atkin *et al*, 1967).

In order to obtain an optimal yield of biomass and protein from any given substrate, it is essential that the major nutrients which are required for balanced growth are, present together in the correct proportions. Some wastes may con-

**TABLE 14.2.   The Yield of Biomass of Different Organisms and
and Different Wastes.**

Substrate	Organism	Reduction (%) C.O.D.	B.O.D.	Yield Biomass (g/l)	Yield Substrate (%)	Batch Time (hr)
Coffee waste waters	Verticillium sp.[1]	70	–	3.4	31C.O.D.	24
Cane blackstrap molasses	Verticillium sp.[1]	–	–	3.4	47CHO	24
Brewery spent grain liquor	Aspergillus niger[2]	–	96	13.0	57T.S.	144
Acid brine	Geotrichum candidum[3]	–	87	13.0	62B.O.D.	96
Carob extract	Aspergillus niger (M1)[4]	–	–	20.0	45T.S.	24
Carob extract	Fusarium sp. (M4)[5]	–	–	20.0	47CHO	20
Cassava	Aspergillus fumigatus[6]	–	–	24.0	46CHO	20
Potato starch	Aspergillus oryzae[7]	91	–	–	46.2CHO	–
Acid brine	Candida utilis[8]	–	93	7.9	–	–
Whey	Saccharomyces fragilis[9]	84	–	24.0	55lac	4

Substrate	Organism	Reduction (%) C.O.D.	B.O.D.	Yield Biomass (g/l)	Productivity (g/l/hr)	Yield Substrate (%)	Dilution Rate (1/hr)
		Continuous Processes					
Confectionery effluent	Candida utilis[10]	74	81	10.12	–	43C.O.D.	0.3
Molasses	Saccharomyces kloeckerianus[11]	–	–	–	1.33	51.4S	1.0
Corn waste (lab.)	Trichoderma viride[12]	96.2	99.2	1.9–2.1	–	50B.O.D.	0.055
Corn waste	Trichoderma viride[12]	92.0	97.5	–	–	50B.O.D.	0.04
Whey	Saccharomyces[13]	–	–	–	–	45–52lac	0.125

NOTE: CHO—carbohydrate; T.S.—total sugars; lac—lactose; S—sucrose.

[1] Espinosa *et al.* 1977.
[2] Hang *et al.* 1975.
[3] Hang *et al.* 1974.
[4] Imrie and Vlitos. 1975.
[5] Imrie and Righelato. 1976.
[6] Reade and Gregory. 1975.
[7] Worgan. 1976.
[8] Hang. 1977a,b.
[9] Wasserman. 1960.
[10] Forage. 1978.
[11] Meyrath. 1975.
[12] Church *et al.* 1972.
[13] Bernstein *et al.* 1977.

tain adequate amounts of nitrogen (Hang, 1977a,b), while in others, nitrogen levels are such that a supplementary source is required. Wasserman (1960) indicated that whey alone will only support some 15 g/l of yeast biomass, whereas with a supplementary source of ammonium sulphate, the yields of biomass may be increased to some 24 g/l. The lower cell yields obtained by Amundson (1967) may indeed be partly attributed to the fact that such a source of nitrogen was not supplied. The source and levels of nitrogen may also affect other growth linked parameters. This is clearly indicated in the data of Shannon and Stevenson (1975b), in which the yields of cell mass, protein, and efficiencies of B.O.D. reduction by *Candida steatolytica* and *Calvatia gigantea* on brewery wastes varied with the type and amount of supplementary nitrogen supplied (Table 14.3). Similarly, Reusser *et al* (1958) demonstrated that the source of nitrogen and its levels may alter the yields of biomass and protein of *Tricholoma nudum*. However, Reade and Gregory (1975) showed that while the carbohydrate conversion efficiencies of *Aspergillus fumigatus* are related to the source of nitrogen, the yields of protein are largely unaffected by the source of nitrogen. The source of nitrogen may also influence the direction of a fermentation indirectly, through a pH effect. This observation may be exploited in providing some measure of control over a fermentation. Where a simple non-aseptic process is required, the use of an ammonium salt may allow culture pH values to remain at levels which minimize the risks of contamination (Church *et al*, 1972; Imrie and Righelato, 1976). However, in general, the effects of the various nutritional and physiological parameters of a process on the behavior of an organism are complex. Consequently, for each combination of organism and substrate, various parameter combinations should be tested out to determine the optimum combination, at both experimental and production scale levels.

SELECTED PROCESSES

The Symba Process

This is a process designed by the Swedish Sugar Company and Chemap Co., for the conversion of liquid wastes which contain starch to biomass. Such wastes are produced during the processing of potato, grain, rice, vegetable, and pasta products (Skogman, 1976). The process involves two species of yeast: *Endomycopsis fibuliger*, which produces the amylase necessary to break down the starch, and *Candida utilis*, which utilizes the resultant sugars. The process is capable of reducing waste liquids with an initial B.O.D. of 10,000 to 20,000 mg/l by approximately 90%. The biomass product contains 96% *Candida* yeast and 4% *Endomycopsis*. This product has been extensively tested as an animal feed. The composition of Symba yeast is given in Table 14.4.

TABLE 14.3. Reducing Sugar Utilization, Yields, Protein content, and B.O.D. Reduction by *Candida Steatolytica* and *Calvatia gigantea* on Brewery Wastes.

Waste	Reducing Sugar: Nitrogen Ratio	Reducing Sugar Utilization		Yield (dry weight basis)				Protein Content		B.O.D. Reduction	
		C. gigantea (%)	*C. steatolytica* (%)	*C. gigantea* (g/l)	*C. gigantea* (g/g sugar)	*C. steatolytica* (g/l)	*C. steatolytica* (g/g sugar)	*C. gigantea* (%)	*C. steatolytica* (%)	*C. steatolytica* (%)	*C. gigantea* (%)
Grain press liquor (GPL)	69.1:1	74.2	83.8	6.25	0.248	6.28	0.282	13.69	21.30	28.2	56.2
GPL + 0.05% N as $(NH_4)_2SO_4$	32.1:1	76.7	90.3	15.19	0.660	6.51	0.283	29.40	28.75	28.1	75.0
GPL + 0.10% N as $(NH_4)_2SO_4$	20.9:1	78.3	86.7	13.03	0.501	7.89	0.336	35.62	40.13	31.2	67.2
GPL + 0.15% N as $(NH_4)_2SO_4$	15.5:1	76.0	73.3	10.17	0.462	5.45	0.239	37.37	34.87	29.7	66.4
GPL + 0.05% N as FSL	28.4:1	79.2	83.0	10.67	0.498	8.08	0.384	33.54	27.75	40.0	55.3
GPL + 0.10% N as FSL	16.0:1	80.4	87.0	9.96	0.498	7.84	0.424	37.94	34.87	41.8	65.1
GPL + 0.15% N as FSL	10.1:1	82.1	82.1	9.23	0.576	6.93	0.433	44.40	34.65	39.2	59.1
Trub press liquor (TPL)	72.0:1	63.6	68.4	27.72	0.749	10.56	0.302	17.50	23.49	45.5	56.1
TPL + 0.05% N as $(NH_4)_2SO_4$	43.5:1	68.1	70.9	39.74	0.883	10.71	0.286	28.56	27.75	50.7	65.2
TPL + 0.10% N as $(NH_4)_2SO_4$	31.2:1	70.0	76.4	36.61	0.872	12.70	0.330	38.27	44.25	54.5	62.1
TPL + 0.15% N as $(NH_4)_2SO_4$	24.3:1	65.4	74.5	15.65	0.382	11.03	0.306	29.70	34.81	50.7	55.3
TPL + 0.05% N as FSL	37.3:1	72.3	81.8	31.44	0.516	10.25	0.301	40.27	37.16	52.1	54.5
TPL + 0.10% N as FSL	22.6:1	71.8	83.9	28.21	0.881	9.06	0.324	44.23	33.60	50.0	63.3
TPL + 0.15% N as FSL	13.9:1	71.0	85.5	25.52	0.945	8.04	0.365	44.52	29.56	48.5	59.8

Source: Shannon and Stevenson. 1975b. (Reprinted from the *Journal of Food Science* 40:826–829. Copyright © by Institute of Food Technologists.)

TABLE 14.4. Composition of Symba Yeast in Food Quality

Protein	48.0%
Fat	3.0%
Carbohydrates	36.5%
Fiber	1.0%
Minerals	5.5%
Water	6.0%
Nucleic acid	4.0%
B vitamins	
Thiamine	145 mg/kg
Riboflavine	80 mg/kg
Pyridoxine	35 mg/kg
Niacin	430 mg/kg
Folic acid	20 mg/kg
Amino acid profile	(g/100 g raw protein)
Arginine	4.6
Aspartic acid	10.3
Cystine	1.0
Glutamic acid	13.8
Histidine	2.0
Isoleucine	4.3
Leucine	7.5
Lysine	6.3
Methionine	1.5
Phenylalanine	5.4
Threonine	5.4
Trytophan	1.3
Tyrosine	4.8
Valine	4.2

Source: Skogman. 1976.

The nucleic acid content (4%) is low for a yeast product and compares favorably with other micro-fungi, such as *Aspergillus oryzae* and *Fusarium semitechtum* (Worgan, 1976). As with many fungi, the level of sulfur amino acids is low.

In the Symba process, the waste effluent is sterilized, then split into two streams, one to a small fermentor, in which the *Endomycopsis* is grown, and the remainder into a larger fermentor, in which both *Candida* and *Endomycopsis* are present. The resultant broth from the *Endomycopsis* fermentor is passed continuously into the larger fermentor, providing a continuous input of amylase activity, which enables the starch to be broken down. When the plant is operating continuously, it is capable of processing 20 m³/hr of effluent containing 2

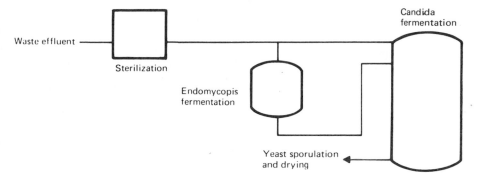

Fig. 14.6. Diagrammatic representation of the Symba Process. *Source:* Skogman. 1976.

to 3% dry matter, and produces dried yeast at 250 kg/hr. It is recommended for effluents which contain 2% or more dry matter (Fig. 14.6).

Whey Fermentation

The production of SCP from whey is currently under considerable investigation in view of the problems associated with the disposal of this waste. Early attempts at the conversion of whey to yeast biomass were described by Wasserman (1960), where a strain of *Saccharomyces fragilis* completely utilized the available lactose in a period of 4 hours and gave a yield of cell mass of 0.55 g/g of lactose utilized. The time required to achieve the maximal yields varied with the size of the inoculum, and Wasserman (1960) considered that an inoculation level equivalent to 30% of the weight of lactose initially present would enable batch times to be minimized. The chosen temperature for yeast growth varied between 31 and 33°C, and although yields were largely unchanged at temperatures as high as 41 to 43°C, bacterial contamination was noted during growth at elevated temperatures. A variation of pH between 5.0 and 5.7 was found to be acceptable for growth, but control of this parameter within these limits was essential to prevent reductions in the yields obtained at higher pH values. The use of a low pH in addition to a high inoculum level also favored the operation of a batch process under non-aseptic conditions. In order to maximize the rates of yeast growth, it was found necessary to supplement the whey with 0.5% KH_2PO_4 and 0.1% yeast extract (which acted as a source of growth factor). Furthermore, a supplementary source of nitrogen in the form of $(NH_4)_2SO_4$ at a level of 0.5% enabled the yields of yeast to be increased from 15.0 to 24 g/l. However, during pilot scale operation, process rates were found to be limited by the rates of oxygen transfer, and Wasserman (1960) concluded that improved substrate conversion efficiencies would probably be obtained through the use of

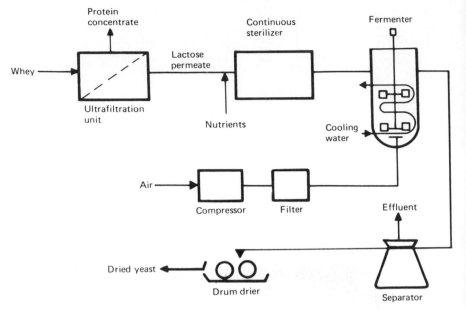

Fig. 14.7. Ultrafiltration/fermentation plant for processing whey. *Source:* Horton *et al.* 1972.

a different fermentor design or, alternatively, by using a semi-continuous or fully continuous system.

Wasserman (1960) indicated that *S. fragilis* will grow well on both supplemented de-proteinized whey and on supplemented whole whey. This observation has proved useful in the production of yeast biomass from lactose permeates, which may be produced through the use of ultrafiltration techniques (Fig. 14.7; Horton *et al*, 1972). Horton *et al* (1972) have described a process whereby whey is ultrafiltered to produce both a lactose permeate as well as a protein concentrate. The permeate is then supplemented with nutrients, sterilized, and fermented continuously in a stirred tank reactor. The yeast is harvested centrifugally to a concentration of 15 to 20% solids and then drum dried. The dried yeast can then be milled and packaged or sold in bulk and is suitable for use as an animal feed.

Bernstein *et al* (1977) recently described a process for whey utilization which may be operated either batchwise, semi-continuously, or fully continuously (Fig. 14.8). Concentrated acid and/or sweet whey is initially diluted with water, raw whey, or condensate water to an appropriate lactose concentration. The use of a "closed loop" system, where fermentor effluent condensate water is recycled to the initial process steps, enables process effluents to be eliminated. The whey is then supplemented with the nutrients, heated to 80°C for 45

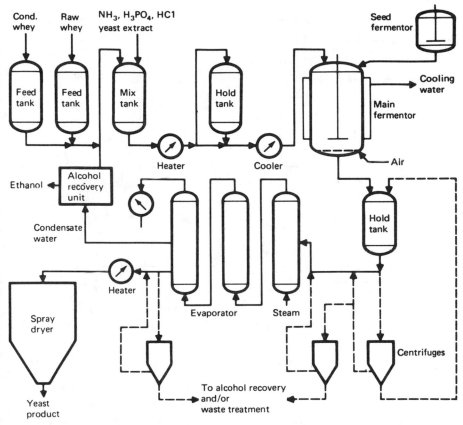

Fig. 14.8. Whey fermentation process of Bernstein. *Source:* **Bernstein** *et al.* **1977.**

minutes, cooled, and then fermented. By the use of a low pH as well as a high cell count and seed size, the necessity for asepsis is eliminated. In batch operation, a yeast seed of 1×10^9 cells/ml, at an inoculum level of 10%, enables the complete lactose utilization to be completed in 8 hours. The maximum specific rate of growth under these conditions is approximately 0.35 hr^{-1}. Fermentation "down times" may be reduced through semi-continuous operation, where 90% of the broth is removed prior to the next batch, and are eliminated completely by fully continuous operation. At a culture dilution rate of 0.125 hr^{-1}, biomass is harvested at a rate of 1250 gal/hr, which approximates to a batch cycle of 10,000 gal/8 hr. When the yeast is harvested centrifugally, a good grade yeast is produced. Alternatively, to reduce costs and simplify the process, the whole fermented material may be concentrated and spray dried. This material is then suitable for use as an animal feed.

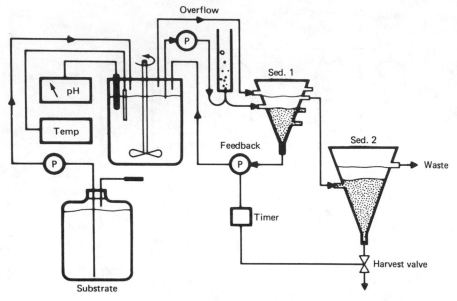

Fig. 14.9. Diagrammatic representation of the biomass recycling process of Meyrath. *Source:* Meyrath. 1975. (Reprinted from *Process Biochemistry* 10:20–22.)

Production of Feed Yeast from Low Grade Wastes

Many effluents contain less than 1% organic matter. The use of these in normal continuous culture leads to a low concentration of biomass in the broth, giving low biomass productivity per unit volume of fermentor. This problem can be resolved by recycling biomass to increase the concentration of biomass in the fermentor. Meyrath (1975) has described a procedure for growing yeast on low concentrations of sugar (2 to 5 g/l), using a recycling technique to increase the productivity of the fermentor. In this process, a highly flocculent strain of *Saccharomyces kloeckerianus* (which could be concentrated by allowing it to settle in funnel shaped sedimentation vessels) was used (Fig. 14.9).

Using a sugar solution containing 2.5 g/l sucrose as molasses, a yeast concentrate of 1.2 g/l was achieved, using the first sedimentation vessel to concentrate and recycle the yeast cells. This concentration was maintained in the fermentor, and additional yeast was harvested, using the second sedimentation vessel. This gave a productivity of 5.3 g/hr yeast from a 3.5l fermentation volume and a yield of 51.4%. It was found that with a fermentor substrate residence time of 1 hour, few problems with contaminating bacteria arose, even when non-sterile substrates were used. In this process, the thermotolerant yeast strains growing at 42°C were used, and the temperature of the fermentor was controlled by adjusting the temperature of the effluent feed.

PRODUCT EVALUATION

Nutritional Quality of the Microbial Biomass

The first requirement of a biomass product is that it is acceptable as a food to the animal or human population for which it has been produced. Animals normally accept fungal biomass in their diet (Smith *et al*, 1975). However, if the product is to be sold for human consumption, then further treatment of the biomass is necessary to make it palatable. Products with a beef-, chicken-, or veal-like flavor have been produced from the mycelium *Fusarium graminearum* (Spicer, 1971). The texture of the mycelium of the filamentous fungi is a valuable property for producing such products (Worgan, 1976).

Since the purpose of feeding microbial biomass to animals is to increase the nutritional quality of the diet, it is also essential that the biomass contains a high level of protein which has a good amino acid proflie. The protein values presented in many publications are simply the nitrogen content determined by digestion and multiplied by 6.25. Using such values, the protein content of filamentous fungi has been determined at 31 to 50%, yeasts at 47 to 53%, algae at 47 to 63%, and bacteria at 72 to 78% (Romantschuk, 1975). However, this is not a satisfactory method of estimating protein content, since much of the nitrogen present in micro-organisms is in the form of nucleic acids and cell wall materials, such as chitin and free amino acids and bases. *Fusarium graminearum* has been studied in detail after being grown in continuous culture at a growth rate of 0.1 hr^{-1}. It contained 9.66% nitrogen on a dry weight basis, of which 65% could be accounted for as protein nitrogen, 7% as free amino acid nitrogen, 15% as RNA nitrogen, 1.5% as free nucleotide nitrogen, and 10% as *N*-acetyl glucosamine nitrogen.

Not all protein has the same nutritional value. The nutritional value of proteins is influenced by the level of certain amino acids—the essential amino acids which must be present in the diet of man. The levels of essential amino acids in protein from a range of filamentous fungi were detailed by Anderson *et al* (1975). Fungal protein compares well with other more conventional foods; however, most fungal protein is deficient in the sulfur amino acids. Of 15 species tested, methionine and cysteine were found to be the limiting amino acids in 13 of them.

Nucleic acids are present in all microbial biomass. These are metabolized to uric acid, which can in excess cause a gout-like condition in man (Kihlberg, 1972). Thus, a high nucleic acid content can present a nutritional problem. An ideal biomass would contain less than 3.3% RNA (Anderson *et al*, 1975), but most fungal products contain higher levels than this. Values of between 3.2 and 4.7% w/w have been reported for filamentous fungi (Worgan, 1976) and the RNA content of yeast can rise to 10% w/w. Products containing a high level of nucleic acids may be blended with low nucleic acid products or may be processed to reduce the level of RNA. Several techniques have been described to reduce the level of RNA in biomass. The level of RNA can be reduced by controlling the fermentation conditions or by treatment of the biomass after it has been harvested (Sinskey and Tannenbaum, 1975). A high protein to RNA ratio has been achieved by growth on a phosphate limited medium, by growth at a low specific growth rate, and by growth at elevated temperatures. Treatments of biomass after it has been harvested include heat shock, which stimulates endogenous RNAase activity, treatment with exogenously applied RNAase, and chemical extraction of the RNA.

The safety of a given biomass product can be influenced by several factors, which can be considered in two categories.

1. Harmful chemicals or micro-organisms present in the substrate, which can become incorporated into the final product. Here, heavy metals, herbicides, and microbial pathogens constitute the most important threat.

2. Toxic materials produced by the organism during the fermentation. Certain fungi produce mycotoxins which are harmful to man and animals (Butler, 1975)—e.g., *Fusarium graminearum* and *Aspergillus flavus*. Other fungi (*Saccharomyces cerevisiae*, *Aspergillus oryzea*, and *Agaricus hisporus*) have been consumed by man throughout history and do not produce toxins. Although mycotoxin production is suppressed in the rapid growth conditions normally used for biomass production. the possibility that conditions suitable for mycotoxin production may arise as a result of faulty operating conditions cannot be ignored. This being so, it is essential that only strains which are genetically incapable of producing mycotoxins be used in biomass processes. Care must also be taken to prevent contamination of the fermentation by mycotoxin producing strains of fungi.

Procedures for carrying out preclinical tests on novel food materials for man have been recommended by the Protein Advisory Group of the United Nations (Tannenbaum and Wang, 1975). These are classified in the following categories of information that are required for the satisfactory evaluation of a new product.

1. *Safety*. Chemical composition, physical properties, content of micro-organisms, toxicological effects on laboratory animals.
2. *Nutritional value.* As determined by chemical and biological assay.
3. *Sanitation.* Consideration of the source of raw materials, from an aesthetic view as well as potential pathogenicity.
4. *Acceptability.* Consideration of the flavor and cultural acceptability of a product.
5. *Technological properties.* Consideration of potential for incorporation into existing products or for incorporation into new products.

These guidelines are designed for products which are intended to be sold as food products for man. As yet, the controls on animal feeds are less stringent;

however, since evidence is available that pathogenic organisms present in animal feeds can pass through the animal into meat products (Rolfe, 1976), it seems probable that similar guidelines will eventually be applied to animal feeds.

Economics of Biomass Production

Introduction. Although there can be little doubt that good quality biomass can be produced from liquid wastes, the success or failure of many of the biomass processes which have been described will depend more on economic factors than on the technological feasibility of the process. There have been several articles in recent years on the cost of production of biomass, and analyses of individual processes have been presented (Imrie and Righelato, 1976; Mateles, 1975; Tomlinson, 1976b; Litchfield, 1977; Moo-Young, 1977).

Moo-Young (1977) carried out a detailed analysis of the cost of several fermentation processes for both waste and hydrocarbon substrates. Perhaps the most valuable feature of this study is the way it illustrates the difficulty of obtaining good comparative data from studies carried out in different geographical locations using different substrates, different fermentation processes, and different scales. The choice of an Israeli process for the production of biomass from molasses in this review was unfortunate, since in Israel molasses is an imported substrate, and thus expensive (Mateles, 1975). This additional expense represents a major percentage of the total cost of the process. The data shown in Table 14.5 have been selected to illustrate how depreciation, raw materials, or labor can be the major cost of a biomass process, depending upon which process is being studied. However, a more meaningful form of analysis for the technologist is one which indicates the cost of individual stages of the

TABLE 14.5.   Analysis of Cost of Four Different Biomass Processes.

		Percent of Total Cost		
	I	II	III	IV
Depreciation	9.1	4.5 (10 yr)	51.5 (10 yr)	19.9 (5 yr)
Raw materials	55.1	61.6	12.2	16.9
Utilities	24.8	9.0	17.7	10.2
Normal operating costs (including labor)	11.0	11.0	18.6	53.0
Financing	—	13.9	—	—
	100	100	100	100

NOTE: I. Pekilo, Finland. Waste starch, liquor. (Moo-Young, 1977).
    II. Israel. Molasses. (Mateles, 1975).
    III. U. K. Potato wastes. (Tomlinson, 1975).
    IV. U. K. Palm Oil. (Imrie and Righelato, 1976).

TABLE 14.6.   Breakdown of Cost of the Development and Operation of
the Individual Steps of the Pekilo Process.

	Investment (%)	Utilities (%)
Supplies and storage	4.9	0.2
Medium preparation	1.6	1.0
Inoculum preparation	1.3	2.6
Fermentation	50.6	76.8
Harvesting	7.6	3.6
Drying	17.0	15.1
Product storage	17.0	0.7

Source: Moo-Young. 1977.

process. The relative costs of both investment and utilities of different stages of the Pekilo process are shown in Table 14.6.

The stages used in this analysis can be considered to be common to all biomass processes and deserve further discussion.

Substrate Supplies and Storage.   Since we are concerned in this chapter with the processing of waste materials, the substrate cost should be zero or even negative if present disposal methods involve expense. However, this does not always represent the situation when a process is operating. Economics of scale require that large quantities of substrate are available at the site of the plant. This may introduce a transport cost into the true cost of the substrate, if effluents from several different food processing plants are to be treated in a single biomass plant. Ideally, the substrate should be available at a steady rate throughout the year; if this is not the case, storage of the substrate or a seasonal shutdown of the plant can introduce additional costs (Tomlinson, 1976b). Finally, it is frequently noted that once a process has been developed to convert a waste material into a useful product, the cost of that waste often increases with market demand.

Choice of Fermentation Process.   Medium preparation represents only 1% of the utility costs of the Pekilo process. The cost of this stage can be kept to a minimum if the process operates using a non-sterile medium. Both yeast and fungal fermentations that operate satisfactorily using a non-sterile procedure have been described (Meyrath, 1975; Imrie and Righelato, 1976). Inoculum preparation also represents a small percentage of the overall cost of utilities. The use of continuous culture should minimize the cost of this stage.

The choice of fermentation apparatus and its mode of operation are influenced by many technical factors; however, economic considerations cannot be ignored.

The high cost of utilities in the Pekilo process can be attributed to the high cost of aeration and agitation in a stirred tank reactor. The overall cost of utilities is much lower in tower fermentors (Table 14.5). Moo-Young (1977) considered that recent developments in tower and air lift fermentors should reduce the cost of utilities in the fermentation stage. These fermentors are also cheaper to build and maintain (Snell and Schweiger, 1949).

A high yield can be obtained in continuous cultures using either a high cell density and a low growth rate or a low cell density and a high growth rate. A high cell density facilitates harvesting by reducing the quantity of water which has to be removed to produce a given quantity of dried product, and cells grown at a slow growth rate have a low nucleic acid content. However, a fast growth rate has been reported to give a higher substrate yield (Humphrey, 1975). High cell densities can be readily achieved when a concentrated substrate is being used, but if the effluent being processed contains low concentrations of substrate (<1%), then some mechanism of biomass recycle is required to achieve a high cell density (Meyrath, 1975).

Continuous culture has generally been the favored method of biomass production, although it has been argued that the efficiency of substrate removal in continuous cultures at dilution rates associated with minimum volume and power requirements is only 60% of that obtained using a semi-continuous or batch process. This has been attributed to the inevitable presence of substrate in the outlet stream of a single stage continuous culture (Inkson et al, 1974). A serial batch process in which 10% of the biomass is left in the fermentor to inoculate the following fermentations has been recommended for the fermentation of semi-solid wastes (Imrie and Righelato, 1976).

Many biomass processes are primarily concerned with pollution control rather than biomass formation. The relationship between cost and B.O.D. reduction (Fig. 14.10) has been analyzed by Tannenbaum and Pace (1976). These calculations indicate that the higher the percentage B.O.D. reduction, the higher the cost. This increase in cost can be minimized by increasing the scale of the process.

Temperature control of the fermentation is a significant cost, particularly in hot climates. Mateles (1975) reported that a saving of 9.2% in the cost of producing microbial biomass from molasses in Israel could be achieved by operating the fermentation at 40°C rather than 30°C.

Harvesting and Drying. Fermentation broths contain between 10 and 20 g/l dry cells. The water present in the fermentation broth is normally removed in two stages: an initial de-watering stage, which increases the percentage of solids to around 30%, and a final dehydration stage (Labuza, 1975). The de-watering stage is readily carried out either by filtration or centrifugation. The cost of centrifugation of yeasts or other fungi is less than 2¢/kg of dry solids. However,

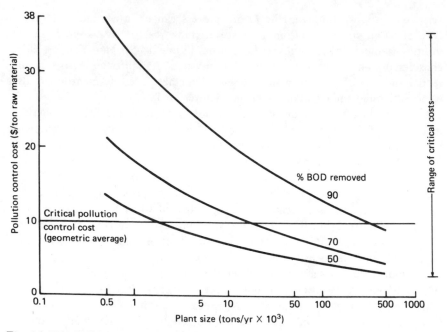

Fig. 14.10. Pollution control costs versus plant size. *Source:* Tannenbaum and Pace. 1976. (Reprinted with permission of Applied Science Publishers.)

the centrifugation of bacteria, which are smaller and less dense, costs approximately four times this amount. Material containing 30% solids is amenable to dehydration by several techniques, including spray drying, tray drying, drum drying, freeze drying, and drying in the sun. Sun drying may be appropriate to small-scale operations in hot climates, but spray drying and drum drying are the preferred techniques in most countries. These have an operating cost of between 0.5 and 1.5¢/l of water removed. Spray drying is the more expensive of the two processes, but it usually produces a better quality product. Tray drying is competitive with these two processes; freeze drying at 4.15¢/l of water removed is too expensive.

One cost that is often ignored, but which has probably deterred the development of biomass processes more than anything else, is the cost of testing the nutritional and toxicological properties of the product. The cost of carrying out such a program for a human food has been conservatively estimated at $500,000 (Tannenbaum and Pace, 1976; Van der Wal, 1976). This must, of course, be spent before any return is obtained on the process.

It is clear that many factors, both technological and non-technological, affect the viability of a biomass process, and that a decision on economic viability can be made only after a detailed analysis of an individual process in a known location. The cost of producing biomass from molasses has been estimated at between 30¢/kg (Moo-Young, 1977) and 55¢/kg (Litchfield, 1977), which gives a protein costs of 54¢/kg and $1.38/kg. The value of the product may be increased if it is fractionated into different cellular components. The extraction of nucleic acids to provide food flavoring compounds has been suggested by Peschard-Mariscal and Viniegra-Gonzalez (1977), and a detailed scheme for the fractionation of yeast into protein, glycan, and yeast extract fractions has been proposed. The properties and potential commercial applications of each fraction were indicated by Seeley (1977).

REFERENCES

Aiba, S., S. Nagai, and Y. Nishizawa. 1976. "Fed batch culture of *Saccharomyces cerevisiae*. A perspective of computer control to enhance the productivity in baker's yeast cultivation." *Biotech. Bioeng.* **18**:1001–1016.

Amundson, C. H. 1967. "Increasing protein content of whey." *American Dairyfood Manufac. Rev.* **29**:22–23, 96–99.

Anderson, C., J. Longton, C. Maddix, G. W. Scammell, and G. L. Solomons. 1975. "The growth of microfungi on carbohydrates." *Single Cell Protein* **II**. S. R. Tannenbaum and D. I. C. Wang. (Eds.). MIT Press, Cambridge, Massachusetts, pp. 314–329.

Anon, 1976. "The runner that was nearly left at the starting gate." *ICI Magazine*, **54**:250–254.

Anon. 1977. "Biotechnology." *Biologist* **24**:11–15.

Atkin, L., L. D. Witter, and Z. J. Ordal. 1967. "Continuous propagation of *Trichosporon cutaneum* in cheese whey." *Appl. Microbiol.* **15**:1339–1344.

Ali, F. A., A. G. Meiring, C. L. Duitschaever, and A. E. Reade. 1975. "Air injection by self-aspirating impeller in aerobic fermentations." *J. Milk Food Tech.* **38**:94–99.

Baidya, T. K. N., F. C. Webb, and U. D. Lilley. 1967. "The utilisation of mixed sugars in continuous fermentation." *Biotech. Bioeng.* **9**:195–204.

Bernstein, S., C. H. Tzeng, and D. Sisson. 1977. "The commercial fermentation of cheese whey for the production of protein and/or alcohol." *Biotech. Bioeng. Symp.* No. 7. E. L. Gaden, Jr. and A. E. Humphrey (Eds.). John Wiley & Sons, New York, pp. 1–9.

Bhavaraju, S. M. and H. W. Blanch. 1975. "Mass transfer in mycelial pellets." *J. Ferment. Tech.* **53**:413–415.

Birch, G. G., K. J. Parker, and J. T. Worgan. 1976. *Food from Waste*. Applied Science Publishers, London.

Blakebrough, N., P. G. Shepherd, and I. Nimmons. 1967. "Equipment for hydrocarbon fermentations." *Biotech. Bioeng.* **9**:77–89.

Blanch, H. W. and S. M. Bhavaraju. 1976. "Non-Newtonian fermentation broths: rheology and mass transfer." *Biotech. Bioeng.* **18**:745–790.

Bonotan-Dura, F. M. and P. Y. Yang. 1976. "The application of constant recycle solids concentration in activated sludge process." *Biotech. Bioeng.* **18**:145–165.

Bu'lock, J. D. 1965. "Aspects of secondary metabolism of fungi." *Biogenesis of Antibiotic Substances.* Z. Vanek and Z. Hostalek (Eds.). Academic Press, New York, pp. 61–71.

Burrows, S. 1970. "Bakers' yeast." *The Yeasts* **III**. A. H. Rose and J. S. Harrison (Eds.) Academic Press, London, pp. 349–420.

Butler, W. H. 1975. "Mycotoxins." *The Filamentous Fungi* **I**. *Industrial Mycology.* J. E. Smith and D. R. Berry (Eds.). Edward Arnold, London, pp. 320–329.

Carter, B. L. A. and A. T. Bull. 1969. "Studies of fungal growth and intermediary carbon metabolism under steady and non-steady state conditions." *Biotech Bioeng.* **11**:785–804.

Carter, B. L. A., A. T. Bull, S. J. Pirt, and B. I. Rowley. 1971. "Relationship between energy substrate utilisation and specific growth role in *Aspergillus nidulans.*" *J. Bact.* **108**:309–313.

Chakravarty, M., S. Begum, H. D. Singh, J. N. Barvah, and M. S. Iyengar. 1973. "Gas hold-up distribution in a gas-lift column." *Advances in Microbiol Engineering. Biotech. Bioeng. Symp.* **No. 4**. P. Sikyta, A. Prokop, and M. Novak (Eds.). John Wiley & Sons, New York, pp. 363–378.

Church, B. D., H. A. Nash, and W. Brosz. 1972. "Use of *Fungi imperfecti* in treating food processing wastes." *Dev. Ind. Microbiol.* **13**:30–46.

Cocker, R. 1975. "The morphological development of *Aspergillus niger* in tower fermentor culture." Ph.D. Thesis, University of Aston in Birmingham, U.K.

Cooney, C. L. and N. Makiguchi. 1976. "Temperature as an engineering parameter in SCP production." *Continuous Culture* 6: *Applications and New Fields.* A. C. R. Deans, D. C. Ellwood, C. G. T. Evans, and J. Melling (Eds.). Ellis Horwood, Chichester, England, pp. 146–157.

Daunter, B. 1972. "The evaluation of the fungi using tower fermentors." M. Sc. Thesis, University of Aston in Birmingham, U.K.

Deindoerfer, F. H. and E. L. Gaden, Jr. 1955. "Effects of liquid physical properties on oxygen transfer in penicillin fermentors." *Appl. Microbiol.* **3**:253–257.

Dunn, I. J. and J. R. Mor. 1975. "Variable-volume continuous cultivation." *Biotech. Bioeng.* **17**:1805–1822.

Edwards, V. H., R. C. Ko, and S. A. Balogh. 1972. "Dynamics and control of continuous microbial propagators to subject substrate inhibition." *Biotech. Bioeng.* **14**:939–974.

Erickson, L. E., S. S. Lee, and L. T. Fan. 1973. "Modelling and analysis of washout in tower fermentation processes." *Advances in Microbial Engineering, Biotech. Bioeng. Symp.* **No. 4.** B. Sikyta, A. Prokop, and M. Novak (Eds.). John Wiley & Sons, New York, pp. 301–330.

Espinosa, R., O. Maldonado, J. F. Menchu, and C. Rolz. 1977. "Aerobic nonaseptic growth of *Verticillium* on coffee waste waters and blackstrap molasses at a pilot plant scale." *Single Cell Protein from Renewable and Nonrenewable Sources. Biotech. Bioeng. Symp.* **No. 7.** E. G. Gaden, and A. G. Humphrey. (Eds.) John Wiley & Sons, New York, pp. 35–44.

Falch, E. A. and E. L. Gaden, 1969. "A continuous multistage tower fermenter **I.** Design and performance tests." *Biotech. Bioeng.* **11**:927–943.

Feichter, A. 1975. "Continuous cultivation of yeast." *Methods in Cell Biology* **11.** D. M. Prescott (Ed.). Academic Press, New York, pp. 97–130.

Fencl, Z. 1969. "Production of microbial protein from carbon sources." *Biotech. Bioeng. Symp.* **No. 1.** E. L. Gaden (Ed.). John Wiley & Sons, New York, pp. 63–70.

Forage, A. J. 1978. "Recovery of yeast from confectionery effluent." *Proc. Biochem.* **13**:8–11, 30.

Freedman, W. and J. F. Davidson. 1969. "Hold-up and liquid circulation in bubble columns." *Trans. Inst. Chem. Eng.* **47**:251–262.

Fu, T. T. and D. W. Thayer, 1975. "Comparison of batch and semi-continuous cultures for production of protein from mesquite wood by *Brevibacterium sp.* JM984." *Biotech. Bioeng.* **17**:1749–1760.

Fukuda, H., Y. Sumino, and T. Kanzaki. 1968. "Scale-up of fermentors." *J. Ferment. Tech.* **46**:829–837.

Green, J. H., S. L. Paskell, and D. Goldmintz, 1976. "Lipolytic fermentations of stickwater by *Geotrichum candidum* and *Candida lipolytica.*" *Appl. Environ.* **31**:569–575.

Greenshields, R. M. and E. L. Smith. 1971. "Tower-fermentation systems and their applications." *Chem. Eng.* **49**:182–190.

Greenshields, R. N. and E. L. Smith. 1974. "The tubular reactor in fermentation." *Proc. Biochem.* **9**:11–17.

Gromov, B. V. 1967. "Main trends in experimental work with algal cultures in the U.S.S.R." *Man, Algae and the Environment.* D. F. Jackson (Ed.). Syracuse University Press, Syracuse, New York, pp. 249–278.

Hang, Y. D. 1976. "Some factors affecting the treatment of sauerkraut waste with yeast." *5th Inter. Ferment. Symp., Berlin.* H. Dellweg (Ed.). p. 350.

Hang, Y. D. 1977a. "Baked-bean waste: a potential substrate for producing fungal amylases." *Appl. Environ. Microbiol.* **33**:1293–1294.

Hang, Y. D. 1977b. "Waste control in sauerkraut manufacturing." *Proc. Biochem.* **12**:27–28.

Hang, Y. D., D. F. Splittstoesser, and R. L. Landschoot. 1973. "Production of yeast invertase from sauerkraut waste." *Appl. Microbiol.* **25**:501–502.

Hang, Y. D., D. F. Splittstoesser, and R. L. Landschoot. 1974. "Propagation of *Geotrichum candidum* in acid brine." *Appl. Microbiol.* **27**:807–808.

Hang, Y. D., D. F. Splittstoesser, and E. E. Woodhams. 1975. "Utilisation of brewery spent grain liquor by *Aspergillus niger.*" *Appl. Microbiol.* **30**:879–880.

Hatch, R. T. 1973. "Experimental and theoretical studies of oxygen transfer in the airlift fermentor." Ph.D. Thesis, MIT Press, Cambridge, Massachusetts.

Hatch, R. T. 1975. "Fermentor design." *Single Cell Protein* **II**. S. R. Tannenbaum and D. I. C. Wang (Eds.). MIT Press, Cambridge, Massachusetts, pp. 46–68.

Herbert, D. 1959. "Some principles of continuous culture." *Recent Progress in Microbiology*. G. Tunevall (Ed.). Stockholm, pp. 381–396.

Herbert, D. 1961. "Theoretical analysis of continuous culture systems." *Continuous Culture of Microorganisms. S.C.I. Monograph* **No. 12.**, *Society of Chemistry and Industry*. London, pp. 21–53.

Herbert, D.; R. Elsworth, and R. C. Telling, 1956. "The continuous culture of bacteria; a theoretical and experimental study." *J. Gen. Microbiol.* **14**:601–622.

Hockenhull, D. J. D. and R. M. Mackenzie. 1968. "Present nutrient feeds for penicillin fermentation on defined media." *Chem and Ind.* **I**:607–610.

Hofsten, B. V. and A. L. Ryden. 1975. "Submerged cultivation of a thermotolerant basidiomycete on cereal flours and other substrates." *Biotech. Bioeng.* **17**:1183–1197.

Horton, B. S., R. L. Goldsmith, and R. R. Zall. 1972. "Membrane processing of cheese whey reaches commercial scale." *Fd. Tech.* **26**:30–35.

Hsu, K. H., L. E. Ericksson, and L. T. Fan, 1975. "Oxygen transfer to mixed cultures in tower systems." *Biotech. Bioeng.* **17**:499–514.

Humphrey, A. E. 1975. "Product outlook and technical feasibility of SCP." *Single Cell Protein* **II**. S. R. Tannenbaum and D. I. C. Wang (Eds.). MIT Press, Cambridge, Massachusetts, pp. 1–23.

Imrie, F. K. E. and R. C. Righelato. 1976. "Production of microbial protein from carbohydrate wastes in developing countries." *Food from Waste*. G. G. Birch, K. J. Parker, and J. T. Worgan (Eds.). Applied Science Publishers, London, pp. 79–97.

Imrie, F. K. E. and A. J. Vlitos. 1975. "Production of fungal protein from carob." *Single Cell Protein* II. S. K. Tannenbaum and D. I. C. Wang (Eds.). MIT Press, Cambridge, Massachusetts, pp. 223–243.

Inkson, M. B., K. C. Phillips, and F. K. E. Imrie, 1974. "Microbial protein from agricultural waste." *Waste Recovery and Recycle Symp. Chem. Eng. Cong.*, Part 1. p. A7.

Kihlberg, R. 1972. "The microbe as a source of protein." *Ann. Rev. Microbiol.* 26:427–466.

Kitai, A., H. Tone, and A. Ozaki. 1969. "Performance of a perforated plate column as a multistage continuous fermenter." *Biotech. Bioeng.* 11:911–926.

Kitai, A. and T. Yamagata. 1970. "Perforated plate column fermenter." *Proc. Biochem.* 5:52–53, 58.

Kosaric, N., H. T. Nguyen, and M. A. Bergougnou. 1974. "Growth of *Spirulina maxima* algae in effluents from secondary waste-water treatment plants." *Biotech. Bioeng.* 16:881–896.

Labuza, T. P. 1975. "Cell collection, recovery/drying for SCP manufacture." *Single Cell Protein* II. S. R. Tannenbaum and D. I. C. Wang (Eds.). MIT Press, Cambridge, Massachusetts, pp. 68–104.

Le François, L., C. G. Mariller, and J. V. Mejane. 1955. "Efectionnements aux procedes cultures iongiques et de Fermentations industrielles, brevet d'invention, France." Patent No. 1,102,200.

Leudeking, E. 1967. "Fermentation process kinetics." *Biochemical and Biological Engineering Science* 1. N. Blakebrough (Ed.). Academic Press, New York, pp. 181–243.

Lilley, V. G. 1965. "Chemical constituents of the fungal cell." *The Fungi* I. G. C. Ainsworth and A. S. Sussman (Eds.). Academic Press, London, pp. 163–175.

Lin, C. H., B. S. Fang, C. S. Wu, H. Y. Fang, T. F. Kuo, and G. Y. Hu. 1976. "Oxygen transfer and mixing in a tower cycling fermentor." *Biotech. Bioeng.* 18:1557–1572.

Lin, L. J. and P. Y. Yang. 1976. "The application of constant recycle solids concentration in completely mixed oxygen activated sludge processes." *Biotech. Bioeng.* 18:1695–1711.

Linek, V., M. Sobotka, and A. Prokop. 1973. "Measurement of aeration capacity of fermentors by rapidly responding oxygen probes." *Advances in Microbial Engineering. Biotech. Biochem. Symp.* No. 4. B. Sikyta, A. Prokop, and M. Novak (Eds.). John Wiley & Sons, New York, pp. 429–453.

Litchfield, J. H. 1968. "The production of fungi." *Single Cell Protein.* R. I. Mateles and S. R. Tannenbaum (Eds.). MIT Press, Cambridge, Massachusetts, pp. 309–329.

Litchfield, J. H. 1977. "Single-cell protein processes." *Adv. Appl. Microbiol.* 22:267–305.

Marshall, K. C. and M. Alexander. 1960. "Growth characteristics of fungi and actinomycetes." *J. Bact.* 80:412–416.

Mateles, R. I. 1975. "Production of SCP in Israel." *Single Cell Protein* II. S. R. Tannenbaum and D. I. C. Wang (Eds.). MIT Press, Cambridge, Massachusetts, pp. 208–222.

Mateles, R. I. and S. R. Tannenbaum, 1968. *Single Cell Protein*. MIT Press, Cambridge, Massachusetts.

Meadows, D. H. 1974. "Limits to growth. A report for the Club of Rome." *Project on the Predicament of Mankind*. Pan Books, London.

Metz, B. and N. W. F. Kasson. 1977. "The growth of molds in the form of pellets— A literature review." *Biotech. Bioeng.* 19:781–799.

Meyrath, J. 1975. "Production of feed yeast from liquid waste." *Proc. Biochem.* 10:20–22.

Miller, D. N. 1964. "Liquid film controlled mass transfer in agitated vessels." *Ind. Eng. Chem.* 56:18–27.

Monod, J. 1942. *Recherches sur la Croissance des Cultures Bacteriennes*. Hermann & Cie, Paris.

Monod, J. 1950. "La technique de culture continue. Theorie et applications." *Ann. Inst. Pasteur.* 179:390–410.

Moo-Young, M. 1977. "Economics of SCP production." *Proc. Biochem.* 12:6–10.

Morris, G. G., R. N. Greenshields, and E. L. Smith. 1973. "Aeration in tower fermenters containing microorganisms." *Advances in Microbial Engineering. Biotech. Bioeng. Symp.* No. 4. B. Sikyta, A. Prokop, and M. Novak (Eds.). John Wiley & Sons, New York, pp. 535–545.

Moser, H. 1958. *The Dynamics of Bacterial Populations Maintained in the Chemostat*. Pub. 614. Carnegie Institution, Washington, D.C.

Mukhopadhyay, S. N. and T. K. Ghose. 1976. "Oxygen participation in fermentation, part 1. Oxygen-microorganism interactions." *Proc. Biochem.* 11:19–27.

Muller, L. L. 1969. "Yeast products from whey." *Proc. Biochem.* 4:21–26.

Murphy, D., D. S. Clark, and C. P. Lentz. 1959. "Aeration in tower-type fermentors." *Can. J. Chem. Eng.* 37:157–161.

Ng, F. M. W. and E. A. Dawes. 1973. "Chemostat studies on the regulation of glucose metabolism in *Pseudomonas aeruginosa*." *Biochem. J.* 32:129–140.

Novick, A. and L. Szilard. 1950. "Experiments with the chemostat on spontaneous mutations of bacteria." *Proc. Nat. Acad. Sci.* (*Washington*) **36**:708–719.

Oswald, W. J. and C. G. Golueke. 1968. "Large scale production of algae." *Single Cell Protein*. R. I. Mateles and S. R. Tannenbaum (Eds.). MIT Press, Cambridge, Massachusetts, pp. 271–305.

Ovaskainen, P., R. Lundell, and P. Laiho. 1976. "Engineering of fermentation plants, part 2. Fermentor design and scale-up." *Proc. Biochem.* **11**, 37–38, 39, 55.

Paca, J. and V. Gregr. 1976. "Design and performance characteristics of a continuous multistage tower fermentor." *Biotech. Bioeng.* **18**:1075–1090.

Pace, G. W. and R. C. Righelato. 1976. "Kinetics of mould growth in a continuous tower fermenter." *Abstracts, 5th Inter. Ferment. Symp., Berlin* **No. 102**.

Pamment, N. B., R. J. Hall, and J. P. Barford, 1978. "Mathematical modelling of lag phases in microbial growth." *Biotech. Bioeng.* **20**:349–381.

Pannell, S. D. 1976. "Studies on continuous tower fermentation." Ph.D. Thesis, University of Aston, Birmingham, U.K.

Peppler, H. J. 1970. "Food yeasts." *The Yeasts* **3**. A. H. Rose and J. S. Harrison (Eds.). Academic Press, New York, pp. 421–462.

Peschard-Mariscal, E. and G. Viniegra-Gonzalez. 1977. "Cost analysis of yeast protein and RNA production by aerobic fermentation of cane molasses." *Single Cell Protein from Nonrenewable Sources. Biotech. Bioeng. Symp.* **No. 7**. E. G. Gaden and A. G. Humphrey (Eds.) John Wiley & Sons, New York, pp. 119–128.

Phillips, D. H. 1966. "Oxygen transfer from mycelial pellets." *Biotech. Bioeng.* **8**:456–460.

Pirt, S. J. 1966. "A theory of the mode of growth of fungi in the form of pellets in submerged culture." *Proc. Royal Soc. B.* **166**:369–373.

Pirt, S. J. 1975. *Principles of Microbe and Cell Cultivation*. Blackwell Scientific Publications, London.

Pirt, S. J. and D. S. Callow. 1960. "Studies on the growth of *Penicillium chrysogenum* in continuous flow culture with reference to penicillin production." *J. Appl. Bact.* **23**:87–98.

Pirt, S. J. and W. M. Kurowski. 1970. "An extension of the theory of the chemostat with feedback of organism and its experimental realisation with a yeast culture." *J. Gen. Microbiol.* **63**:357–366.

Pohland, F. G. and J. W. Hudson. 1976. "Aerobic and anaerobic microbial treatment alternatives for shellfish processing waste waters in continuous culture." *Biotech. Bioeng.* **18**:1219–1247.

Pollard, R. and H. H. Topiwala. 1976. "Heat transfer coefficients and two-phase dispersion properties in a stirred-tank fermentor." *Biotech. Bioeng.* **18**:1517–1535.

Powell, O. and J. R. Lowe. 1964. "Theory of multistage continuous culture." *Continuous Cultivation of Microorganisms (Proc. 2nd Symp.).* I. Malek, K. Beran and J. Hospodka (Eds.). Czechoslovak Academy of Sciences, Prague, p. 45.

Prescott, S. C. and C. G. Dunn. 1959. *Industrial Microbiology*, 3rd ed. McGraw-Hill, New York.

Prokop, A., L. E. Erickson, J. Fernandez, and A. G. Humphrey. 1969. "Design and physical characteristics of a multistage continuous tower fermentor." *Biotech. Bioeng.* **11**:945–966.

Prokop, A. and M. Sobotka. 1975. "Insoluble substrate and oxygen transport in hydrocarbon fermentation." *Single Cell Protein* II. S. R. Tannenbaum and D. I. C. Wang (Eds.). MIT Press, Cambridge, Massachusetts, pp. 127–157.

Radovcich, N. A. and R. Moissis. 1962. "The transition from two-phase bubble flow to slug flow." *MIT Report* No. 7-7673-22. Department of Mechanical Engineering, MIT, Cambridge, Massachusetts.

Reade, A. E. and K. F. Gregory. 1975. "High-temperature production of protein-enriched feed from cassava by fungi." *Appl. Microbiol.* **30**:897–904.

Reith, T. 1970. "Interfacial area and scaling-up of gas-liquid contactors." *Brit. Chem. Eng.* **15**:1559–1563.

Reith, T., G. Renken, and B. A. Israel. 1968. "Gas hold-up and axial mixing in the fluid phase of bubble columns." *Chem. Eng. Sci.* **23**:619–629.

Reusser, F., J. F. T. Spencer, and H. R. Sallans. 1958. "*Tricholoma nudum* as a source of microbiological protein." *Appl. Microbiol.* **6**:5–8.

Righelato, R. C. 1975. "Growth kinetics of mycelial fungi." *The Filamentous Fungi* 1. *Industrial Mycology*. J. E. Smith and D. R. Berry (Eds.). Edward Arnold, London, pp. 79–103.

Righelato, R. C. and R. Elsworth. 1970. "Industrial applications of continuous culture: pharmaceutical products and other products of processes." *Adv. Appl. Microbiol.* **13**:399–417.

Righelato, R. C., A. P. J. Trinci, S. J. Pirt, and A. Peat. 1968. "The influence of maintenance energy and growth rate on the metabolic activity, morphology and conidiation of *Penicillium chrysogenum*." *J. Gen. Microbiol.* **50**:399–412.

Robinson, C. W. and C. R. Wilke. 1973. "Oxygen adsorption in stirred tanks: a correlation for ionic strength effects." *Biotech. Bioeng.* **15**:755–782.

Rolfe, E. J. 1976. "Food from waste in the present world situation." *Food from*

Waste. G. G. Birch, K. J. Parker, and J. T. Worgan (Eds.). Applied Science Publishers, London, pp. 1–7.

Rolz, C. 1975. "Utilization of cane and coffee processing by-products as microbial protein substrates." *Single Cell Protein* **II**. S. R. Tannenbaum and D. I. C. Wang (Eds.). MIT Press, Cambridge. Massachusetts, pp. 273–313.

Romanktschuk, H. 1975. "The Pekilo Process: protein from spent sulphite liquor." *Single Cell Protein* **II**. S. R. Tannenbaum and D. I. C. Wang (Eds.). MIT Press, Cambridge, Massachusetts, pp. 344–356.

Ryder, D. N. and C. G. Sinclair. 1972. "Model for the growth of aerobic microorganisms under oxygen limiting conditions." *Biotech. Bioeng.* **14**:787–798.

Ryu, Y. and A. G. Humphrey. 1972. "A reassessment of oxygen-transfer rates in antibiotics fermentations." *J. Ferment. Tech.* **50(6)**:424–431.

Sanchez-Marroquin, A. 1977. "Mixed cultures in the production of single cell protein from agave juices." *Single Cell Protein from Renewable and Nonrenewable Sources. Biotech. Bioeng. Symp.* No. 7. E. G. Gaden and A. G. Humphrey (Eds.). John Wiley & Sons, New York, pp. 23–34.

Schaftlein, R. W. and T. W. F. Russell. 1968. "Two-phase reactor design." *Ind. Eng. Chem.* **60**:12–27.

Seeley, R. D. 1977. "Fractionation and utilisation of bakers' yeast." *Master Brewers' Association of America., Tech. Quart.* **14**:35–39.

Shannon, L. J. and K. E. Stevenson. 1975a. "Growth of fungi and BOD reduction in selected brewery wastes." *J. Fd. Sci.* **40**:826–829.

Shannon, L. J. and K. E. Stevenson. 1975b. "Growth of *Calvatia gigantea* and *Candida steatolytica* in brewery wastes for microbial protein production and BOD reduction." *J. Fd. Sci.* **40**:830–832.

Sideman, S., O. Hortacsu, and J. Fulton. 1966. "Mass transfer in gas-liquid contacting systems." *Ind. Eng. Chem.* **58**:32–47.

Sinclair, C. G. and D. N. Ryder. 1975. "Models for the continuous culture of microorganisms under both oxygen and carbon limiting conditions." *Biotech. Bioeng.* **17**:375–398.

Sinskey, A. J. and S. R. Tannenbaum. 1975. "Removal of nucleic acids in SCP." *Single-Cell Protein* **II**. S. R. Tannenbaum and D. I. C. Wang (Eds.). MIT Press, Cambridge, Massachusetts, pp. 158–179.

Skogman, H. 1976. "Production of Symba yeast from potato wastes." *Food from Waste*. G. G. Birch, K. J. Parker, and J. T. Worgan (Eds.). Applied Science Publishers, London, pp. 167–179.

Smith, E. L. and R. N. Greenshields. 1974. "Tower fermentation systems and their application to aerobic processes." *Chem. Eng.* **52**:28–34.

Smith, R. H., R. Palmer, and A. E. Reade. 1975. "A chemical and biological assessment of *Aspergillus oryzae* and other filamentous fungi as protein sources for simple stomached animals." *J. Fd. Agric.* **26**:785–795.

Snell, H. L. and L. B. Schweiger. 1949. "Production of citric acid by fermentation." U.S. Patent No. 2492667.

Solomons, G. L. 1972. "Improvements in the design and operation of the chemostat." *J. Appl. Chem. Biotech.* **22**:217–228.

Solomons, G. L. 1975. "Submerged culture production of mycelial biomass." *The Filamentous Fungi* **1**. *Industrial Mycology*. J. E. Smith and D. R. Berry (Eds.). Edward Arnold, London, pp. 249–264.

Solomons, G. L. and G. O. Weston. 1961. "The production of oxygen transfer rates in the presence of mould mycelium." *J. Biochem. Microbiol. Tech. Eng.* **3**:1–6.

Spicer, A. 1971. "Protein production by microfungi." *Topical Sci.* **13**:239–250.

Standing, C. N., A. G. Frederickson, and H. M. Tsuchiya. 1972. "Batch and continuous-culture transients for two substrate systems." *Appl. Microbiol.* **23**: 354–359.

Stanton, W. R. 1977. "The microbes, potential in resource economy." *Biotechnology and Fungal Differentiation*, *FEMS Symp.* **No. 4**. J. Meyrath and J. D. Bu'lock (Eds.). Academic Press, London, pp. 195–215.

Steel, R. and W. D. Maxon. 1966. "Dissolved oxygen measurements in pilot and production-scale Novobiocin fermentations." *Biotech. Bioeng.* **8**:97–108.

Steel, R. 1969. "Systems for high solid processes." *Fermentation Advances*. D. Perlman (Ed.). Academic Press, New York, pp. 491–514.

Sucuru, G. A., R. S. Engelbrecht, and E. S. K. Chian. 1975. "Thermophilic microbiological treatment of high strength waste waters with simultaneous recovery of single cell protein." *Biotech. Bioeng.* **17**:1639–1662.

Taguchi, H. 1971. "The nature of fermentation fluids." *Adv. Biochem. Eng.* **1**: 1–30.

Tannenbaum, S. R. and G. W. Pace. 1976. "Food from waste, an overview." *Food from Waste*. G. G. Birch, K. J. Parker, and J. T. Worgan (Eds.). Applied Science Publishers, London, pp. 8–22.

Tannenbaum, S. R. and D. I. C. Wang. 1975. *Single Cell Protein* **II**. MIT Press, Cambridge, Massachusetts.

Tempest, D. W. 1970. "The continuous cultivation of microorganisms **1**. Theory of the chemostat." *Methods in Microbiology* **2**. J. R. Norris and D. W. Ribbons (Eds.). Academic Press, New York, pp. 259–276.

Thanh, N. C. and R. E. Simard. 1973. "Biological treatment of wastewater by yeasts." *J. Water Poll. Con. Fed.* **45**:674–680.

Tomlinson, E. J. 1975. "Utilisation of strong organic effluents." *Technical Report* **TR 15**. *Water Research Centre*. P.O. Box 16, Medmenham, Marlow, Bucks, SL7 2HD.

Tomlinson, E. J. 1976a. "Production of SCP from strong organic waste waters from food and drink processing industries **I**. Laboratory cultures." *Water Res.* **10**:367–371.

Tomlinson, E. J. 1976b. "Production of SCP from strong organic waste waters from food and drink processing industries **II**. Practical and economic feasibility of non-aseptic batch culture." *Water Res.* **10**:372–376.

Topiwala, H. H. and G. Hamer. 1973. "A study of gas transfer in fermentors." *Advances in Microbial Engineering. Biotech. Bioeng. Symp.* **No. 4**. B. Sikyta, A. Prokop, and M. Novak (Eds.). John Wiley & Sons, New York, pp. 547–557.

Trinci, A. P. J. 1969. "A kinetic study of the growth of *Aspergillus nidulans* and other fungi." *J. Gen. Microbiol.* **57**:11–24.

Trinci, A. P. J. 1972. "Culture turbidity as a measure of mould growth." *Trans. Brit. Mycol. Soc.* **58**:467–473.

Van der Wal, P. 1976. "Nutritional and toxicological evaluation of novel feed." *Food from Waste*. G. G. Birch, K. J. Parker, and J. T. Worgan (Eds.). Applied Science Publishers, London, pp. 256–263.

Van Uden, N. 1969. "Kinetics of nutrient limited growth." *Ann. Rev. Microbiol.* **23**:473–486.

Van Uden, N. and A. Madeira-Lopes. 1976. "Yield-maintenance relations of yeast growth in the chemostat at supraoptimal temperatures." *Biotech. Bioeng.* **18**: 791–804.

Wasserman, A. E. 1960. "The rapid conversion of whey to yeast." *Dairy Eng.* **77**:374–379.

Weiner, B. A. and R. A. Rhodes. 1974. "Fermentation of feedlot waste filtrate by fungi and streptomycetes." *Appl. Microbiol.* **28**:845–850.

Whitaker, A. and P. A. Long. 1973. "Fungal pelleting." *Proc. Biochem.* **8**:27–31.

White, J. 1954. "Baker's yeast." *Yeast Technology*. Chapman & Hall, London, pp. 27–41.

Wood, B. J. B. W. and Yong Fook Min. 1975. "Original food fermentations." *The Filamentous Fungi* **1**. *Industrial Mycology*. J. E. Smith and D. E. Berry (Eds.). Edward Arnold, London, pp. 265–280.

Worgan, J. T. 1968. "Culture of higher fungi." *Prog. Ind. Microbiol.* **8.** D. J. D. Hockenhull (Ed.). J. & A. Churchill, London, pp. 73–139.

Worgan, J. T. 1972. "Protein production by microorganisms from carbohydrate substrates." *The Biological Efficiency of Protein Production.* J. G. W. Jones (Ed). Cambridge University Press, Cambridge, England, pp. 339–761.

Worgan, J. T. 1976. "Wastes from crop plants as raw materials for conversion by fungi to food or livestock feed." *Food from Waste.* G. G. Birch, K. J. Parker, and J. T. Worgan (Eds.). Applied Science Publishers, London, pp. 23–41.

Yoshida, F. and K. Akita. 1965. "Performance of gas bubble columns: volumetric liquid-phase mass transfer coefficient and gas hold-up." *J. Amer. Inst. Chem. Eng.* **11**:9–13.

Yoshida, P., T. Yamane, and K. Nakamoto. 1973. "Fed-batch hydrocarbon fermentation with colloidal emulsion feed." *Biotech. Bioeng.* **15**:257–270.

15

Fertilizers and energy from anaerobic fermentation of solid wastes

R. S. Holdom,* Ph.D. and B. Winstrom-Olsen

Department of Applied Microbiology, University of Strathclyde, Glasgow, Scotland.

INTRODUCTION

In recent years, there has been a growing awareness of the implications of the finite nature of our resources, especially of gas, coal, and oil (Meadows *et al*, 1972; Mesarovic and Pestel, 1975). Many people have investigated alternative sources of energy, ones that do not have finite limitations (wind, sun, waves, and water.) Photosynthetically derived materials are an important organic store of solar energy but are not generally in a sufficiently dry state to have an attractive calorific value, nor are they in a form which suits the energy using technology on which our present society is founded. A process which can overcome these difficulties with a minimum of technological progress is the production of methane from organic materials by anaerobic digestion.

*Present address: Director Research and Consultancy Services, UMIST, Manchester, U.K.

ANAEROBIC DIGESTION AS MICROBIAL PROCESS

Anaerobic digestion of organic material is a much researched field, serviced by a vast amount of literature, which crosses several disciplines and spans a period of 70 years or so. Nevertheless, it is currently enjoying a revival of interest and is being "rediscovered" or developed by groups all over the world as a means of converting wastes into energy, as an anti-pollution technology, and as an integral process in the production of chemical raw materials from renewable starting materials (Table 15.1). It is a remarkable fermentation in that solid and liquid

TABLE 15.1.   Recent Evidence of Renewed Interest in Anaerobic Digestion Processes.

Reference	Comment
General:	
Bailey. 1976.	U.N. World Bank study.
Wise. 1977.	Commercial scale. Acetate intermediate.
Fraser. 1976.	Commercial Energy Plantation (decisions).
Boersma et al. 1974.	Integrated system using waste heat.
Reviews/bibliographic:	
Goluke and McGauhey. 1976.	Alternatives to digestion covered.
Freeman and Pyle. 1977.	Third World applications—annoted bibliography.
Tietjen et al. 1975.	Survey, 16 papers including Pyrolysis, Composting, etc.
Systems analysis/modeling:	
Benemann et al. 1977.	Micro-algae used.
Pfeffer and Liebman. 1976.	Mathematical simulation and costings.
Ostrovski et al. 1976.	Modeling, small versus centralized digesters.
Kispert et al. 1974.	Computer model, costings.
Use in different countries:	
Parikh and Parikh. 1976.	Rural energy and fertilizer, India.
Troughton and Cousins. 1976.	Energy farming, New Zealand.
Kang. 1975.	8,000 biogas units, Taiwan.
Smil. 1977.	Chinese experience, household systems.
Hansford. 1976.	Energy balance for West Germany.
Allison et al. 1976.	Energy center, Sri Lanka.
Clements and Soderstrom. 1976.	Tropical/subtropical units.
Fertilizer:	
Parikh and Parikh. 1976.	
Webber and Bastiman. 1967.	Poultry manure/grass dressing.
Webber. 1961.	Organic manures compared.
Davies. 1970.	Cow slurry on grassland.
Special applications:	
Anon. 1977.	Brewery wastes.
Diaz and Trezek. 1977.	Municipal solid wastes.
Abeles. 1977.	Plug-flow farm systems.
Hawkes et al. 1976.	Chicken litter.

materials of all kinds can be accommodated by the complex mixed species culture in a digester, and, with a few exceptions, methane gas is always produced. The highly interactive state of the wide range of species in the bacterial population confers stability to the fermentation in the face of disruption by process upsets of various kinds. A suitable culture is readily obtained from animal manures, and virtually any sealed container with provision for gas take-off can become a fermentation vessel. Attention to a few simple rules guarantees successful production of biogas (typically 64% methane, 36% CO_2).

While the basic fermentation is easy to accomplish, it is difficult to:

- Scale up to a size which suits farms and small industries;
- Convert the process into commercially viable units;
- Operate the total system with a net energy surplus; and
- Manipulate the microbial system towards higher net energy production.

Kinds of Wastes

There are three basic wastes: industrial, domestic, and agricultural (Table 15.2). Each of these occurs in solid and liquid forms or as a blend of the two (slurry). The decision as to which of the available technologies is appropriate to use in each case follows no coherent plan. If it did, materials less than about 14% moisture would probably be pyrolyzed (Tietjen *et al*, 1975), especially if they were of high calorific value but poor biodegradability, such as tyres and plastics. Liquids of biological oxygen demand (B.O.D.) less than 8000 mg/l would be aerobically processed by one of the many types of conventional sewage plants. Solid materials already containing a high percentage of water (e.g., vegetable and agricultural wastes) are best slurried for anaerobic digestion. Slurries much over 10% solids are difficult to pump and will need dilution, preferably with a high B.O.D. liquid (e.g., silage liquor, abattoir, or meat factory effluent). For anaerobic digestion, the higher the B.O.D., the greater the net energy output.

Current State of Anaerobic Technology

The use of our various organic wastes in small local schemes could begin immediately. Already, in California, about ten sanitary landfills (the primary method of solid waste disposal in the United States) are extracting the methane which is automatically produced in the landfill by natural fermentation. The processes are operated by government agencies and private firms. One of these, Reserve Synthetic Fuels, Inc., Palos Verdes, is selling pipeline quality landfill gas to the Southern California Gas Company (Hekiman *et al*, 1976). However a major feature of the technology of biogas production is that it is manageable by the ordinary man. For this reason, a range of small-scale systems has appeared in recent years, most of them designed for use on farms.

TABLE 15.2. Waste Organic Materials Suitable for Anaerobic Digestion.

Type of Waste	Region	Annual Production (million tonnes)	B.O.D. (mg/l)	Reference
Farm manure	U.S.A.	2000	16–58,000	—
	U.K.	—	—	—
Crop residues:				
Cane bagasse	World	70	—	—
Straws (three types)	U.K.	1	—	—
Potato and pea haulms	U.K.	1.6	—	—
Sugar beet tops	—	—	—	—
Forestry wastes	U.S.A.	400	—	—
Shavings and sawdust	U.K.	1	—	—
Other wastes:				
Bracken	U.K.	1	—	—
Seaweed	U.K.	1	—	—
Silage liquor		—	25,000	Smith, 1977.
Industrial wastes:	U.S.A.	110	—	—
	U.K.	56	—	—
Sewage	U.K.	16	350	Bolton and Klein. 1971.
Cheese whey (lactose)	U.K.	0.4	1200	Shabi, F. A. and Cannon, M. C., 1975.
	U.S.A. (gal)	1000	—	—
Sulfite waste liquor	U.S.A. (gal)	12,000	20,000–100,000	Berger. 1958.
Animal collagen	U.S.A.	1	—	—
Spent wash from:				
Beet molasses distillery	—	—	60,000–100,000	Basu and Leclerc. 1972.
Cane sugar distillery	—	—	18,000–37,000	Sen and Bhaskaran. 1962.
Process wastes* from:				
Meat factory	—	—	470–5700	Shabi and Cannon. 1975.
Vegetable canning	—	—	240–1100	Dickinson. 1971.
Fruit canning	—	—	140–970	Dickinson. 1971.
Citrus processing	—	—	60–400	Dickinson. 1971.
Urban wastes:				
Domestic sewage sludge	U.K.	20	—	—
Domestic refuse	U.K.	20	—	—

*Not economical for gas production unless used as the aqueous phase in slurry-making with solid wastes, or unless already warm (30 to 40°C).

Farm-Scale Technology

The amount of energy readily available from digestion of animal wastes is not large on a national scale. For example, it has been calculated that in the United Kingdom, only 2% of the national gas consumption (0.3% of the total energy consumption) could be provided from the digestion of presently available animal wastes (Meynell, 1976; Best, 1975).

However, for an individual investor who uses digestion, the energy yield is more significant. For example, a 60,000 bird poultry unit could provide its own power needs plus a small surplus for an investment of six times the unit's annual electricity bill (Savery, 1972). This would require a plant of 300 cubic meters fed with a slurry of 10% total solids operating with a 20 day retention time in the digester (Hobson, 1977). Also, an intensive dairy unit could produce sufficient surplus gas to be sold to nearby energy consumers (Gaddy et al, 1974).

Tyrrell (1978) has prepared a realistic costing of a digester as an integral part of a 5,000 pig rearing unit (Table 15.3). This shows that if good use can be made of the biogas and the fertilizer liquid, a considerable cash surplus is available after one year, and "pay-back time" on the original outlay can be as short as four years. Thereafter, savings of over £10,000/annum can be realized.

In countries of low energy intensity (Slesser, 1974), the advantages of methane production are more pronounced. In India gross energy consumption is 6×10^9 j/annum per capita, compared to the United Kingdom, 140×10^9 j/annum per capita (Meynell, 1976). The success of Gobar Gas plants demonstrates the usefulness of the process (Finlay, 1976). In addition, nutrients in the waste need not be lost from the system and could play an important role in agriculture (Department of Environment, 1972). This concept of nutrient conservation is exceedingly important for a favorable economic assessment of methane production. Use of the concept in "energy accounting" is discussed later in this chapter, but the agricultural impact is illustrated by the following facts. In the United Kingdom, about 970,000 tonnes of organic nitrogen are applied as fertilizer to cropland every year. Very nearly the same amount of nitrogen (800,000 tonnes) is produced as animal wastes. Half this tonnage represents the potential for recycle by technological means (Cooke, 1976)—400,000 tonnes of nitrogen excreted by housed animal stock; 400,000 by grazing stock. This potential is improving all the time as a result of intensification of animal rearing and the amalgamation of farms. For example, over the period 1960 to 1974, there was a ten-fold increase in the number of large stock farms. We now have in the United Kingdom 14 million cattle housed in units of >200 animals, 8 million pigs in units of >5,000 and 30 million poultry in units of >20,000 (Agricultural Research Council, 1976). Farm-scale digestion becomes economical with such units because of the concentrated availability of the waste, but also for another reason. The B.O.D. (pollution index) of an intensive pig farm can approach 30,000 mg/l of run-off (compare sewage, at 250 mg/l B.O.D.). Thus, one farm

TABLE 15.3. Costing* for Methane Digester System, 5000 Pig Unit.

Debits	Capital Cost (£)	Annual Charge (£)
Digester (conventional tank)	13,875	—
Amortization over 20 yr	—	793
Interest @ 10%, 1st yr	—	1387
Pumps, stirrers, boilers, holding tanks, etc.	4625	—
Amortization over 3 yr	—	1542
Interest, 1st yr	—	462
Electricity generator, heat exchanger and hut	6500	—
Amortization over 4 yr	—	1622
Maintenance costs	—	1000
Interest, 1st yr	—	650
Total	25,000	7456

Credits

Assuming the gas yield is sufficient for an average production of 100 kw/day from the generator (optimistic for all-year running of digester), and allowing

5.36p/unit electricity on first 233 units/$\frac{1}{4}$ yr
2.95p/unit electricity on second 233 units/$\frac{1}{4}$ yr
2.75p/unit electricity on remaining units/$\frac{1}{4}$ yr

Potential electricity savings	=	£9660
Estimated fertilizer savings/sales =		£4000
		£13,660/annum
less annual costs		7456
1st yr surplus		6204.

Pay-back Time

After deduction of charges each yr:

1st yr	25,000 – 6204	–18,796
2nd yr	– 8146	–10,650
3rd yr	– 6483	– 4167
4th yr	–13,243	+ 9076
Thereafter	c. – 1300	c. +12,000

Pay-back time approximately 3.5 yr.

Source: Tyrell. 1978.
*No allowance for maintenance labor charges, tax relief, grant aid.

unit can now pollute at a rate equivalent to that of a small town of 10,000 population. If this pollution load is treated by a conventional aerobic sewage plant, all that is achieved is a redistribution of the B.O.D. into different solids fractions (i.e., microbial cells). Large amounts of fossil fuel are consumed in the process, 96.8% of the nitrogen is lost as volatiles, and the small amount of nitrogen left over is usually present as nitrate (NO_3), undesirable for land spreading (Hobson *et al*, 1977). Without taking account of the gas produced by the anaerobic process, the energy costs of running an aerobic plant are some 700 times those of an anaerobic plant and increase in proportion to the throughput. Running costs of the anaerobic plant decrease because large-scale digesters have lower heat losses. Hobson's figures for experimental pilot plants show that if the value of the excess gas from the digester (after deduction of the amount needed to heat the digester—about 45% of the total gas) is accounted for, the digester is cheaper overall than the aerobic plant above 4m^3/day throughput. This allows for the costs of equipment to burn the excess gas for some useful purpose. At throughputs above about 8m^3/day, the anaerobic system comes into increasing credit balance. Further credits are the increased availability of nitrogen (all organic nitrogen forms are reduced to NH_4^+) and preservation of phosphorus, potassium, and micro-nutrient levels in the effluent. On a national scale, the energy equivalent of sparing manufactured fertilizer is 1 million tonnes coal equivalent (about £100 million worth). From another aspect, of the total nitrogen in circulation in a year, only about 50% appears in harvested crops and about 10% is known to be leached from the soil. Of the remaining 40%, an unknown amount (probably about one-tenth) is volatilized in the soil and all the rest can be thought of as "unavailable" through inefficient coupling of recyclable materials into our agricultural practice. Furthermore, of the total nitrogen in the system, 55% is derived by natural means (nitrogen fixation), 37% derives from applied fertilizer, and 8% enters as imported animal feeds (this percentage is growing annually). It is especially important, therefore, that imported nitrogen is retained in the system, preferably in its reduced form (NH_4^+), which is available to plants.

Quantitative data on the best use of the digester effluent as a crop fertilizer are difficult to find. However, Tyrell (1978) reports a better harvest of grass dressed with digester effluent than with raw slurry. On one farm (100 sows, 40 cattle), it is claimed that 178 l/ha of digester effluent applied after each grazing or cropping now replaces all the inorganic fertilizer on the 45 ha farm and substitutes for about £1,000 worth of organic fertilizer.

Advantages of Anaerobic Digestion

Anaerobic technology reduces the "biological turnover time" of carbon energy forms from millions of years to about 10 or 15 years. It also has a low impact on the environment and is sustainable far into the future (Troughton and Cousins, 1976; Mesarovic and Pastel, 1975). Such features have recently attracted the United Kingdom Department of Energy to study the potential for growing "energy crops" on marginal or non-arable land. In the United States, it has been proposed by Klass (1974) that it may be feasible to grow crops on a large scale to provide a "methane economy". This would require the selection of suitable crops with a high yield in terms of mj/ha/hr and the cultivation of massive areas

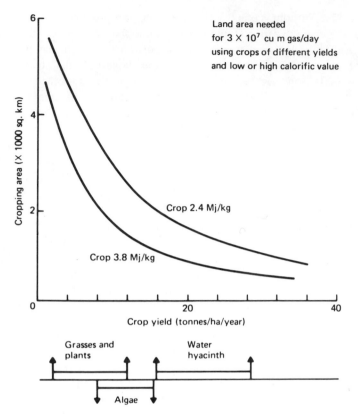

Fig. 15.1. Energy crops for biogas production. *Source:* Klass. 1974.

of land and/or water (Fig. 15.1), but the crop need not be a conventional food species. For example, 65% of the current gas consumption in the United States could be provided by an energy plantation of area 260,000 square kilometers using water hyacinth of 3.8 mj/kg dry material. Marine algae, in particular, seem to have an assured future as a raw material for digestion. Kelp (*Macrocystis pyrifera*) can be grown in currently non-productive sea water using deep water nutrients in a process developed by United States Navy scientists (Leese, 1976) and a potential solution to the high cost of storing methane gas is to ferment the algae to acetate as the storage material, the acetate being converted as required to methane (Wise, 1977).

Energy Accounting

An alternative to economic analysis is now described which allows an optimum design to be established for each set of local conditions where the nature and quantity of the wastes, climate, and capital available may vary.

Fig. 15.2.   Rear view of two stage biogas plant under construction, showing 2 × 5 m³ tanks, and large insulated hot water accumulator as energy store. The plant is now totally enclosed by polythene wrapped straw bales and roofed (as a normal barn). Slurry holding tanks are outside. The shed houses gas boiler and control instruments.

In realizing economic advantages of running a small biogas system, it is imperative to design the system as a true net energy producer. Some schemes proposed in the literature are net energy consumers when all system inputs are accounted for. The system boundary should include any energy consuming or producing activity in the processing of the waste from its point of origin up to the ultimate distribution and consumption of end products (gas, fertilizer, soil conditioner). Unless a wide system boundary is drawn in this way, erroneous economics result.

In the worked example below, energy analysis accounts for the energy debits and credits based on measurements (or calculations) for a farm system in Perthshire (Figs. 15.2 and 15.3).

For this analysis, no account has been made of the "energy values" which could be attributed to soil conditioner, reduction of health risks, reduced pollution load on the biosphere, or energy costs of materials used to construct the plant (including insulation). Such factors are important but beyond the scope of the present remit.

The range limits to the data used in the analysis are listed in Table 15.4 and are derived from the farm-scale system and from a careful screening of the published

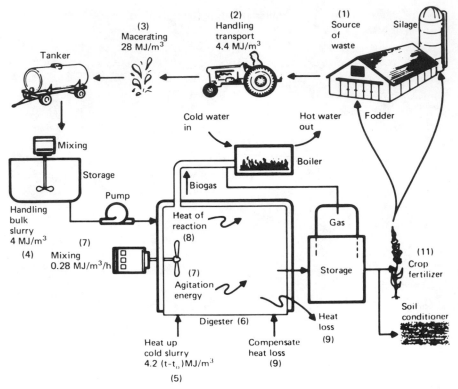

Fig. 15.3. Energy accounting sheet for a farm system, with general notes for adaptation to alternative systems.

[1] The animal waste has its own "energy value" if it serves some useful purpose other than as a substrate for digestion. Here it is taken as a resource of negative financial value because of the pollution and disposal problem and is rated at zero energy value. In cases where there is an alternative use for the material (e.g., specially grown energy crops which also possess some food value) it becomes a question of comparing the usefulness to man of the end products.

[2] Uses existing farm tractor and attachments, including slurry tanker/trailer.

[3] Necessary in this case to break up the bedding straw, but more typically may be omitted, or can be a less energy intensive process.

[4] Sedimented slurry in the storage tank must be mixed prior to pumping. This is best achieved with a short burst of mixing, using a separate propeller-mixer (energy intensive, short time) rather than with the slurry loading pump used on a recycle principle (less energy intensive, long time, ineffective).

[5] t_0 = fermentation temperature (typically 35°C), t = ambient temperature. Preheating can be achieved in several ways, but heat exchangers block up too readily when a high-solids slurry is used. This farm system uses hot water to warm the digester. The water is heated by electricity at start-up (use 4.2 in calculation) and biogas subsequently (use 5.6 in calculation if gas 67% CH_4).

[6] On this site, there are two vessels arranged in multi-stage cascade, total capacity 10m³. A single large vessel is more usual but twin vessels are an advantage when a process upset occurs or when special availability of a waste requires a high feed rate for a few days (e.g., seasonal silage liquor).

work of others. Important variables affecting the optimization of the analysis are given in Table 15.5.

"True net energy yield" was determined for typical ranges of digester size, ambient temperatures, and retention time within the digester. A computer was used to process these interactive parameters. Hundreds of two-dimensional graphs can be obtained, a few of which are now used to illustrate the general results.

A very important point of process design is shown in Fig. 15.4. It is normally thought that a digester should be loaded at the maximum rate before instability sets in (short retention time) but the results of the author's work show that a retention time of 30 days and unfavorable temperature (0°C) gives more *net* energy than a digester running at 10 days retention time (r.t.) at the most favorable of ambient temperatures. The reason for this is the higher percentage destruction of solid materials retained in a digester for 30 days as against 10 days, especially if the mixed waste contains a substantial portion of cellulose and other slowly degraded materials. For entirely liquid wastes with readily converted soluble compounds, this result would not hold and a 10 day retention time would be better. In a new reactor, to be described shortly, a short liquid residence time and a long solids residence time are combined in one system. Also, digester size is critical. For a 10 cubic meter (9,000 liter) insulated digester, 26% of the energy debits are attributable to heat loss, 25% to heating cold slurry. For a 100 cubic meter digester, these figures are 34 and 5%, respectively (Table 15.4). Thus, there is every incentive to build a large digester, perhaps as a cooperative venture with pooled wastes.

Substitution of Primary Fossil Fuel

If the energy value of the fertilizer in the effluent is considered about equal to the calorific value of the gas, the energy economics are improved (Line 2, Fig.

---

[7] Agitation consumes power, but apart from efficiency losses in the drive motor, this all emanates as useful heat, which partly compensates heat losses.

[8] Heat of reaction (fermentation) is about 5% of the calorific value of the biogas produced (67% CH_4) and also partly compensates for heat losses.

[9] Heat loss $Q = 4.63 \times 10^{-6}(t - t_0)V^{2/3}$ mj/°C/sec/m^2.

[10] The most efficient use of gas is direct burning, although it can be used to drive a gas engine generator to make electricity. It cannot be liquified at normal temperatures, and filling of gas cylinders (3000 psi) is too energy intensive to be worthwhile. On this site, it is mostly burned as it is made, and the energy is stored in a hot water accumulator for space heating. This minimizes the size of gas storage vessels (expensive) and some of the hot water is available to warm the digester. If wastes are available when gas is least wanted (e.g., grass clippings throughout the summer), they can be conveniently silaged until the peak energy demand occurs. Silage acids are themselves an ideal substrate for the biogas fermentation, and meanwhile, they prevent biodegradation and consequent loss of carbon as CO_2, nitrogen as ammonia.

[11] The energy value of the "fertilizer" is approximately equivalent to the calorific value of the gas produced (Table 15.4). Organic forms of nitrogen in the wastes are reduced during digestion to NH_4^+.

TABLE 15.4.   Data from Experimental Farm Plant and Published Work, Used in Energy Accounting of Biogas Systems.

Animal	Volume of Biogas[1] per Volatile Solid	Biogas per Animal (mj)	Fertilizer Equivalent per Animal (mj)	Fertilizer Equivalent per Biogas Energy	Average Excreta per kg Animal per Day (g)					
					Wet Manure	Total Solid	Volatile Solid	Total Nitrogen	P_2O_5	K_2O
Cattle	0.094–0.31	–	–	–	–	–	–	–	–	–
Dairy	–	13.4	11.7	0.9	84	7.9	5.3	0.23	0.34	0.12
Beef	–	11.9	15.0	1.1	66	9.5	4.7	0.32	0.18	0.29
Pig	0.37–0.50	4.0	3.3	0.8	74	8.9	5.4	0.51	0.42	0.40
Poultry	0.31–0.62	0.22	0.14	0.6	62	14	9.6	0.74	0.60	0.30

Energy Accounting of a Plant[2]

Digester Volume (m^3)	Biogas Produced (mj/hr)	Total Energy Before Digestion (mj/hr)	Input Plant Operation (mj/hr)	Percentage of Total Energy Input			
				Handling Before Digestion	Heating of Input Sludge	Compensation for Heat Loss	Stirring of Digester Content
10	8.3	2.1	4.2	8	25	26	41
100	83	21	25	11	34	5	50

[1] CH_4 65%, CO_2 35%, saturated, calorific value 25 mj/m³.
[2] Assumptions: Ambient temperature 10°C, fermentation 35°C, 30 day retention time, 0.8 digester volumes/day biogas, slurry strength 2% (w/v) digestible solids (the minimum practical). Gas output proportional to slurry strength up to 10% solids, assuming no changes to suspending fluid.

TABLE 15.5.   Parameters Important in Energy Accounting of Anaerobic Digestion Toward Optimum Net Energy Output.

Parameter	Remark
Liquid detention time	Ideally short to reduce plant size.
Solids detention time	Ideally long enough for 100% destruction of biodegradable solids.
Solids concentration in feed	Directly affects gas yield.
Alkalinity and pH	Crucial to process stability.
Temperature	Crucial to energy accounting in temperate climate.
Solids recycle rate	Optional, but increases gas production rate at expense of pumping energy.
Mixing	Essential for high rate digestion; severe problems with strong slurries.
Additional nutrients	Stimulate gas production and raise fertilizer value of effluent.

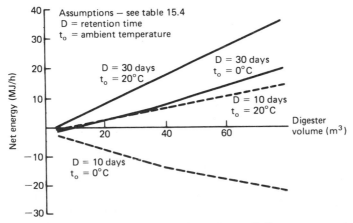

Fig. 15.4. Net energy yield from different sizes of digester operating at the normal extremes of ambient temperature and retention time. Assumptions: see Table 15.4. D = retention time, t_0 = ambient temperature.

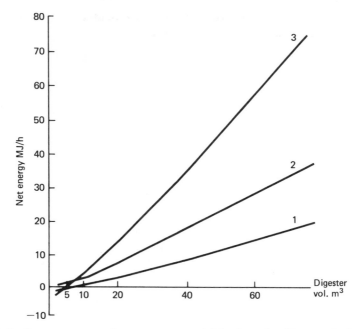

Fig. 15.5. Improvements in net energy yield when fertilizer and electricity equivalent of biogas are considered. Assumptions: see Table 15.4.

Line 1. Retention time 30 days, t_0 0°C, only. Gas accounted for as energy credit.

Line 2. Energy value of fertilizer salts (including N, P, and K) also credited. Calculated as gross energy requirement (GER)—see Winstrom-Olsen *et al*, 1977.

Line 3. Energy spared by using site produced biogas instead of electrical energy at GER = 14.0 mj/kw · hr.

15.5). If gas is used to substitute electricity, then according to the rules of systems analysis, instead of using the calorific value (C.V.) of the produced biogas, the saving in primary fuel (coal, oil, etc.) used in making and distributing the electricity should be credited. If this is done, the biogas plant acquires even more favorable energy economics for the country as a whole (Line 3, Fig. 15.5) and shorter "pay-back time" (1 year) for the individual farmer.

Planning the Size of a Digester—Cooperative Ventures

Since net energy yields are better from a larger digester than from a small one, it is important to list the magnitude of all available fermentable sources reasonably local to the site and build as large a digester as capital available will permit. However, the energy used to collect wastes must not negate the energy value of the extra gas produced. Thus, there is an optimum distance for collecting material for delivery to a central plant. Some models will be used to show the principles for a farm cooperative. The models do not account for the possibility of collecting material during the course of inter-farm trips made for other essential purposes, in which case even more favorable energy economics result. The models apply equally well to other collection systems. For example, it is envisaged that a central plant, currently running in a Glasgow park on grass clippings and feeding gas to existing greenhouses, will be scaled up and will receive other organic wastes from the Glasgow area. Some of these wastes (e.g., fruit market, abattoir) have to be removed from the source and represent a free resource. Other wastes (e.g., silage liquor of high B.O.D.) can be collected in season from local sites as a service to the community. If the volume of biogas generated is large enough, additional investment in greenhouses would then be feasible.

The mutual position of the sources of waste are, in the models, made to satisfy the following idealization. The sources are placed mainly along one road; however, at some points, there will be side roads complicating the pattern of travel. The real situation is likely to be a combination of the following two models.

Linear model 1. The sources occur at equal intervals along one road (Fig. 15.6a).

Square model 2. The sources approximate to a quadratic pattern (Fig. 15.6b).

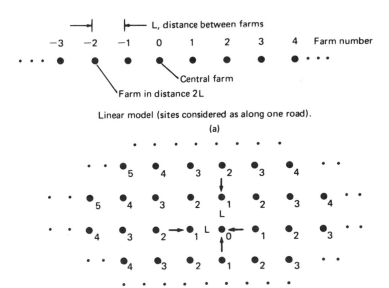

Linear model (sites considered as along one road).

(a)

Square model (sites requiring complex routing of collection vehicle).

(b)

Fig. 15.6. Assumptions in modeling of the collection of wastes from neighboring sites for use at central site.

(*a*) Linear model (sites considered as along one road).
(*b*) Square model (sites requiring complex routing of collection vehicle).
(*c*) Formulae used.

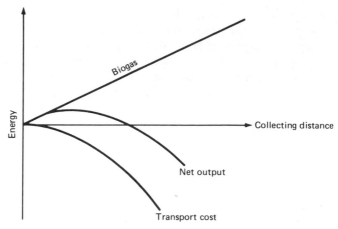

Fig. 15.7. Collection of wastes for use at a central site—general form of energy yield curves.

Formulas are now derived for the total available voltile solid (*VS*) within a distance of *NL*, where *L* is the distance between the farms and *N* is the number of farms in one direction from the center point. *P* is the percentage of digestible solid in the waste. *VSC* and *VSO* are the digestible materials available from the "central" and "other" sources, respectively (Fig. 15.6c). *VSO/P* is measured in cubic meters. The energy cost of transportation is proportional to the volume of material and the distance *L*.

The gross energy produced is linearly related to the volume of material collected (*VSC* + *2N VSO*), the actual value depending on the concentration of digestible material (*P*). The *net* energy follows a curve passing through an optimum, the general form appearing as shown in Fig. 15.7. It is tedious to calculate these relationships for each case, and therefore the models were rewritten as a computer program into which were fed a large number of values chosen from within a realistic band of data (see Table 15.4). The program allowed rapid optimization of net energy yields from any particular combination of data, such as would be provided by a new "client" wishing to adopt the technology. As an example, we wish to know the best size of digester for a cooperative venture in which a 4% slurry is made from the blend of wastes from local sites, each producing 10 kg solid material per day (this is usually in the form of a slurry). Two inter-site distances (1 and 4 kilometers) were chosen, real distances varying between the two values. Also, the usual extremes of ambient temperature for the United Kingdom's climate (0° and 20°C) were put in as constraints. The results are shown in Fig. 15.8. Conclusions can be drawn from several viewpoints and can be expressed with different emphasis, as follows:

1. At any normal ambient temperature between 0° and 20°C, having to travel

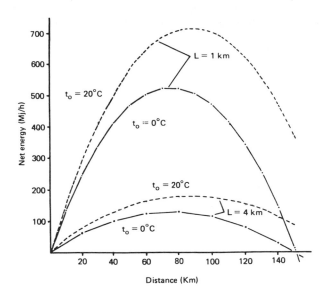

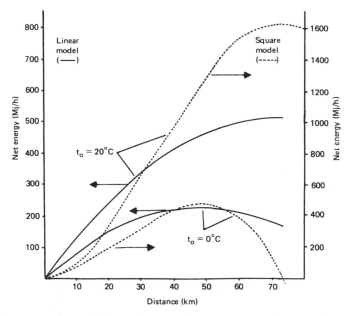

Fig. 15.8. Optimum collection distance for transport of wastes from outlying sources to a central biogas plant operating between two extremes of ambient temperature.

(*a*) Using a linear model (Fig. 15.6a) and two different intersite distances (1 and 4 km).

(*b*) Comparing linear and square models (left and right vertical scales, respectively).

In both cases the slurry strength is 4% solids.

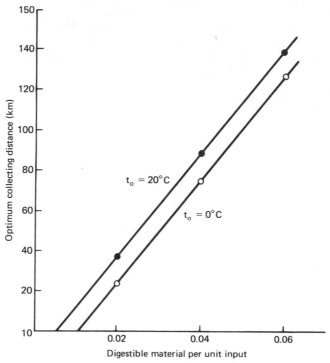

Fig. 15.9. Influence of slurry solids content on the distance which can be traveled in collecting wastes from several sites.

4 kilometers between sites instead of 1 kilometer hardly changes the *optimum* distance which can be traveled before net energy yield begins to fall, but drastically affects (500% less) the absolute value of net energy yield. The latter value mainly affects "pay-back" time on investment and can be improved if the wastes include a liquid fraction of high B.O.D.

2. If 4 kilometers between sites is unavoidable, little is to be gained by collection of material from other farms and alternative methods of increasing the capacity and productivity of a plant should be considered. (The exception is where transport of the waste is already essential and the energy debit of transport could be overlooked.)

3. When collection points are within 1 kilometer of each other, it is worth collecting wastes up to 80 to 90 kilometers of the total journey distance. Ambient temperature becomes critical to the overall energy output. (If a source of waste heat were available, this constraint would be removed.)

Fig. 15.9 illustrates another important result. If the strength of the slurry can be increased, then *either* the distance between sites can be greater, or, for a given inter-site distance, collection can usefully cover a wider area. The slurry cannot exceed about 10% solids for engineering reasons, but one site may provide a solid material (e.g., pulverized local authority waste) which allows a weak slurry blend to be upgraded. Note that the line does not pass through the origin; i.e., some materials are such poor substrates that even if they are immediately local to the plant they are not worth including in the slurry.

Process Improvements—New Reactor Design

Sometimes a slurry has to be diluted to make it more manageable when being pumped or stirred. In such cases, the diluent should not be water but a high B.O.D. liquid such as silage liquor or soluble wastes from a food factory. These soluble nutrients are converted to methane more quickly than are solid particles and so the ideal residence time in the digester is much less for liquids than solids. It is wasteful of digester capacity if liquids are retained unduly long, and incomplete gasification occurs if solids leave the reactor too soon with the spent liquid. A design of reactor which overcomes this problem and offers several other advantages is a special version of the tower fermentor (Fig. 15.10). Typically, this can operate on a short hydraulic retention time (3 to 10 days) simultaneously with a longer solids retention time (15 to 30 days, depending on the physical size and digestibility of the solids). When operated and *fed continuously*, the tower automatically selects for floc-forming microbes, where the flocs consist variously of substrate solids, microbes in mixed synergistic association with one another, or both. For grass clippings mixed with brewery effluent, 100% degradation of biodegradable solids occurred within an 18 day solids residence time, while liquid throughputs were higher (10 day residence time), and gas yields were maximized. In operational terms, the slurry mixture fed to the digester (grass and effluent liquor) could be weak (c.4% solids) and therefore readily pumped by simple, cheap, centrifugal pumps, while within the tower, the solids level held at a steady 7%. Other advantages of the tower reactor are listed below.

1. Microbial biomass flocs are retained longer than in a tank reactor; higher productivity per unit volume of reactor results.

2. No impressed mixing is required if the height to diameter ratio of the tower is chosen carefully (typically 15 : 1 for small pilot plant units up to 1,000 l; 11 : 1 for larger). Agitation of the contents of the tower is important, of course, but is achieved automatically in the following way. As gas forms (CO_2 only in start-up, CO_2 and CH_4 thereafter), the bubbles tend to remain associated with particles which thereby move upwards with the bubble. As the hydrostatic head in the tower decreases, the gas bubble expands and usually breaks free. The particle falls back under gravity and is retained, having been reduced in size as part of the digestion process. Eventually, the particles are so small as to pass out of the top of the fermentor, or they consist of a non-digestible residue. The latter either pass out of the fermentor as described or, if they are large enough, accumulate at

Fig. 15.10. Pilot plant tower fermentor, forerunner to a 10 m^3 scaled up version operating on grass clippings in a Glasgow park.

the base of the tower and must be removed from time to time (these residues de-water readily and make an ideal peat substitute or soil dressing with some residual fertilizer value in the form of the microbial biomass trapped within).

3. The gas bubbles not associated with particles also play a fundamental role in mixing the tower contents; the superficial gas velocity up the tower has to be a certain minimum for mixing to occur. This velocity is dependent, among other things, on the viscosity of the liquid and/or slurry mix in the *steady state* condition of the fermentation. A tower fermentor must therefore be brought to its

maximum productivity level over a period of 6 to 8 weeks. Extremely helpful in monitoring gas output on a continuous recorder chart is a gas meter system with analogue output described by Holdom *et al* (1976). The chart recording allows early detection of "trends" so that over-feeding, under-feeding, toxicity problems, etc. can be avoided.

4. The stability of a tower fermentation is high because of (*a*) the higher biomass concentration in the reactor, and (*b*) the slight stratification (inhomogeneity) of biochemical function over the length of the tower. Thus, slurry feeds of pH outside the essential range for methane producing bacteria have been readily accommodated (e.g., pH 3.0 brewery effluent) and equipment malfunctions (e.g., pump failure or over-run) and process upsets relating to the quality of different batches of slurry are buffered to a large extent.

A LOOK TO THE FUTURE

As energy costs rise, it will make greater economic sense to invest capital in biogas technology. The life of the main plant should be very large (20 to 25 years) and although technical improvements to the plant may be expected as time passes, these will not undermine the original investment. Knowledge of the mixed culture system (one of the most complicated known) sufficient to make yield improvements will not accrue for ten years or so, if at all. There are sound biochemical reasons to suggest that the thermodynamic efficiency of the fermentation cannot be improved on to any great extent and we must turn to process design and to control systems technology for yield improvements. New and cheaper construction materials for digesters and gas storage vessels are needed. Micro-processor technology could provide cheaper controllers of temperature, agitation, and feeding rates. A cheap analyzer of the methane content of biogas is desirable so that a process control loop linking feed rate to calorific value of gas can be installed. This will force the culture to evolve toward maximum gas production at maximum feed rate, a situation of minimum stability, where sophisticated control technology is necessary, but need not be expensive.

Finally, a new patented system for retaining biomass in a fermentation vessel (Atkinson, 1974) may allow gas production from dilute wastes to become feasible. In this system, the microbial culture colonizes knitted steel or plastic mesh balls, the weight of which ensures that they are retained indefinitely in the continuous open flow reactor housing them (Fig. 15.11). The mesh may particularly suit the synergistic or syntrophic associations which are known to occur in anaerobic digestion and for which conventional fermentors are not ideal.

In conclusion, there will soon be a complete range of anaerobic technologies to suit most kinds of biodegradable matter. It is to be hoped that "technology transfer" through commercial channels can soon take place.

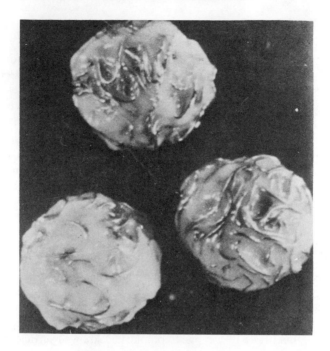

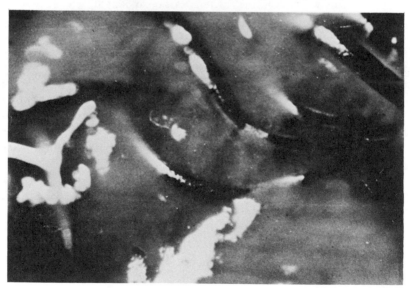

Fig. 15.11. Wire mesh particles (approximately 1 cm in diameter) colonized with biomass. The weight of these particles ensures that they are retained in the reactor vessel. (see Atkinson, 1974). A ×25 magnification shows the microbial film and knitted mesh.

ACKNOWLEDGMENT

This chapter would not have been possible without major contributions from R. Welborn and R. Cocker. The generous support of the Wolfson Foundation is also gratefully acknowledged.

REFERENCES

Abeles, T. P. 1977. "Design and engineering considerations in plug flow farm digesters—A preliminary analysis." *Inst. Gas. Tech. Symp. on Clean Fuels from Biomass and Wastes*. Orlando, pp. 417–424.

Agricultural Research Council. 1976. *Report: Studies of Farm Livestock Wastes*. State House, High Holborn, London.

Allison, J. H., S. R. Southerland, and C. E. Gordon. 1976. *An Energy Centre in Sri Lanka. SAE Report* **769011**. Sept. 12–17.

Anon. 1959. *Anaerobic Digestion of Industrial Wastes, Ministry of Technology Notes on Water Pollution No.* **4**. (HMSO Public).

Anon. 1976. "Methane production through bioconversion of agricultural residues." *Sharing the Sun Conference* **7**. K. W. Boer (Ed.). Pergamon Press, Elmsford, New York, pp. 119–129.

Anon. 1977. "Energy recovery in brewery industry," *J. Env. Eng. Div-Asce.* **103**:445–459.

Atkinson, B. 1974. *Biochemical Reactors*. Pion Press, London.

Bailey, D. G. F. 1976. "The bio-gas system: serious energy force or environmental fad?." *Nat. Res. Forum, U.N. World Bank* **1**:55–63.

Basu, A. K. and E. Leclerc. 1972. "Mesophilic digestion of beet molasses distillery waste water." *Adv. Poll. Res. 1973., Proc. 6th Int. Conf. Assoc. Water Poll. Res.* S. H. Jenkins (Ed.). Pergamon Press, New York, pp. 581–590.

Benemann, J. R., B. L. Koopman, D. C. Baker, J. C. Weissman, and W. J. Oswald. 1977. "A systems analysis of bioconversion with microalgae." *Inst. Gas. Tech. Symp. on Clean Fuels from Biomass and Wastes*. Orlando, pp. 101–127.

Berger, H. F. 1958. "Anaerobic digestion of pulp and paper waste." *Biological Treatment of Sewage and Industrial Waste* **2**. J. McCabe and W. W. Eckenfelder (Ed.). Reinhold, New York, pp. 136–144.

Best, P. 1975. "Turning droppings into gas." *Poultry International* **5**:6.

Boersma, L., J. R. Miner, and H. K. Phinney. 1974. *Animal Waste Conversion Systems Based on Thermal Discharge. NTIS Report* **PB-240 113**.

Bolton, R. L. and L. Klein. 1971. *Sewage Treatment. Basic Principles and Trends*. Butterworths, London, p. 31.

Clements, A. C. and K. G. Soderstrom. 1976. "Self sufficient energy integrated design and construction method for low cost self help housing programs." *Sharing The Sun Conference* **4**. K. W. Boer (Ed.). Pergamon Press, Elmsford, New York, pp. 262–273.

Cooke, G. W. 1976. "The role of organic manures and legumes in crop production." *Energy Use and Brit. Agric. Proc. 10th Ann. Conf. Reading Univ. Agric. Club*. D. M. Bather and H. I. Day (Eds.). Reading University Agricultural Club, University of Reading, U.K., pp. 15–18.

Davies, M. T. 1970. "Experiments on the fertilizing value of animal slurries." *Experimental Husbandry* **19**:49–60.

Department of Environment. 1972. *Agricultural Use of Sewage Sludge, Notes on Water Pollution, No.* **57**. H. M. Government.

Diaz, L. F. and G. J. Trezak. 1977. "Biogasification of a selected fraction of municipal solid wastes." *Compost Sci.* **18**:8–13.

Dickinson, D. 1971. *Practical Waste Treatment and Disposal.* Applied Science Publishers, London.

Finlay, J. H. 1976. *Operation and Maintenance of Gobar Gas Plants.* United Mission Economic Development Agency, Butwal, Nepal.

Fraser, M. D. 1976. *Solar NG: Large-Scale Production of Sng by Anaerobic Digestion of Specially Grown Plant Matter. SAE Report* **769016**. pp. 83–91.

Fraser, M. D. 1977. "The Economics of Sng Production by Anaerobic Digestion of Specially Grown Plant Matter." *Inst. Gas Tech. Symp. on Clean Fuels from Biomass and Wastes*. Orlando, pp. 425–432.

Freeman, C. and L. Pyle. 1977. *Methane Generation by Anaerobic Fermentation: An Annotated Bibliography*. Intermediate Technology Report, Imperial College, London.

Gaddy, J. L., E. L. Park, and E. B. Rapp. 1974. "Kinetics and economics of anaerobic digestion of animal waste." *Water, Air and Poll.* **3**:161–169.

Ghosh, S. and D. L. Klass. 1977. "Two-phase anaerobic digestion." *Inst. Gas. Tech. Symp. on Clean Fuels from Biomass and Wastes*. Orlando, pp. 373–403.

Golueke, C. G. and P. H. McGauhey. 1976. "Waste materials." *Ann. Rev. Energy* **1**:257–278.

Hansford, G. S. 1976. "The power requirements for waste disposal." *Chemsa* **2**:70–73.

Hawkes, D., R. Horton, and D. A. Stafford. 1976. "The applications of anaerobic digestion to producing methane gas and fertilizer from farm wastes." *Proc. Biochem.* **12**:32–37.

Hekiman, K. K., W. J. Lockman, and J. H. Hirt. 1976. "Methane as recovery from sanitary land-fills." *Waste Age* **7**:2–9.

Hobson, P. N. 1977. *Notes on Anaerobic Digestion and Anaerobic Digesters Based on Experimental Plant in Aberdeen*. Rowett Inst., Bucksburn, Aberdeen.

Holdom, R. S., V. L. Larsen, and M. J. Spivey. 1976. "Gas flow measurements in fermentations: an inexpensive system with analogue out-put." *J. Appl. Bacteriol.* **41**:255-261.

Hungate, R. E. 1976. "Suitability of methanogenic substrates, health hazards and terrestrial conservation of plant nutrients." *Inst. Training and Res. Seminar on Microbial Energy Conversion.* pp. 339-345.

Kang, E. K. Y. 1975. "The development of marsh gas production from hog waste in Taiwan, ROC." *10th Intersoc. Energy Conversion Conf.* Delaware, pp. 828-834.

Kispert, R. G., L. C. Anderson, D. H. Walker, S. E. Sadek, and D. L. Wise. 1974. *Fuel Gas Production from Solid Waste. NTIS Report* **PB-238 068**.

Klass, D. L. 1974, "A perpetual methane economy—is it possible?." *Chem. Tech.*, March, pp. 161-169.

Klass, D. L., S. Ghosh, and J. R. Conrad. 1976. "The conversion of grass to fuel for captive use." *Inst. Gas Tech. Symp. on Clean Fuels.* pp. 229-253.

Lecuyer, R. P. and J. H. Marten. 1976. "An economic assessment of fuel gas from water hyacinths." *Inst. Gas Tech. Symp. on Clean Fuels.* pp. 267-287.

Leese, T. M. 1976, "The conversion of ocean farm kelp to methane and other products." *Inst. Gas Tech. Symp. on Clean Fuels.* pp. 253-262.

Loll, U. 1976. "Engineering, operation, and economics of bio-digestion." *Inst. Training and Res. Seminar on Microbial Energy Conversion.* pp. 361-379.

Meadows, D. H., D. L. Meadows, J. Randers, and W. W. Behrens, 1972. *The Limits to Growth*. A report for the Club of Rome's project on the predicament of Mankind. Earth Island, London.

Mesarovic, M. and E. Pastel. 1975. *Mankind at the Turning Point*. The 2nd Report to the Club of Rome. Hutchinson, London.

Maynell, P. J. 1976. *Methane: Planning a Digester*. Prism Press, Dorchester, U.K.

Ostrovski, C. M., N. Peters, and J. L. Sullivan. 1976. "A feasibility study of bio-gas production in individual farms in S. Western Ontario." *Sharing the Sun Conference* 7. K. W. Boer (Ed.). Pergamon Press, Elmsford, New York, pp. 129-146.

Oswald, W. J. 1976. "Gas production from micro algae." *Inst. Gas Tech. Symp. on Clean Fuels.* p. 311.

Parikh, J. K. and K. S. Parikh. 1976. "Potential of bio-gas plants and how to realise." *Inst. Training and Res. Seminar on Microbial Energy Conversion.* pp. 555-592.

Pfeffer, J. T. and J. C. Liebman. 1976. "Energy from refuse by bioconversion, fermentation, and residue disposal processes." *Resource Recovery and Conservation* 1:295–314.

Savery, C. W. 1972. "Methane recovery from chicken manure digestion." *J. Water Poll. Contr. Fed.* 44:2349–2352.

Sen, B. P. and T. R. Bhaskaran. 1962. "Anaerobic digestion of liquid molasses distillery wastes." *J. Water Poll. Contr. Fed.* 34:1015–1021.

Shabi, F. A. and M. C. Cannon. 1975. "Characteristics and treatment of dairy and meat effluents." *Effluent Water Treat. J.* 15:130–140.

Slesser, M. 1974. "The energy basis for development." *Chem. Eng.*, April, pp. 231–236.

Smil, V. 1977. "China claims lead in biogas supply." *Energy Intnl.* 14:25–28.

Smith, J. E. 1977. "Inventory of energy use in wastewater sludge treatment and disposal." *Ind. Water. Eng.* 14:20–27.

Tietjen, C., A. E. Hassan, J. R. Fischer, G. R. Morris, E. J. Kroeker, T. B. Abeles, and G. M. Wong-Chong. 1975. "Energy reclamation from agricultural wastes." *Energy, Agric. and Waste Management Conf.* pp. 247–485.

Troughton, J. H. and W. J. Cousins. 1976. "Prospects for energy farming." *New Zealand Energy J.* 49:190–195.

Tyrell, M. A. 1978. Personal communication.

Webber, J. 1961. "An experiment to compare bulky organic manures." *Expt. Hort.* 5:53–65.

Webber, J. and B. Bastiman 1967. "Experiments testing poultry manure as a source of nitrogen for grasses." *Experimental Husbandry* 15:11–18.

Winstrom-Olsen, R. S. Holdom, R. Cocker, and R. Welborn. 1977. "Energy analysis of anaerobic biogas fermentation of agricultural effluents." *FEMS Microbiol. Letters* 1:153–156.

Wise, D. L. 1977. "Biomass progress and plans." *ERDA/AGA/NCA/EPRL/4th Energy Tech. Conf.* Washington, March 14–16, pp. 434–455.

Index

Index